Guide to Research
Techniques in Neuroscience

Guide to Research Techniques in Neuroscience

Matt Carter
Jennifer C. Shieh
Stanford University, School of Medicine,
Stanford

AMSTERDAM • BOSTON • HEIDELBERG • LONDON
NEW YORK • OXFORD • PARIS • SAN DIEGO
SAN FRANCISCO • SINGAPORE • SYDNEY • TOKYO

Academic Press is an Imprint of Elsevier

Academic Press is an imprint of Elsevier
30 Corporate Drive, Suite 400, Burlington, MA 01803, USA
525 B Street, Suite 1900, San Diego, California 92101-4495, USA
84 Theobald's Road, London WC1X 8RR, UK

Library of Congress Cataloging-in-Publication Data
Carter, Matt, 1978-
    Guide to research techniques in neuroscience / Matt Carter, Jennifer C. Shieh.
        p. ; cm.
    Includes bibliographical references.
    ISBN 978-0-12-374849-2
    1. Neurosciences—Research—Methodology. I. Shieh, Jennifer C., 1981- II. Title.
    [DNLM: 1. Nervous System Physiological Phenomena. 2. Diagnostic Imaging—methods.
    3. Laboratory Techniques and Procedures. WL 102 C324g 2010]
    RC337.C37 2010
    616.85′20072—dc22

                                                                                2009021267

British Library Cataloguing-in-Publication Data
A catalogue record for this book is available from the British Library.

ISBN: 978-0-12-374849-2

For information on all Academic Press publications
visit our Web site at www.elsevierdirect.com

Printed in Canada

10   9 8 7 6 5 4 3 2 1

Working together to grow
libraries in developing countries

www.elsevier.com | www.bookaid.org | www.sabre.org

ELSEVIER   BOOK AID
              International   Sabre Foundation

Contents

The story of why we decided to write this book can essentially be summarized as: we tried to find a book just like it, couldn't find one anywhere, and ultimately decided the book would be so useful that we would write it ourselves.

When we were advanced undergraduates and beginning graduate students, we found the number of techniques used in neuroscience research daunting. In no other biological field are students required to learn biochemistry, molecular biology, genetics, electrophysiology, microscopy, histology, behavioral assays, and human imaging technologies, not to mention the basic biological and physical properties that allow these methodologies to work. It is true that each neuroscientist only practices a handful of these general techniques. However, all neuroscientists must learn to understand and appreciate one another's research, even if they don't perform the specific experiments themselves.

For example, consider the study of the auditory system. Auditory research can be performed in humans using functional magnetic resonance imaging (fMRI) and other methods of whole-brain imaging. Auditory research can also be performed in animal models, from standard laboratory animals (mice, rats, flies) to animals that have unique auditory properties that make them particularly well suited for auditory research (bats, barn owls). These animal models can be studied using *in vivo* electrophysiology while they are anesthetized or while performing behavioral tasks. Additionally, the cells responsible for hearing and interpreting auditory stimuli can be examined *in vitro*—for example, by recording electrical signals from isolated hair cells. Histological and biochemical techniques can highlight the specific ion channels and other proteins that provide auditory cells their physiological properties. Genetic methods can be used to investigate the role of molecules that make hearing possible or cause auditory deficits. Thus, anyone interested in the auditory system, or any other specific subfield of neuroscience, must be prepared to read about and analyze the contributions of studies using a wide spectrum of techniques.

As students new to neuroscience research, we were overwhelmed with the expectation that we understand all of these methods. Moreover, we were surprised by the number of advanced students who seemed to pretend that they understood all these techniques when they really only understood the specific techniques they used in their own studies. Sometimes, these students, and even some postdocs and faculty, would confess to us, "I should have learned that long ago, but I just don't really understand it." We would hear scientists who use electrophysiology and fMRI techniques sometimes say, "Who cares about a blob on a gel?" Likewise, we would hear geneticists and molecular biologists confess, "I don't know what the squiggly lines in electrophysiology studies mean," or "I don't know why fMRI experiments are so hard—can't you just put someone in a scanner and turn on the machine?" In short, it became obvious that there should be a guide to neuroscience techniques. So we tried to find one.

Finding a guide to neuroscience techniques turned out to be more difficult than we imagined. During our first trips to the annual Society for Neuroscience conference, we tried to find a simple techniques book among the hundreds of books being promoted by different publishers. Amazingly, there was no single book that aimed to explain the breadth of neuroscience techniques at a basic level. There were detailed books written about single techniques (*PCR Methods* or *Making Transgenic Animals*) and recipe books filled with protocols, but no book addressed techniques as a subject themselves. Even the almighty Internet did not provide thorough insight into standard techniques. On the Internet, you can find protocols for doing a western blot and learn that it is a technique that measures the amount of protein in a sample. But you cannot find information about *other* ways a scientist can measure protein expression or why you might choose a western blot over these other methods. There are no example figures from the literature, nor are there suggestions of what to look for in these figures. Our dream book would contain a description of western blots, a comparison of western blots with similar methods, and an example of what the data looked like in the literature.

No such book existed, and we wanted one. With the encouragement of faculty members and students at Stanford, we decided to write this book. We also decided that the information would make an excellent seminar course. Along with a talented student colleague of ours, Saul Villeda, we created a 9-week techniques course called "Understanding Techniques in Neuroscience." This class surveys neuroscience techniques and provides examples from the literature. We wrote a 110+-page course reader to accompany the class and quickly decided that this text should also form the basis of a future book that could exist independently of the class.

The class was amazingly well received. The first year, our class had 15 students. Through word-of-mouth, the class doubled in size the next year. The third year, over 100 students attended our lectures, including undergraduates, graduate students, postdocs, and occasionally a faculty member. We believe the number of people taking the course demonstrates that neuroscientists are interested in learning about one another's methods, as well as the need for formalized education about neuroscience techniques.

Therefore, we decided to adapt our course reader into this book. It is the book that we so desperately wanted when we were beginning neuroscience students, and we are happy that it is now possible to find at Society for

Neuroscience meetings. We learned a tremendous amount of information while researching and writing this book, and we hope that you find it helpful for your own education.

We would like to first thank the editorial staff at Elsevier/Academic Press for making this book a reality. From Susan Lee, who initially saw the potential of our course reader, to Melissa Turner and Mica Haley, who managed to get the final product with only minor hair-pulling.

This book would not be possible without the guidance and help of the incredible faculty at Stanford. In particular, Bill Newsome provided tremendous encouragement and support throughout the writing process. As promised, he will receive a fine bottle of Woodford Reserve upon the successful publication of this book. We must also acknowledge the understanding of our research advisors Luis de Lecea and Sue McConnell. As promised, they will receive our theses without further delay. Finally, we thank Mehrdad Shamloo and Mehrdad Faizi at Stanford's Behavioral and Functional Neuroscience Laboratory and Jon Mulholland and Lydia-Marie Joubert of the Cell Sciences Imaging Facility for their expert advice.

Our teaching partner Saul Villeda researched many of the methods that are detailed in this book and helped prepare the original course reader. Saul is an amazing teacher, and it has been our pleasure to work with him over the years, both as colleagues and friends.

We also received substantial guidance and support from the Stanford neuroscience community. So many students and postdocs spent their valuable time and energy to help us understand techniques, edit our chapters, and provide encouragement as we finished the book. Special thanks to Raag Airan, Björn Brembs, Brittany Burrows, Laurie Burns, Kelsey Clark, Emily Drabant, Mary Hynes, Pushkar Joshi, Rachel Kalmar, Jocelyn Krey, Dino Leone, Jie Li, Scott Owen, Georgia Panagiotakos, Chris Potter, Elizabeth Race, Victoria Rafalski, Andreas Rauschecker, Magali Rowan, Rory Sayres, Bob Schafer, Jana Schaich Borg, John Wilson, and Sandra Wilson, who provided substantial suggestions to improve the content and text. We completed this book with our neuroscience colleagues in mind, and we hope that all future neuroscience students find this book helpful to their education and throughout their careers.

Finally, we thank our significant others, Vishal Srivastava and Alison Cross. Thank you *so* much for giving us time, love, and encouragement.

Matt Carter
Jennifer C. Shieh
Stanford, California, 2009

Understanding how the brain works is perhaps the greatest challenge facing contemporary science. How mental life is rooted in the biology of the brain is intrinsically fascinating, and researchers from remarkably diverse scientific backgrounds are being drawn increasingly to the field of neuroscience. Psychologists, molecular biologists, physiologists, physicists, engineers, computer scientists, and more are all contributing importantly to the richness of modern neuroscience research. The brain remains the most complex and enigmatic entity in the known universe, and all relevant scientific techniques and perspectives must ultimately be brought to bear to solve its mysteries.

The quickening pace and diversity of neuroscience research pose substantial challenges, however, for new students and for established researchers seeking to enter the field for the first time. The techniques used to measure and manipulate the nervous system are dizzying in their scope and complexity, and some of the important new approaches are accompanied by clouds of jargon that are all but impenetrable to outsiders. Fortunately for those of us who are perplexed, a concise, no-nonsense guidebook has now arrived. The *Guide to Research Techniques in Neuroscience* presents the central experimental techniques in contemporary neuroscience in a highly readable form. It is all there—from brain imaging to electrophysiology to microscopy to transgenic technologies. Matt Carter and Jennifer Shieh take us on a very practical cook's tour of research techniques, all the while providing concise overviews of exactly where and how particular techniques fit into the grand scheme of the basic questions being asked of the nervous system.

As a neurophysiologist, I read quickly through the electrophysiology and brain imaging chapters, noting with appreciation the efficient presentation of basic information that should be in the intellectual repertoire of any aspiring neuroscientist. But I read avidly the chapters on cloning, gene delivery, and use of transgenic animals—the exotica (to me!) of molecular biological approaches made clean, simple, and pleasingly devoid of specialist vocabulary. I rather suspect that most neuroscience professionals will have the same experience reading this book, encountering high-yield veins of useful information interspersed with much they already know. For the student or postdoc just entering the field, however, the vast majority of the book will be very high-yield.

This book is an essential resource for anyone—from the beginning graduate student to the seasoned faculty member—who could use an efficient guidebook at his or her side while reading papers written by *any* of their colleagues, irrespective of the level of analysis, from small molecules to spike trains. It may not render the mysterious raster plots of systems lectures immediately clear, but it will orient you to the key types of information you want to take home from that talk. This book is a little gem! Read it, rely on it, and pass it on to others.

William T. Newsome, Ph.D.
Professor, Department of Neurobiology, Stanford University
Investigator, Howard Hughes Medical Institute

The human mind has been studied for thousands of years, but the human brain, as well as the brains of other species, has only been studied for about a century. Only 150 years ago, the ability to study the nervous systems of humans and other animals was limited to direct observation and by examining the effects of brain damage in people and other organisms. With the advent of histology came the ability to visualize and differentiate between neurons based on morphology. The great neuroscientist Santiago Ramón y Cajal used a method called Golgi staining to visualize the morphology and architecture of neurons and their circuits throughout the brain. Cajal used the Golgi stain to propel the field of neuroscience into its modern state.

In the history of neuroscience, each leap forward in knowledge has been based on a leap forward in techniques and technology. Just as Ramón y Cajal used Golgi staining to greatly advance our understanding of the structure of the nervous system, scientists throughout the twentieth century used more and more advanced techniques to contribute to our understanding of the function of the nervous system: Eccles, Hodgkin, and Huxley used intracellular recording technology to investigate the ionic basis of membrane potentials; Hubel and Wiesel used extracellular recording technology to investigate how information is processed and recorded in the visual system; Neher and Sakmann used patch-clamp technology to investigate the physiology of single ion channels. In the latter half of the twentieth century, the explosion of molecular biology techniques and methods of genetically manipulating model organisms allowed neuroscientists to study individual genes, proteins, and cell types. Technology has progressed so far in the past 100 years that the Golgi stain itself seems to have been reinvented through powerful technologies (Chapter 11) that allow investigators to turn specific neurons different colors to further investigate the structure and connectivity of the nervous system.

The modern neuroscientist now has hundreds of techniques that can be used to answer specific scientific questions. This book contains 14 chapters that provide an overview of the most commonly used techniques. Although there are dozens of techniques that seem very different at first glance, many of them attempt to study the nervous system in the same way. For example, transcranial magnetic stimulation (Chapter 1), physical lesions (Chapter 3), pharmacological inhibition (Chapter 3), optogenetic inhibition (Chapter 7), and genetic knockdown or knockouts (Chapter 12) are all attempts to test the effect of a *loss-of-function* of some aspect of the nervous system on another aspect of the nervous system. For each level of investigation (whole brains to individual genes), research strategies can be similar even if the techniques used are very different.

LEVELS OF INVESTIGATION

Something immediately obvious to all students of neuroscience is that the nervous system is exceptionally complicated and can be examined at multiple levels of investigation. The basic functional unit of the nervous system is the neuron. The human brain is composed of approximately 100 billion neurons that are connected into circuits via approximately 100 trillion synapses. Neural circuits are organized into anatomical structures and larger networks of neurons that can integrate information across modalities from many different parts of the brain. These networks process sensory information from the external and internal environment and provide the neural basis of cognition—learning, memory, perception, decision making, emotion, and other higher-order processes. The final output of the nervous system is a behavior composed of a coordinated motor action. This behavior can either be extremely simple, such as a motor reflex, or incredibly complicated, such as dancing, typing, or playing a musical instrument. Behavior is usually defined not just by what an organism does, but what it *chooses* to do. Therefore, except in rare circumstances of lesion or disease, cognition and behavior are inseparably linked, and in animals other than humans, behavior is used as a read-out of animal cognition.

Just as one can start with a neuron and scale up toward circuits, cognition, and behavior, a scientist can also scale down and examine the components that make up a neuron. A neuron is itself defined as having a cell body (soma), axon, and dendrites. These neuronal components contain subcellular specializations that make the neuron unique among other cell types. Specialized organelles in a neuron, such as vesicles containing neurotransmitters, provide the cell with the ability to signal to other neurons. Specialized cytoskeleton processes allow a neural process to extend great distances throughout the brain and body. Several proteins provide neurons with their intercellular signaling abilities and physiological characteristics. For example, biosynthetic enzymes produce neurotransmitters, while other proteins serve as receptors for these signaling molecules. One of the most important types of proteins in the nervous system form ion channels, the transmembrane structures that allow neurons to become electrically active under certain conditions. All of these proteins are the products of genes, the functional units of an organism's genome. The human genome contains approximately 30,000 genes, with each neural subtype expressing its own subset of these genes.

The complexity of the nervous system is awesome in scope. It is amazing that a mutation in a single gene, such as a gene that codes for a transmembrane ion channel, can produce effects that alter the electrical properties of a neuron, in turn altering the normal firing patterns of a neural circuit and thus causing an abnormal behavior.

A neuroscientist can approach the study of the nervous system through any of these levels of organization. The 14 chapters of this book provide a guide to the types of experiments that can be performed at each level. However, irrespective of technique, the basic scientific approach one can use to study the nervous system is consistent from level to level, whether the subject is human cognition or axon guidance in cell culture. Next we will examine the basic approaches to designing experiments in the nervous system.

METHODS OF STUDYING THE NERVOUS SYSTEM

There are four general methods of studying the nervous system: (1) *examining case studies*—identifying interesting events that have occurred naturally and using these events to develop hypotheses that can be tested in future experiments; (2) *screens*—searching for anatomical structures, neurons, proteins, or genes that could play a role in a subject of interest; (3) *description*—using techniques that allow a scientist to observe the nervous system without manipulating any variables; and (4) *manipulation*—testing hypotheses by determining the effect of an independent variable on a dependent variable. Each of these four methods is described in detail here.

Examining Case Studies

A **case study** is an example of an event that happened to a subject (most often a human or group of humans) that demonstrates an important role for an aspect of the nervous system. The circumstances surrounding the event are usually nonrepeatable and cannot be precisely recreated in a laboratory setting. Such demonstrations are, therefore, not true experiments in that no variables are deliberately controlled by a scientist. However, these events can often reveal substantial information about an aspect of neural function that was previously unknown.

For example, consider the case of Phineas Gage, a railroad worker who was involved in an accident in 1848 that caused an iron rod to pass through his skull. The rod entered the left side of his face, passed just behind his left eye, and exited through the top of his head, completely lesioning his frontal lobes. This is an amazing event, not only because Gage survived (and lived for another 12 years), but also because it informed scientists about the function of the frontal lobe of the brain. The event allowed investigators to retrospectively ask the question "What is the effect of removing the frontal lobe on consciousness and behavior?" According to Gage's friends, family, and co-workers, he was "no longer Gage." He retained the ability to learn, remember, sense, and perceive his environment, to execute motor functions, and to live a fairly normal life, but it seemed to people who knew him that his personality had changed completely. After the accident, Gage was less polite, erratic, unreliable, and offensive to others. He wound up losing his job at the railroad, not because of any physical or mental incapacity, but because he was simply so disrespectful and offensive that people could not stand to work with him.

This case study is not a true experiment; no scientist decided to test the removal of the frontal lobe on personality. But the incident, and others like it, allows neuroscientists to form hypotheses based on naturally occurring events. Because of Gage's story, neuroscientists could hypothesize about the contribution of the frontal lobe to human personality. Future experiments could test these hypotheses on animal models (that share certain human personality traits) and even attempt to identify neural circuits that contribute to human behaviors.

Screens

A screen is a method that allows an investigator to determine what nuclei, neurons, or genes/proteins may be involved in a particular biological process. Such experiments are not necessarily driven by a hypothesis, but the experiment identifies candidates that can form the basis for future hypothesis-driven research. For example, a neuroscientist who wants to identify genes involved in body weight regulation may compare gene expression profiles in central feeding centers of the brain in both fed and starved animals: genes expressed in starved animals relative to fed animals may be important for generating the motivation to eat.

Screens can be performed at multiple levels of investigation. When a cognitive neuroscientist places a human subject into an fMRI scanner and examines which brain areas show increased activation in response to a specific stimulus or task, the scientist is essentially performing a screen of different brain regions. When a fly geneticist examines thousands of mutagenized flies for deficiency in a behavioral task, the scientist is attempting to identify genes necessary for that behavior to occur. Such genes can then be tested in future experiments. Thus, screens can be performed to identify interesting molecules or entire brain regions.

Description

Descriptive science is the act of simply observing properties of the nervous system without manipulation. This type of research is usually the first step in acquiring knowledge about a newly discovered gene, protein, or neuronal subtype. For example, an investigator could describe the sequence of a gene and where in the brain the gene is expressed. Likewise, in the case of proteins, an investigator could describe the amino acid sequence of the protein and where in the brain the protein is expressed. Neurons can be described in terms of what genes/proteins they express, their morphology, how many neurons make up a population of neurons, and their electrophysiological properties.

It is important to note that just because a study may be descriptive does not mean that it is necessarily easier than other types of experiments. Observation and description form the foundation for understanding the relationship between structure and function, as well as providing insight about what elements to manipulate in future experiments. Thus, descriptive neuroscience plays just as important a role in modern research as when Ramón y Cajal observed the structure of neurons 100 years ago.

Manipulation

Manipulating an aspect of the nervous system or environment and examining the effect this perturbation has on a separate aspect of the nervous system is the only way to test a hypothesis in neuroscience. A manipulation experiment tests *the effect of X on Y*. The variable that is manipulated, *X*, is referred to as the **independent variable**. The part of the system that is measured, *Y*, is referred to as the **dependent variable**.

Two of the most common types of manipulation experiments are loss-of-function and gain-of-function experiments. In a **loss-of-function** (also called "**necessity**") experiment, a part of the nervous system is diminished or removed in an attempt to determine if it is *necessary* for a certain process to occur. The following questions are all loss-of-function questions:

- Is a normal copy of the gene *Fezf2* necessary for the proper development of the cerebral cortex?
- Is the receptor for the hypocretin neuropeptide necessary for normal sleep/wake transitions in mammals?
- Is electrical activity in the medial geniculate nucleus required for auditory-driven spikes in auditory cortex?
- Can human patients with damage to the cerebellar vermis perform as well as healthy controls on a verbal-memory task?

In all of these experiments, an aspect of the nervous system is partially or totally disrupted, whether it is a gene, a protein, electrical activity, or an entire brain structure. The independent variable is the loss of the structure, and the dependent variable is the effect on another aspect of the nervous system. Sometimes, a good follow-up for a loss-of-function experiment is a **rescue experiment** in which the aspect of the nervous system that is lost is deliberately returned. For example, if it is found that a certain line of fruit flies lacks a gene that is necessary for proper development of an eye, the specific gene can be reintroduced to the flies in transgenic experiments to see if the functional gene can rescue the aberrant phenotype.

In a **gain-of-function** (also call "**sufficiency**") experiment, an aspect of the nervous system is increased relative to normal. This may include an increased expression of a gene or protein, an increase in electrical activity in a brain region, or an increase of a particular neurotransmitter in the extracellular medium. The aspect of the nervous system that is increased is the independent variable, and the effect on another part of the nervous system is the dependent variable. The following questions are all gain-of-function questions:

- Can an increase in the gene *TrpA1* cause mice to be hypersensitive to cold temperatures?
- Can an introcerebroventricular injection of Neuropeptide Y cause an increase in feeding behavior in rats?
- Can electrical microstimulation of the lateral geniculate nucleus increase spike frequencies over time in area V4 of visual cortex?
- Can stimulation of the motor cortex in human subjects using transcranial magnetic stimulation (TMS) cause motor behaviors?

In both loss-of-function and gain-of-function experiments, it is important not to overstate the conclusions of the experiments. For example, consider a loss-of-function experiment in which a mouse that lacks a gene is unresponsive to painful stimuli. The investigator could conclude that this gene is necessary for proper performance on an assay for pain detection. However, an inappropriate conclusion would be that this gene regulates pain detection. Perhaps this gene is responsible for normal development of the spinal cord and the mouse lacks all peripheral sensation. Alternatively, this gene may code for a protein that is necessary for normal development of the thalamus; if improper development of the part of the thalamus that receives information about painful stimuli causes the stimuli not to reach somatosensory cortex, the animal will not perform normally on the pain-detection task. Careful controls are necessary to reach appropriate conclusions.

UNDERSTANDING TECHNIQUES IN NEUROSCIENCE

We hope that these 14 chapters serve as a useful guide to studying the nervous system. There are two important points to keep in mind while reading these chapters: (1) All of the techniques described throughout this book depend on the principles just mentioned. For each level of investigation that a scientist may choose to study, the same four general types of methods exist: examining case studies, screens, description, and manipulation. The same principles that are used to study brain activity in awake, human subjects are used to study genes and proteins in tissue samples. The methods may vary, but the principles remain the same. (2) It is also important to remember that techniques and methods should never be the guiding force behind doing experiments in research. Ideally, experiments should be performed in order to answer an interesting question, not the other way around. A technique should not be used for its own sake but because it is the *best* technique available to answer a particular research question. Therefore, we hope this book answers your questions about what techniques are available in modern neuroscience research and answers your specific research questions as well!

Chapter 1

Whole Brain Imaging

After reading this chapter, you should be able to:
- Compare the relative strengths and limitations of different structural and functional brain imaging techniques
- Explain the physical and physiological basis of MRI/fMRI technology
- Describe the components of functional brain imaging experimental design: formulating a hypothesis, choosing task paradigms, performing the experiment, acquiring and analyzing data, and constructing figures

Techniques Covered:
- **Structural techniques:** cerebral angiography, computerized tomography (CT), magnetic resonance imaging (MRI), diffusion MR imaging
- **Functional techniques:** functional magnetic resonance imaging (fMRI), positron emission tomography (PET), single-proton emission computerized tomography (SPECT), electroencephalography (EEG), magnetoencephalography (MEG), optical imaging
- **Techniques used to investigate the necessity and sufficiency of a specific brain region for a cognitive function:** transcranial magnetic stimulation (TMS) and case studies

Modern brain imaging technology can seem like magic. The ability to produce detailed images of the human brain without physically penetrating the skull is a technological marvel that has saved thousands of lives and allowed scientists to study the structure of the brain throughout development, disease, and aging. Furthermore, the ability to image neural activity in the brain during cognition has provided scientists the opportunity to correlate activity in distinct brain regions with specific mental operations, a truly remarkable achievement. Indeed, colorful figures depicting activity in the human brain dazzle scientists and nonscientists alike.

Of course, brain imaging technology is not magic. The technology that produces detailed images of the brain depends on complex physics, expensive equipment, and skilled technicians. As with all scientific experiments, brain imaging studies must be well designed, the data accurately analyzed, and the results carefully interpreted. The purpose of this chapter is to explain the ostensible magic of whole brain imaging technology and provide insight into how experiments are designed and interpreted.

Whole brain imaging technology can essentially be divided into two categories: structural and functional. Structural techniques produce images of the anatomical architecture of the brain, whereas functional techniques produce images of the physiological processes that underscore neural activity. This chapter will survey both classifications of techniques and describe how they can be used in modern neuroscience research. We will focus on MRI and fMRI technology due to the widespread use of these techniques in the neuroscience literature. After reviewing these techniques, we will survey the essential components of a functional imaging experiment: forming hypotheses, choosing appropriate task paradigms, performing experiments, acquiring and analyzing data, and producing figures for publication.

STRUCTURAL BRAIN IMAGING TECHNIQUES

Structural brain imaging techniques are used to resolve the anatomy of the brain in a living subject without physically penetrating the skull. These techniques can be used in combination with **functional brain imaging** techniques to correlate neural activity in specific anatomical regions with behavioral or cognitive functions. Structural techniques can also be used to measure anatomical changes that occur over time, such as a decrease in brain mass that occurs with aging or with the progression of disease. Most often, these techniques are used in clinical neuroscience and neurology to diagnose diseases such as tumors and vascular disorders.

Brain imaging technologies take advantage of the different composition of distinct brain regions and use these differences to form the basis of an image (Figure 1.1). Neural cell bodies contain many biomolecules, including proteins and carbohydrates. Axons and fiber tracts are relatively fatty due to the insulation provided by myelin. Cerebrospinal fluid (CSF) in the ventricles and surrounding the brain is essentially a saline solution. The microanatomy and composition of individual neural structures cause distinct regions of the brain to appear different when examined by the naked eye. For example, when looking at slices of the brain with the naked eye, brain tissue mostly composed of cell bodies appears gray compared with other areas, and thus is referred to as "gray matter." Brain tissue mostly composed of axons and fiber tracts appears white, and thus is referred to as "white matter." Often, the most informative structural images of the brain show the contrast between gray and white matter. Therefore, the ultimate goal of structural imaging technologies is to differentiate between

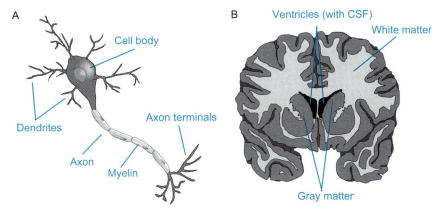

FIGURE 1.1 The composition of the brain. (A) A microscopic view of a neuron. Each neuron is composed of a cell body, dendrites, and an axonal process. The cell bodies and dendrites are rich in proteins and carbohydrates. Axons are surrounded by a myelin insulation made of fats. **(B)** A macroscopic view of the brain. Gray matter is rich in cell bodies and therefore in proteins and carbohydrates. White matter is composed of axon tracts and is therefore rich in fatty myelin. CSF in the ventricles is a saline solution. To produce an image, brain imaging technologies must differentiate among proteins/carbohydrates, fats, and saline.

proteins and carbohydrates (cell bodies), fat (axon tracts), and salt water (CSF), as this contrast reveals the most information about brain architecture.

Until the early 1970s, there was no technology that could differentiate between these substances within the brain. Conventional **X-ray** technology is essentially useless for this purpose. During an X-ray procedure, an X-ray beam is passed through an object and then onto a photographic plate (Figure 1.2A). Each of the molecules through which the beam passes absorbs some of the radiation, so only the unabsorbed portions of the beam reach the photographic plate. X-ray photography is therefore only effective in characterizing internal structures that differ substantially from their surroundings in the degree to which they absorb X-rays, such as bone in flesh (Figure 1.2B). By the time an X-ray beam passes through the relatively soft consistency of the brain (not to mention the relatively hard consistency of the skull!), little information about individual brain structures can be discerned (Figure 1.2C). Therefore, in the 1960s and 1970s there was strong motivation to discover better ways of imaging the brain. The techniques described in the following sections represent 30–40 years worth of innovation in technology ultimately designed to show contrast within the soft tissue of the brain.

Cerebral Angiography

A **cerebral angiogram** is an enhanced X-ray that uses dyes to make up for the relatively poor soft-tissue contrast of conventional X-rays. A radio-opaque dye that absorbs X-rays better than surrounding tissue is injected into an artery that

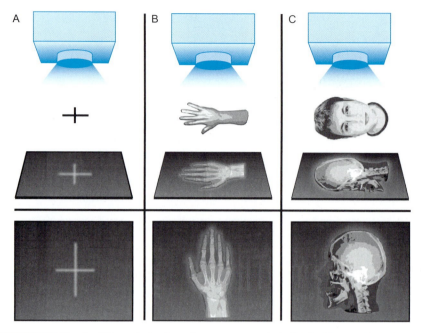

FIGURE 1.2 Standard X-ray technology alone cannot produce detailed images of the brain.
(A) During an X-ray procedure, an X-ray beam is passed through an object and onto a photo-graphic plate. Only the unabsorbed portions of the beam reach the plate, creating an image. (B) The contrast between the soft consistency of skin and muscle compared with the hard consistency of bone is sufficient to form an X-ray image. However, (C) the contrast between the soft consistency of different tissues within the brain is insufficient to form an X-ray image.

delivers blood to the brain. This substance heightens the contrast between the cerebral circulatory system and surrounding brain tissue during an X-ray (Figure 1.3A). Thus, the most prominent aspect of the central nervous system imaged in a cerebral angiogram is the brain vasculature. Angiograms can show vascular damage and indicate the presence of a tumor or aneurysm (Figure 1.3B).

Computerized Tomography (CT)

Another method that improves upon conventional X-ray technology to image the brain and body is **computerized tomography** (the "CT scan"—sometimes also called computerized axial tomography, or CAT scan). A patient or subject lies with his or her head positioned in the center of a cylinder (Figure 1.4A). A narrow beam of X-rays is aimed through the person's head and hits a detector on the opposite side. The beam and detector rotate in a slow arc, taking many individual X-ray scans at the same **axial** plane. As mentioned previously, a single X-ray scan would supply little information about the structure of the brain. However, multiple scans taken from different angles can combine to provide information

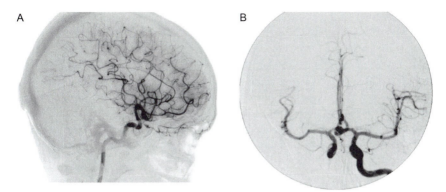

FIGURE 1.3 **Cerebral angiography.** (**A**) A cerebral angiogram depicting the vasculature of the right hemisphere of the brain. (**B**) A cerebral angiogram indicating the presence of a brain aneurysm. (A, B: Reprinted from Nolte, J. and Angevine, J. B., (2007). *The Human Brain in Photographs and Diagrams,* 3rd ed. with permission from Mosby/Elsevier: Philadelphia. Courtesy of (A) Dr. Joachim F. Seeger and (B) Dr. Raymond F. Carmody.)

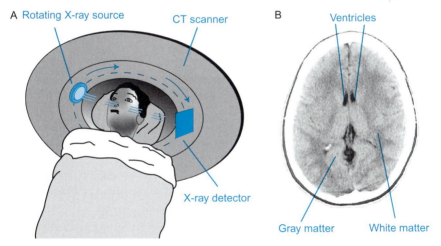

FIGURE 1.4 **Computerized tomography (CT).** (**A**) During an imaging session, a narrow beam of X-rays is slowly rotated around a subject's head to hit a detector on the opposite side. Signals from around the head are combined into a computer program that constructs a composite picture based on the various X-ray angles. (**B**) A modern CT scan can distinguish between gray and white matter, differentiate ventricles, and depict structures with a spatial resolution of millimeters. (B: Reprinted from Nolte, J. and Angevine, J. B., (2007). *The Human Brain in Photographs and Diagrams,* 3rd ed. with permission from Mosby/Elsevier: Philadelphia. Courtesy of Dr. Raymond F. Carmody.)

about small differences in radiodensity between different brain structures. These data are entered into a computer algorithm that constructs a composite picture based on the X-ray scans from all the different angles. With this information, a "slice," or tomogram (*tomo* means "cut" or "slice"), can be generated. Typically, 8–10 images of axial brain sections are obtained for analysis.

The quality of a CT scan depends on the width of the X-ray beam (narrower is better), the sensitivity of the X-ray detector, and the ability of the computer to construct an image from the data. Modern CT scans can distinguish between gray matter, white matter, and ventricles with a spatial resolution of millimeters (Figure 1.4B). They are particularly useful for identifying fluid boundaries in human patients, such as when blood collects on the brain surface in a hematoma, or detecting hard objects in soft tissue, such as a tumor or calcification. CT scanners are faster, cheaper to operate, and less prone to motion artifacts than MRI scanners; therefore, they tend to be the first tool used to diagnose a patient.

Magnetic Resonance Imaging (MRI)

Magnetic resonance imaging (MRI) technology produces highly detailed structural images of the brain and body. The resolution of a modern MR image is far superior to a CT image, typically less than a millimeter. Thus, MRI technology has largely superseded computerized tomography as *the* method of imaging the brain in both clinical and research settings. The technology that makes MRI possible is complex, but it is necessary to understand in order to fully appreciate MR images of the brain.

The Electromagnetic Basis of MRI Technology

As the name suggests, magnetic resonance imaging takes advantage of the magnetic properties of neural tissue to produce an image. Most often, MRI utilizes the magnetic properties of hydrogen protons, as they are highly abundant in the fluids and organic compounds of the brain and body. The main function of an MRI scanner is to artificially excite these hydrogen protons and then measure their relaxation properties over time.

An MRI scanner is composed of a long tube-like chamber, where a subject is placed, surrounded by electric coils hidden within the MRI apparatus (Figure 1.5). As current passes through the coils in a clockwise rotation, a magnetic field is produced longitudinal to the patient, in the direction of the feet to head. The purpose of putting the subject in a magnetic field is to affect the hydrogen protons in the subject's tissues. Protons can be thought of as miniature magnets: they spin about an axis and their spinning positive charge induces a tiny magnetic field (Figure 1.6A). Normally, the magnetic fields of individual protons orient in random directions (Figure 1.6B). However, when a subject is placed inside the strong magnetic field of an MRI machine, the magnetic fields of individual protons align in the axis of the field. Some protons align parallel to the magnetic field, toward the subject's head, while others align in the opposite, "antiparallel" direction, toward the subject's feet (Figure 1.6C). It is *slightly* more energetically favorable for the protons to orient in the parallel rather than antiparallel direction, so there is a net magnetic field vector in the parallel direction (Figure 1.6D).

There is one more important detail to know about the protons in a magnetic field: they do not simply stay stationary, aligned parallel (or antiparallel) to the

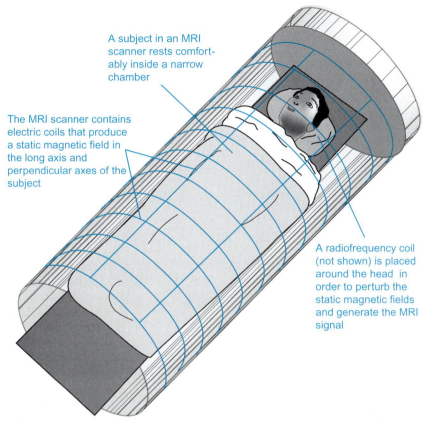

A subject in an MRI scanner rests comfortably inside a narrow chamber

The MRI scanner contains electric coils that produce a static magnetic field in the long axis and perpendicular axes of the subject

A radiofrequency coil (not shown) is placed around the head in order to perturb the static magnetic fields and generate the MRI signal

FIGURE 1.5 A human subject/patient in an MRI scanner. The subject lies inside a chamber surrounded by electric coils. Current passing through these coils induces a strong magnetic field.

magnetic field lines. Instead, they **precess** around their axis, spinning like a top (Figure 1.7). The frequency with which they spin is dependent on the strength of the external magnetic field (generated by the MRI machine). The larger the external magnetic field, the higher the precession frequency. The strength of a magnetic field is measured in Tesla (T). In the literature, you will see that most conventional MRI machines create external magnetic fields at 1.5–3 Tesla. Newer, more powerful MRI scanners create fields at 7 Tesla. Higher magnetic field strengths increase the signal-to-noise ratio and give higher contrast and spatial resolution. However, these powerful magnets are also more expensive and more likely to cause physiological discomfort in subjects, such as nausea or dizziness.

Generating an Image

Prior to the beginning of an imaging session, there is a net magnetic field vector in the parallel direction, longitudinal to the subject's body (Figure 1.8A). In

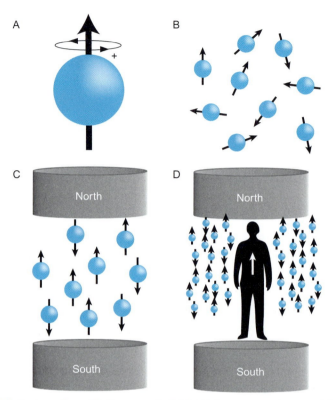

FIGURE 1.6 **Protons align with the magnetic field.** (**A**) As a proton spins around its axis, the rotating positive charge induces a magnetic field. Therefore, a proton can be thought of as a tiny magnet. (**B**) Before an external magnetic field is applied, protons orient in random directions. (**C**) However, in the presence of a strong magnetic field, protons align in either a parallel or antiparallel direction. (**D**) In an MRI scanner, protons aligned in the parallel direction are directed toward the subject's head, while protons aligned in the antiparallel direction are directed toward the subject's feet. Slightly more protons are aligned in the parallel direction, so there is a net magnetic force toward the subject's head.

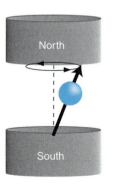

FIGURE 1.7 **Protons precess around an axis in a pattern that can resemble a spinning top.**

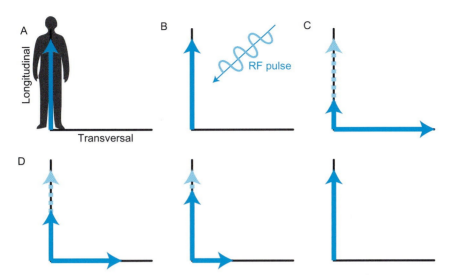

FIGURE 1.8 A radiofrequency pulse alters the net magnetic field. (A) Before the RF pulse is applied, there is a net magnetic force in the longitudinal direction toward the subject's head. **(B, C)** After the RF pulse is applied, the net longitudinal field decreases, as some protons reorient to the antiparallel orientation. Also, a new net magnetic field is created in the transverse direction. **(D)** After the RF pulse is switched off, the longitudinal field increases to normal and the transverse field decreases until the magnetic field returns to the state before the RF pulse was applied.

order to collect data for an MR image, the subject is briefly exposed to pulses of electromagnetic energy, referred to as **radiofrequency (RF) pulses** (Figure 1.8B). Applying an RF pulse *that has the same frequency as the proton precession frequency* causes two effects: (1) Some protons in the parallel phase pick up energy, reverse polarity to the antiparallel phase, and therefore decrease the net longitudinal magnetization; (2) Some protons get in sync and start to precess in phase. Their vectors now add up in a direction that is transverse to the external magnetic field (perpendicular to the subject's body). Thus, a new transversal magnetization is established (Figure 1.8C).

After the RF pulse is switched off, the high-energy nuclei begin to relax and realign (Figure 1.8D). Eventually, the longitudinal magnetization increases to its original value, while the transversal magnetization decreases to zero. The time (in milliseconds) required for a certain percentage of the protons to realign in the longitudinal direction is termed **T1**. The transversal relaxation time is termed **T2**.

Both the information acquired in the longitudinal direction (T1), as well as the information acquired in the transversal direction (T2), are measured by an antenna inside the MRI scanner. Recall that the goal of brain imaging technology is to resolve differences in tissues made of proteins and carbohydrates, fat, and salt water. What makes MRI technology useful for producing images of the brain is that these substances exhibit different values for both T1 and T2 relative to each other. For any given point in time during the relaxation phase, the T1 white matter signal is stronger than that of gray matter, and the gray matter signal is stronger than that of CSF (Figure 1.9A). These differences in signal intensity are exactly opposite for a T2 measurement: the CSF signal is strongest, followed by gray matter and then white matter (Figure 1.9B). By examining the contrast in signal from different points in space, it is possible to differentiate between different substances and form an image.

An image of the brain formed from T1 data is referred to as a **T1-weighted** image, while an image formed from T2 data is said to be **T2-weighted**. These images appear different from one another because of the differences in signal intensity among substances in T1- vs. T2-weighted images (Figures 1.9C–D). As a rule of thumb, if the CSF is black, you are looking at a T1-weighted image, as CSF has the lowest relative T1 signal intensity. If, on the other hand, the CSF appears bright white, you are looking at a T2-weighted image, as CSF has the highest relative T2 signal intensity. In the literature, most anatomical data are presented as T1-weighted images, as these images usually show better contrast between brain structures. However, this is not always the case—for

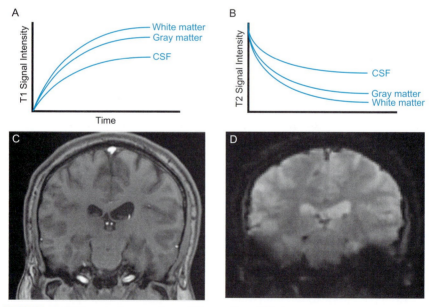

FIGURE 1.9 Different substances have different T1 and T2 time constants. (A) For any given longitudinal magnetization relaxation time (T1), the white matter signal intensity will be greater than that of CSF. **(B)** On the other hand, for any given transversal magnetization relaxation time (T2), the CSF signal intensity is greater than that for white matter. **(C)** Therefore, white matter will be bright and CSF dark on a T1-weighted image, while **(D)** white matter will be dark and CSF bright on a T2-weighted image. (C, D: Courtesy of Dr. Rory Sayres.)

example, lesions of white matter that occur due to the rupturing of blood vessels are more easily detectable on a T2-weighted image. Therefore, T2-weighted images may be optimal when examining patients following trauma or stroke.

Selecting a "Slice" of Brain to Image

How does an investigator select a **slice** to examine? Recall that an RF pulse will only excite protons with the same precession frequency as the frequency of the pulse. The precession frequency varies with the strength of the external magnetic field, so to select a single slice of the brain to image, an additional magnetic field is applied to the external magnetic field at a gradient (Figure 1.10A). Because the field strength is not equal at all planes in the longitudinal direction, an RF pulse at a specific frequency will only affect the protons, and thus the

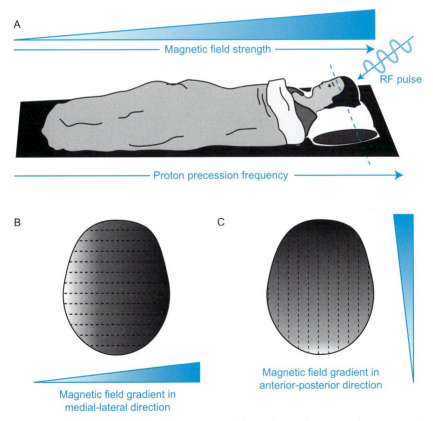

FIGURE 1.10 How a "slice" is selected in MRI. (A) To select a slice to examine, an external magnetic field is applied at a gradient. An RF pulse will only excite protons in a particular slice, as the pulse can only excite atoms with the same precession frequency. To measure the signal intensity for each point within the slice, two additional gradients are applied, one in the medial-lateral direction **(B)** and one in the anterior-posterior direction **(C)**.

signal, at a specific plane. This is the slice of the brain that will be presented as a two-dimensional image.

In order to measure signal from individual points within the slice, two additional magnetic gradients are applied in the other two axes (Figure 1.10B, C). Therefore, each point in space will have its own unique magnetic signature. Each point has a specific volume and is termed a **voxel**, a three-dimensional version of a pixel that represents a cubic volume of brain space. The resolution of each voxel is determined by the values of the gradients applied to the subject. With greater magnetic field strengths, more dramatic gradients can be established. This is why a 3 T MRI scanner can produce an image with higher **spatial resolution** than a 1.5 T scanner.

MRI has a number of features that have made it an especially valuable research tool for both diagnostic and research studies.

- It is entirely noninvasive, as no substances need to be injected into human subjects. Sometimes, a contrast agent is injected that enhances the visibility of water-rich regions. However, for most brain imaging procedures, no additional contrast is necessary.
- Slices of the brain can be obtained from any angle. CT scanners can only image slices of the brain based on the axis of rotation of the X-ray emitter and detector within the apparatus.
- By varying the gradient and RF pulse parameters, MRI scanners can be used to generate images that highlight certain brain regions and provide contrast between specific kinds of neural tissue.
- Finally, no X-ray radiation is applied to the subject in the scanner.

Even with these relative advantages, MRI has not completely replaced CT imaging. CT imaging is better for visualizing bony or calcified structures in the head, and also remains the imaging technique of choice for subjects who might not be able to enter the high magnetic field (for instance, due to a pacemaker), or subjects with claustrophobia. Additionally, CT scanners have much lower operating costs than MRI scanners, making their use relatively less expensive.

Diffusion Magnetic Resonance Imaging (diffusion MRI)

Diffusion MRI is an application of MRI that is used to examine the structure of axon fiber tracts in the brain. Traditional MRI images present white matter as a homogenous structure (as in Figures 1.9C–D). In reality, fiber tracts originate from various sources, radiate in different orientations, and travel to distinct regions of the brain. Diffusion MRI gives investigators the opportunity to visualize these different white matter pathways and study the complexities of axonal architecture (Figure 1.11).

The term **diffusion** refers to the fact that water molecules, like all other molecules, randomly move through a medium over time. For example, in a glass of water, any particular molecule of water will move around randomly in any

FIGURE 1.11 Diffusion tensor imaging. An example of an image of white matter tracts in the corpus callosum. (Reprinted from Van Hecke, W. et al. (2008). On the construction of an inter-subject diffusion tensor magnetic resonance atlas of the healthy human brain. *Neuroimage* 43(1):69–80, with permission from Elsevier).

direction, limited only by the walls of the container. This type of diffusion is referred to as isotropic—diffusion along all directions. However, water molecules within brain tissue diffuse quite differently due to the physical environment of the brain. They tend to diffuse most rapidly along parallel bundles of fibers with coherent orientations. This type of diffusion is referred to as anisotropic—diffusion that is not equal in all directions but instead tends to move randomly along a single axis.

Diffusion MRI is able to measure the anisotropic diffusion of water along fiber bundles, highlighting the connectivity between brain regions. The physics and mathematics behind this technology are too complex for this text, but the goal is to use MRI technology to analyze the magnitude and direction of the diffusion of water molecules for each voxel of tissue, thus creating a three-dimensional image of fiber tracts. There are various kinds of diffusion MRI methods, the most commonly used called diffusion tensor imaging (DTI).

Diffusion MRI can provide information about which areas of the brain are connected, but it is *not* able to determine the direction of this connectivity (which endpoint is the source and which endpoint is the target). However, it is possible to combine diffusion MRI with functional MRI, and the combination of these methods may reveal information about functional connectivity between brain structures.

FUNCTIONAL BRAIN IMAGING TECHNIQUES

Functional brain imaging techniques are used to measure neural activity in the central nervous system without physically penetrating the skull. The ultimate goal of these techniques is to determine which neural structures are active during certain mental operations. Although these techniques cannot demonstrate that a brain region *causes* certain actions or is *the* specific structure that regulates a cognitive process, they can demonstrate other useful properties. Functional brain imaging techniques can show that activity in specific brain regions is often correlated with a particular stimulus, emotional state, or behavioral task. They can show that information is represented in certain places within the brain, that information may be present in the brain without being consciously represented, and that diseased brains may process information abnormally compared to healthy brains. Thus, these techniques have allowed neuroscientists the opportunity to study the neural basis of cognition, emotion, sensation, and behavior in *humans*, a feat that cannot be achieved by most other techniques in this book.

Functional Magnetic Resonance Imaging (fMRI)

Functional magnetic resonance imaging (fMRI) uses the same physical principles as MRI to produce high-resolution representations of neural activity over time. As with MRI technology, fMRI detects signals from excited hydrogen protons in a magnetic field. Recall that different substances within the brain exhibit different T1 and T2 values following stimulation with an RF pulse. In a structural imaging experiment, these values are used to measure differences in signal intensity between gray matter, white matter, and CSF. fMRI technology takes advantage of the signal intensity of another substance within the brain: **hemoglobin**, the protein in the blood that carries oxygen to cells. On a T2 image, oxyhemoglobin (the oxygen-carrying form of hemoglobin) has a relatively stronger magnetic resonance signal than deoxyhemoglobin (the oxygen-depleted form of hemoglobin). Thus, fMRI allows an investigator the opportunity to examine changes in the oxygenation-state of hemoglobin over time.

The ability to examine changes in oxygen metabolism over time in the brain is useful because it serves as an indirect measure of neural activity. Active neurons will consume more oxygen compared to when they are at rest (Figure 1.12). Initially, this activity decreases the levels of oxyhemoglobin and increases levels of deoxyhemoglobin. However, within seconds, the brain microvasculature responds to this local oxygen depletion by increasing the flow of oxygen-rich blood to the active area. This is referred to as the **blood oxygen level-dependent (BOLD) effect**, and it forms the basis of the fMRI signal. The precise physiological nature of the relationship between neural activity and the BOLD effect is the subject of much current research.

fMRI depends on T2-weighted images because the contrast in signal intensity between deoxyhemoglobin and oxyhemoglobin is greatest on these kinds of images. In a typical experiment, a T2-weighted image of the brain is obtained prior to the presentation of a stimulus. After a stimulus is presented or the subject performs a task, additional T2-weighted images are obtained. As neurons become active due to the stimulus or task, the BOLD effect causes a relative increase in oxyhemoglobin to the microenvironment, and thus an increase in T2-weighted signal. The amount of T2 signal is compared between

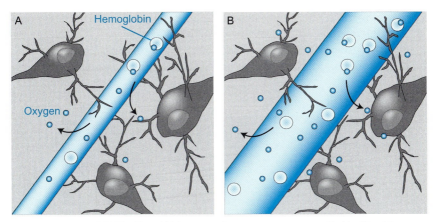

FIGURE 1.12 The BOLD effect. (A) A group of neurons at rest are supplied by blood from capillaries. **(B)** When these neurons become active, they increase their metabolic demand for oxygen. The microvasculature responds by supplying more oxygen-rich blood to the local area. This relative increase in the oxygenated form of hemoglobin causes an increase in the T2 signal.

the prestimulus and poststimulus time points and color coded to depict the signal intensity. This data is typically superimposed over a T1-weighted image that more clearly depicts the underlying anatomy of the brain. The end result is a colorful statistical representation of neural activity superimposed on an anatomical image of the brain—a depiction of the BOLD response over time (Figure 1.13). It is important to understand that fMRI data in a two-dimensional figure is actually *four*-dimensional: the *x, y, z* coordinate planes for each voxel in space, as well as the fourth dimension of time (the "before and after" time points during which the stimulus is presented).

Although fMRI technology provides a powerful tool to study the neural basis of cognition, there are significant limitations. One of the biggest challenges of fMRI research is that the actual T2 signal change for a given voxel of brain space, before and after BOLD changes, can be as low as 0.2%. This change is very difficult to detect, especially given that the noise of the system can be as high as 0.3–0.4%. Therefore, fMRI stimuli must be repeated several times for a single subject, and a series of statistical tests must confirm the presence of a reproducible signal. Another significant limitation is the **temporal delay**: it can take 6–10 seconds after the presentation of a stimulus for oxygenated blood to flow to an active region, so there can be a long time delay between the stimulus or task and the measurement of neural activity. The **temporal resolution**, the ability to resolve neural activity into discrete events, is about 4–8 seconds, relatively poor compared to other techniques such as electroencephalography or magnetoencephalography (see below for descriptions of EEG and MEG). Finally, fMRI cannot identify the neurochemistry of neural events, such as which neurotransmitters or neuromodulators mediate a change in neural activity. Inferences about the neurochemical make-up of neural activity must be based on prior knowledge of brain anatomy or other forms of whole brain imaging, such as PET (see the following section).

Even with these relative disadvantages, fMRI remains a powerful technique for correlating neural activity with mental operations. A well-designed

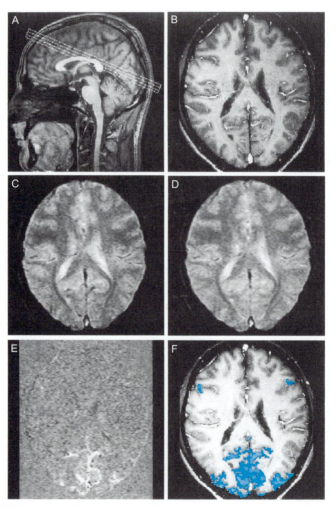

FIGURE 1.13 A simple fMRI experiment. This experiment examines the difference in BOLD signal intensity between subjects exposed to light or no light. **(A)** First, the investigators select a slice to examine. **(B)** A T1-weighted image is obtained that provides structural data. T2-weighted images are used for the actual data collection in the conditions of **(C)** light and **(D)** no light. **(E)** The data analysis compares BOLD signal intensity between **(C)** and **(D)**. **(F)** This result is superimposed on the structural data to produce an image suitable for publication. (A–F: Reprinted with kind permission of Springer Science + Business Media and Jens Frahm from Windhorst, U. and Johansson, H. (eds.), 1999. *Modern Techniques in Neuroscience Research,* Ch. 38: Magnetic Resonance Imaging of Human Brain Function, p. 1064, Fig. 5.)

experiment can reveal much about the human brain and allow an investigator to noninvasively examine the physiology of cognition. See the last part of this chapter for a thorough discussion of the design of a functional imaging experiment.

BOX 1.1 fMRI Experiments in Animals

The main motivation to develop functional brain imaging technology was to non-invasiveley study neural activity in humans. However, this technology can also be used with other animals. Functional imaging studies using animal subjects have been used for examining the physiological basis of fMRI, investigating animal models of neurological disorders, and exploring the basic mechanisms of perception, behavior, and cognition. Indeed, rat and mouse subjects pioneered the early development of fMRI technology. Dogs, cats, and even songbirds have also been utilized in fMRI experiments. Anesthetized and awake nonhuman primates, such as macaque monkeys, have been used as research subjects in fMRI experiments since the late 1990s.

The major benefits of using animal subjects include the ability to validate the use of animal models and to bridge human fMRI experiments and animal electro-physiology experiments. Studies using these two different techniques complement each other, providing knowledge about a specific field of neuroscience, such as the physiology of the visual system. Being able to use both fMRI and electrophysiology in the same animal (and even at the same time) greatly aids our understanding of the functional activity of individual neurons and entire brain regions in the same experiment. Furthermore, fMRI studies in primates that screen neural activity across the entire brain can inform future electrophysiological studies about brain regions containing neurons of interest to an investigator.

However, fMRI studies using primates present additional limitations and challenges compared to traditional human studies. The horizontal position of most MRI scanners is not ideal for primate studies, so vertical scanners have been created to accommodate the special chambers that support and brace a conscious primate. The animal's head must be fixed in place with a head post so there is no head movement during an experiment. The animal must also be acclimated to these experimental conditions so it is comfortable in its environment and can perform the task. A further challenge is that during the actual experimental sessions, a primate may lose motivation to perform a task or attend to a stimulus. Unlike an electrophysiology experiment in which an animal occasionally receives a juice reward (see Chapter 2), juice must be consistently provided so that the animal stays attentive and completes the scanning session.

MRI scanners with higher magnetic field strengths allow high-resolution, detailed imaging in small animals such as rats and mice. Obviously, these animals cannot perform complicated cognitive tasks and most likely need to be anesthetized during an imaging session. However, the benefit of doing fMRI in rodents is the possibility of injecting psychoactive drugs during an imaging session or performing lesion studies. Thus, it is possible to study the global effects of a pharmacological agent, such as a receptor antagonist, or the loss of a brain region over time. Furthermore, a scientist can follow up functional imaging data with histological studies after the completion of an experiment.

Positron Emission Tomography (PET)

Positron emission tomography (PET) provides a representation of neural activity but no information about brain structure. This technology was developed in the 1970s and 1980s as a novel method of functional imaging, but has largely been superseded by fMRI technology for most cognitive experiments. In a PET experiment, an unstable **positron-emitting isotope** is injected into a subject's carotid artery (a neck artery that feeds the ipsilateral cerebral hemisphere). As the isotope decays, it emits a **positron**, an antimatter counterpart of an electron. When a positron comes into contact with an electron, an annihilation event occurs, resulting in a pair of gamma photons that move in opposite directions (Figure 1.14A). These photons pass through the body and can be measured by a gamma-detecting device that circles the subject's head. The detector identifies a pair of gamma photons that arrive at opposite sides of the subject's head at the same time (within a few nanoseconds; Figure 1.14B). As the detector rotates around the subject's head, these signals can be used to derive the source of the annihilation events within the subject (Figure 1.14C).

A PET experiment can use a variety of positron-emitting isotopes. One of the most commonly used is **fluorodeoxyglucose (FDG)**, a radioactive form of glucose. As metabolically active neurons require an increase in the uptake of glucose from the blood, the presence of FDG can be used as an indirect marker of neural activity. In addition to FDG, radioactive water can be injected into the brain's circulatory system. Because there is an increase in blood flow to active areas of the brain, the PET scan will indicate the areas in which blood flow is increased during activity.

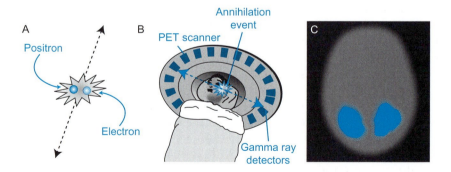

FIGURE 1.14 PET imaging. (A) An unstable positron-emitting isotope will decay over time and emit a positron. When a positron comes into contact with an electron, an annihilation event occurs and a pair of gamma photons are released. **(B)** If this unstable isotope is injected into a subject, the annihilation event can be detected by a PET scanner. This scanner consists of a series of gamma ray detectors arranged around the subject's head. **(C)** Unstable isotopes for many metabolic substances, such as glucose or neuropeptide metabolic proteins, can be imaged in a PET experiment. The increase in signal over time can map sites of metabolic activity for a specific neural stimulus or task.

A particularly useful aspect of PET imaging is the ability to use positron-emitting isotopes that can bind to specific receptors in the brain. For example, a radioactive ligand that binds to serotonin receptors can indicate the locations and binding potential of these receptors in the brain, providing information about the relative metabolism of serotonin in human subjects. The ability to image the metabolism of specific bioactive molecules makes PET imaging unique among functional imaging techniques and provides a utility that fMRI cannot.

There are a number of limitations to PET. Compared to fMRI, PET has about the same temporal resolution (4–8 seconds), a lower spatial resolution, and cannot generate anatomical data. PET images are often combined with CT or MRI images to present functional data in an anatomical context. The cost of doing PET is very expensive. Because most isotopes used for PET studies have extremely short half-lives, positron-emitting isotopes must be synthesized on site using a room-sized device called a **cyclotron**, which itself is expensive to purchase and maintain. The half-life of FDG is 110 minutes, so this compound cannot be ordered and delivered from a remote location. Finally, PET requires the injection of these radioactive substances into subjects. Therefore, multiple PET-imaging sessions are not recommended for a single subject. Because of these limitations, most modern functional imaging experiments utilize fMRI technology instead of PET, with the exception of those experiments that focus on the metabolism of bioactive substances, such as neurotransmitters.

Single-Proton Emission Computerized Tomography (SPECT)

Single-proton emission computerized tomography (SPECT) imaging is very similar to PET imaging, producing functional images of neural activity but no structural data. Like PET, a radioactive probe is injected (or inhaled) into the circulatory system. The probes bind red blood cells to be carried throughout the body. Because blood flow is increased in active brain structures, the radioactive signal is used to assess an increase in neural metabolism. As the label undergoes radioactive decay, it emits high-energy photons that can be detected using a gamma camera. The camera is rapidly moved around the head of a subject to collect photons from many different angles, permitting a three-dimensional reconstruction.

Although SPECT is very similar to PET in concept, it is not as costly because the radiolabeled probes are usually commercially available and do not require an on-site cyclotron. Therefore, SPECT can be thought of as a cheaper alternative to PET. The disadvantages to using SPECT as an imaging technique are similar to the disadvantages of using PET: the technology has a relatively low spatial resolution compared to fMRI (about 8 mm), and radioactive substances must be injected into a subject.

Electroencephalography (EEG)

Electroencephalography (EEG) is a measure of the gross electrical activity of the brain. It is not truly a brain *imaging* technique, as no meaningful images of the brain can be produced using this technique alone. However, EEG is noninvasive and can be used to ascertain particular states of consciousness with a temporal resolution of milliseconds. EEG combined with other imaging techniques, such as fMRI, can provide an excellent temporal and spatial representation of neural activity.

To produce an electroencephalogram, several disk-shaped electrodes, about half the size of a dime, are placed on the scalp (Figure 1.15). The scalp EEG reflects the sum of electrical events throughout the head. These events include action potentials and postsynaptic potentials, as well as electrical signals from scalp muscles and skin. An interesting way to think about EEG technology is to imagine what it would be like to place a microphone above a large crowd of people, such as above Times Square during New Year's Eve. In this analogy, it is impossible to make out the signal from an individual person in the crowd. However, it is possible to determine when a meaningful event occurs, such as when the crowd cheers at midnight. Likewise, it is impossible to record the electrical activity from a single neuron with EEG, but it is possible to ascertain when a meaningful event occurs in the brain, such as when a subject detects a salient stimulus in an experiment. Thus, combining the temporal resolution of EEG with the spatial resolution of fMRI provides a powerful method to detect the precise timing and location of neural activity within the brain.

Magnetoencephalography (MEG)

Magnetoencephalography (MEG) measures changes in magnetic fields on the surface of the scalp that are produced by changes in underlying patterns

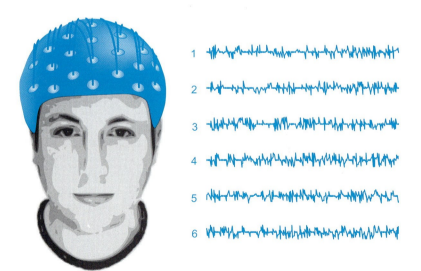

FIGURE 1.15 A human subject wearing electrodes for EEG recordings. The electrodes measure the global electrical activity of billions of neurons in the brain. Each electrode placed on the scalp records a unique trace of activity based on its location on the scalp.

of neural electrical activity (Figure 1.16). About 50,000 neurons are required to produce a detectable signal with MEG, a number that may seem large but is actually much smaller than what is required for an EEG signal, which may require millions of neurons. MEG offers relatively poor spatial resolution but excellent temporal resolution compared with PET, SPECT, and fMRI. Therefore, MEG can be thought of as a compromise technique: it offers excellent temporal resolution and much better spatial resolution compared with EEG, but not as good spatial resolution as other imaging techniques. In particular, MEG in combination with fMRI allows for excellent temporal and spatial resolution of neural activity. Unfortunately, MEG is a very expensive technique, requiring a room that can obstruct magnetic fields from outside sources; even something as small as a coffee machine in a neighboring building can be detected if the room is not adequately insulated.

Optical Imaging

Optical imaging techniques produce images of neural activity by measuring changes in blood flow and metabolism from the surface of the brain. Rather than detecting changes in the magnetic or electrical properties of neurons, as in fMRI, EEG, or MEG, optical imaging detects changes in light reflectance from the surface of the brain due to changes in the amount of blood flowing to neural tissue. In animal preparations or during human surgery, light is shined on an exposed portion of the brain. Any light that is reflected off the surface is detected by a sensitive camera and recorded by a computer. When neurons are more active, changes in the blood volume, blood oxygenation, and the

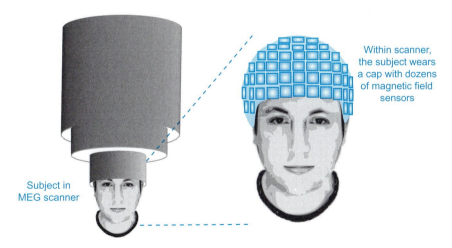

Within scanner, the subject wears a cap with dozens of magnetic field sensors

Subject in MEG scanner

FIGURE 1.16 A magnetoencephalography setup. MEG offers excellent temporal resolution and better spatial resolution than EEG, although not as good spatial resolution as PET or fMRI. MEG equipment is expensive and requires a room with strong insulation.

light-scattering properties of neural tissue (resulting from ion and chemical movements) all cause small (0.1–3.0%) changes in the reflectance of light from the brain's surface. For each experiment, the investigator images a baseline amount of light reflectance, and then compares this baseline parameter to changes in light reflectance that arise due to the presentation of a stimulus. Optical imaging technologies allow for spatial resolutions of <1 mm and temporal resolutions of 2–8 seconds. These technologies have been used to produce high-resolution functional maps of visual cortex in both animals and humans.

Earlier in this chapter, we defined functional imaging techniques as methods that allowed investigators to measure changes in neural activity over time without physically penetrating the skull. Obviously, many optical imaging techniques are an exception to this definition, as the brain surface must be exposed to allow light to penetrate and reflect back to a camera. However, **diffuse optical imaging (DOI)** and **near-infrared spectroscopy (NIRS)** are noninvasive alternatives that utilize the same basic principles as invasive optical imaging but record light reflectance through the scalp (Figure 1.17). The signal is much weaker than invasive optical imaging, as light must pass through the superficial layers of the head to the brain, and then from the brain to optical electrodes, known as **optrodes** or **optodes**, placed on the surface of the scalp. However, these techniques are sensitive enough to detect large changes in neural activity and can be useful in clinical applications as an alternative to fMRI or PET because of their low cost and portability. For example, long-term monitoring of

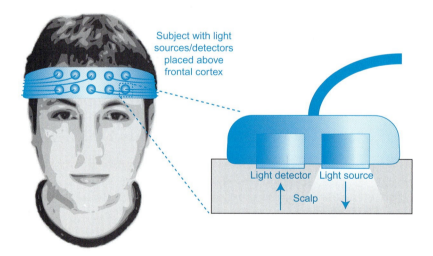

FIGURE 1.17 Optical imaging. Light is shined onto the surface of the brain. A portion of this light is reflected off the brain and detected by multiple optrodes. Changes in neural activity produce changes in the amount of light that is absorbed and reflected by the brain. Therefore, optical imaging can be used to indirectly detect changes in neural activity. These techniques can be either invasive (the skull is opened to reveal the surface of the brain) or noninvasive.

cerebral oxygenation in a patient following a stroke is a potential use of DOI that would be practically impossible with fMRI or PET.

Table 1.1 compares the spatial and temporal resolutions, cost, and invasiveness of the various functional brain imaging techniques described previously. As mentioned before, many imaging laboratories now combine multiple techniques to make up for the inadequacies of any single technique. For example, fMRI offers excellent spatial resolution and is noninvasive, but it does not offer good temporal resolution of neural activity during experiments since the BOLD effect occurs over a period of 6–10 seconds. Therefore, some laboratories combine fMRI with MEG or EEG, in which signals can be detected within a fraction of a second.

So far, this chapter has focused almost exclusively on the technology of structural and functional brain imaging. Of course, the utility of these technologies depends entirely on how they are used: the research hypotheses, experimental designs, and methods of data analysis. In the last part of this chapter, we will examine the approaches that can be taken in the design and analysis of functional imaging experiments.

FUNCTIONAL IMAGING EXPERIMENTAL DESIGN AND ANALYSIS

Functional imaging technologies are distinct from other techniques described in this book in their utility in human **cognitive neuroscience**, the field of neuroscience dedicated to elucidating the neural basis of thought and perception. Cognitive operations routinely studied in the literature include attention, learning and memory, executive function, language, emotion, and higher-order sensory processing, such as the enjoyment of art or music. Determining the neural basis of cognitive processes, especially processes that seem unique to humans, absolutely depends on brain imaging techniques.

TABLE 1.1 Comparison of Functional Imaging Techniques

	Spatial Resolution	Temporal Resolution	Cost	Invasiveness
fMRI	<1 mm	2–8 s	Expensive	Noninvasive
PET	~4 mm	1 min	Very expensive	Radioactive injection
SPECT	~8 mm	2–8 s	Expensive	Radioactive injection
EEG	~1 cm	~1 ms	Moderate	Noninvasive
MEG	~1 mm	~1 ms	Very expensive	Noninvasive
Optical imaging	<1 mm	10–100 ms	Inexpensive	Can be invasive

In addition to answering scientific questions about cognitive neuroscience, fMRI has added important information to other neuroscience fields, such as sensory or motor **systems neuroscience**. These studies identify regions of the human brain that are active in the presence of certain sensory stimuli or motor actions. Some studies reproduce findings in humans that were previously demonstrated in other animals, as well as add important insights that enhance our understanding of these systems. For example, fMRI research has demonstrated that visual pathways in humans are anatomically consistent with visual pathways found in other mammals. However, our understanding of the visual system has been greatly enhanced by the fact that human subjects can report experiences that other animals cannot. For example, activity in visual cortex increases not only when human subjects are exposed to visual stimuli, but also when subjects are told to *imagine* visual stimuli. Furthermore, fMRI research has demonstrated distinctions in how different types of visual scenes are represented in distinct regions of cortex. Thus, fMRI contributes meaningfully to our understanding of both cognitive as well as systems neuroscience.

Unfortunately, the conception and design of functional imaging experiments are not always well understood by scientists who do not use these methods. This is probably because these technologies are unlike most other techniques used in neuroscience, as well as the fact that they are usually only performed by the dedicated researchers who use them regularly. A common complaint among brain imaging specialists is the misconception that you can simply "throw a human subject into a scanner, tell him or her to look at some stimulus, and then publish the results." Like any other technique, whole brain imaging experiments must be carefully designed and interpreted, sometimes more than nonspecialists may appreciate. The next section describes some of the scientific considerations in the design and execution of a functional imaging experiment. We will focus on fMRI, as this technique dominates the field, but the same principles can be applied to other functional imaging techniques as well.

Planning the Experiment

Before any human subject is placed in a scanner, an investigator conceives and designs an experiment months or even years in advance. There are many practical considerations that must be taken into account in the design of the experiment that may affect the ambitions of the investigator. Once these limitations are understood, the investigator designs the experiment to answer a specific question or test a specific hypothesis. Finally, the investigator designs a proper task paradigm and tests efficacy of the stimuli so that the results are appropriate and accurate. We review these considerations below.

Practical Considerations

Perhaps the most important practical consideration of an fMRI experiment is its cost. The actual scanning machine, as well as its support apparatus, costs millions of dollars and is almost always shared by multiple labs. Routine maintenance of the scanner is also expensive, with specialized technicians on call to fix any potential physical problems that may arise. Most research institutions charge an individual lab based on the number of hours the scanner is used. These institutions may also charge money based on the time of day, with scanner use during the day costing more than scanner use late at night. It is not unusual for scanner time to cost between $100 and $1000 per hour, depending on the institution's operating costs. Therefore, an fMRI experiment must be well designed so that scanner time is not wasted. fMRI is too expensive to simply "play around with" and put humans in the scanner for no defined purpose. Finally, scanner time is usually reserved well in advance due to the number of scientists who wish to perform experiments.

Another important practical concern of fMRI is that a subject must keep his or her head completely still during a scanning session. Even small movements can disrupt the magnetic field, causing artifacts in the images. Also, if a subject moves during a scan, the acquired images will not line up with each other, making them uninterpretable. This means the subject will be unable to perform certain activities while inside the scanner, such as speaking, exercising, or moving facial muscles. However, subjects can be trained to perform simple hand movements to complete a task or respond to a stimulus, such as pressing a button on a hand device. Some specialized devices have been invented to deliver olfactory or gustatory stimuli to subjects.

A practical concern for an investigator who wishes to study sleep, attention, or the auditory system is that an fMRI scanner can be very noisy. The RF pulse sequences are loud and last the entire duration of the experiment. Therefore, many subjects do not sleep normally in a scanner, and the changes in attention due to a surprising stimulus may actually be due to the surprise of a noisy scanner turning on and off. Auditory stimuli can also be very difficult to work with, for obvious reasons. These limitations are not insurmountable, but require extra planning by the scientist to make sure that a sporadically loud environment does not bias the results.

Structure of a Functional Imaging Experiment

Like any other experiment in neuroscience, functional imaging experiments examine the effect of an **independent variable** on a **dependent variable** (see the Introduction for a more comprehensive distinction). The independent variable is the experimental variable that is intentionally manipulated by the researcher and is hypothesized to cause a change in the dependent variable. In a functional imaging experiment, the independent variable can be a stimulus, task, or even a difference in the subjects being tested such as age, gender, or disease state. The dependent variable is the quantifiable variable measured by the researcher to determine the effect of the independent variable. This variable is different for each brain imaging technique. In the case of fMRI research, the dependent variable is the BOLD signal intensity for a particular part of the brain. In the case of PET or SPECT technology, the dependent variable is the intensity of gamma radiation. All of these dependent variables are used as proxies for neural activity. Although the end result of a functional imaging experiment is usually a pretty picture, it is important to realize that the overall goal of functional imaging experiments is to test an independent variable on signal intensity.

The specific hypotheses for a functional imaging experiment are usually framed in the following manner:

- Activity in brain region X is correlated with stimulus/task Y.
- For a given stimulus or task, activity in brain region X precedes activity in brain region Y.
- Activity in brain region X is higher under condition Y than condition Z.
- Activity in brain region X is higher in human population Y than human population Z.
- Activity in brain region X changes across time as a subject learns a task.

These hypotheses depend upon the investigator knowing of the existence of "brain region X" and having a good rationale for studying that region. fMRI can also be used to screen the entire brain to find brain regions that increase neural activity in response to a stimulus/condition in the first place. However, the amount of data generated during an fMRI experiment is quite large, and processing fMRI data can be easier if the investigator has an idea of where to look.

It is also important to recognize whether a functional imaging study is **between subjects** or **within subjects**. In a study that is between subjects, the independent variable is the different populations of subjects, such as male vs. female, old vs. young, healthy vs. diseased, genotype A vs. genotype B, and so on. These differences in subjects can potentially create problems in analysis. For example, it is known that the BOLD response can change with age; therefore, comparing a younger group of subjects with an older group of subjects using fMRI may require additional controls to ensure that changes in brain activation are not simply due to changes in the BOLD effect. In a study that is within subjects, each subject participates in all experimental conditions and the independent variable is something other than the identity of the particular subject.

Task Paradigms

After developing a specific research hypothesis, an investigator determines an appropriate **task paradigm**, a strategy for presenting stimuli to subjects during an experiment. These paradigms are usually divided into two major categories: **blocked designs** or **event-related designs**.

In a blocked design, a subject is presented with two or more categories of stimuli that alternate every 1–2 minutes. For example, consider an experiment that features two categories of stimuli, X and Y. Category X might represent something like "smiling faces" while category Y could represent "neutral faces." In the first block of stimuli, the subject is presented with different images made up entirely of category X: X1, X2, X3, X4, and so on. A typical block may last between 10–60 seconds, with each individual stimulus presented from 1–10 seconds. After the end of the block, there may or may not be a brief period when no stimuli are presented. Then, a second block of stimuli are presented made up entirely of category Y: Y1, Y2, Y3, Y4, and so on. These blocks of stimuli are repeated in cycles, alternating X and Y, until a sufficient amount of data is collected from a subject. Some blocked designs contain a third category, alternating between blocks of X, Y, and Z.

Blocked designs are useful for obtaining relatively high signal-to-noise information from active brain regions. Because stimuli of the same experimental category are presented repeatedly for 1–2 minutes, the hemodynamic BOLD response has sufficient time to increase and a strong signal is produced. These paradigms are often chosen for experiments involving the determination of which brain regions are active for a given stimulus or task. They are also good for neural processes that last a relatively long time, such as changes in cognitive or emotional state. A blocked design is not as useful for examining a change in brain activity during relatively short events that last 2–5 seconds, such as when a subject is required to make a decision inside a scanner.

In an event-related design, stimuli are presented as isolated, individual events of short duration. This paradigm is useful for discrete tasks in which a subject must recognize an event, detect a novel stimulus, or make a decision, all processes that take place in short time intervals. The hemodynamic properties of the BOLD effect may return to baseline between events, or for fast-paced stimuli, the BOLD response may not return to baseline and so the effects of two stimuli on neural activity can be examined within a small time window. For example, stimulus A may indicate the nature of stimulus B, and the BOLD signal can be measured for B with and without the presence of A. Event-related designs are not usually used in experiments that affect a subject's emotional state over time, as these states usually last over relatively longer time periods.

Blocked designs and event-related designs are not mutually exclusive, and indeed some studies take advantage of both designs and employ what is referred to as a **mixed design**. The relative advantages and disadvantages of all three task paradigms are compared in Table 1.2.

Pilot Experiments

Because MRI scanner time is so expensive, it is often necessary to test the delivery of stimuli or characterize the behavior of subjects during a task *before* any imaging experiments take place. These pilot experiments take place outside of the MRI facility and demonstrate the efficacy of stimulus delivery in producing an appropriate human response. Statistical analysis of how subjects perform their tasks is sometimes necessary to gain approval before the actual experiments take place in the scanner. Only after the investigator characterizes a cognitive task in terms of the efficacy of the stimuli and behavior of human subjects will it be appropriate to begin to associate neural correlates with that behavior.

TABLE 1.2 Relative Advantages and Disadvantages of Different fMRI Task Paradigms

	Advantages	Disadvantages
Blocked	• More statistical power for detecting subtle differences across different conditions • Tend to be simpler to implement and analyze • Good for examining state changes	• Grouping and predictability of stimuli may confound results • Information about the time course of the activation response is lost within a block • Not applicable to certain types of tasks (e.g., novelty)
Event-related	• Reduce confounds of predictable stimulus order, since stimuli can be presented randomly • Can sort trials after the experiment according to specific behavioral outcomes • Useful for examining temporal characteristics of responses • Flexible analysis strategies	• More complex design and analyses than blocked design • Lower signal-to-noise ratio than block designs • Must perform longer scan runs to compensate for loss in statistical power
Mixed	• Can compare short-term transient activity with long-term sustained activity	• Most complicated analyses

Planning an Experiment that Uses Human Subjects

Once an experiment is designed, a lab must be given permission to use humans as research subjects by an **Institutional Review Board (IRB)**. An IRB is a committee of about 10 individuals, typically composed of physicians and research faculty, as well as non-faculty members such as a nurse, minister, graduate student, or lawyer. The purpose of the IRB is to review studies that use human subjects and determine if the study meets ethical, safety, and scientific obligations. An IRB typically meets anywhere from 2–12 times a year to review and vote on the submitted applications. Additionally, any research project involving human subjects must meet the standards of the Health Insurance Portability and Accountability Act's (HIPAA) "Privacy Rule," which regulates the protection, security, and confidentiality of private health information.

Conducting the Experiment

After an experiment is designed, discussed, and approved, a scientist proceeds with recruiting human subjects and collecting data in the scanner. These scientists must be familiar with the scanner technology and experimental procedures before any actual data collection begins. The difficulty in recruiting human subjects and the high cost of scanner time makes it essential that experiments are conducted properly and without any mistakes.

Working with Human Subjects

Though the exact number varies based on the experimental design, usually about 10–20 subjects are necessary for a typical imaging experiment to reach an appropriate, statistically significant conclusion. Recruiting human subjects can be relatively easy or difficult depending on the type of subjects needed. If the only requirement for a subject is that he or she is an average, healthy adult, then recruiting subjects can be relatively straightforward. At academic institutions, undergraduates and graduate students are often targeted for recruitment, with compensation given in the form of extra-credit for a course or cash payment. When subjects must represent a particular population of individuals (e.g., individuals with posttraumatic stress disorder, elderly subjects, or individuals diagnosed with depression) it can be much harder to recruit a necessary number of subjects. Consider a study in which the hypothesis is that depressed individuals have a significantly different level of activity in a particular brain region than nondepressed individuals. In this case, the investigator must find individuals who have been formally diagnosed with depression. Additionally, because psychotropic drugs can confound the results, it is necessary that none of these subjects received medication after their diagnosis. Therefore, the investigator must identify 10–20 subjects who have been diagnosed with depression but chose not to undergo pharmacological treatment (or who have not yet started treatment). In studies like this, recruiting subjects can be the bottleneck in the time it takes to complete an experiment. Investigators might need to formally recruit individuals from other institutions, such as nursing homes, hospitals, or secondary schools. It might be necessary to formally collaborate with clinical neurologists to recruit patients with various mental health disorders.

Once a subject is recruited, the investigator schedules a specific session time in the scanner. Prior to the start of an experiment, the investigator must inform the subject about the nature of the experiment (although details of the experiment can be withheld for experimental purposes). The subject must sign a release form to indicate that he or she is a willing participant in the study and that any data gathered can be used for publication. Finally, the subject must complete a brief survey about his or her personal health in order to ensure that there are no confounding variables such as history of mental health disease, cardiovascular disorders, or heavy use of alcohol or other drugs. The subject is also requested to remove any metal objects before entering the scanner, such as loose change, belt buckles, earrings, and so on. Often the investigators ask the subjects to change into a disposable gown to ensure that there are no metal objects placed inside the scanner.

During the experiment, the investigator must ensure that the subject is as comfortable as possible inside the scanner. To reduce head movement, the subject's head is surrounded by soft padding to hold it firmly in place inside a head coil. The investigator provides the subject with a squeeze ball that, when pressed, can stop the scans at any time if the subject becomes uncomfortable. For experiments that require the subject to make a choice or respond to a stimulus, the investigator also provides a nonmetallic handheld device with buttons that the subject can press to provide feedback. Visual stimuli can be projected through goggles or to a mirror inside the scanner to allow for video presentations from a nonmetallic surface.

After the experiment is complete, an investigator usually debriefs the subject on the experiment and pays the subject for his or her time. Depending on the experiment, follow-up sessions can be necessary and these are also scheduled well in advance.

Data Acquisition

In a typical experiment, it is common for about 10–20 human subjects to participate (twice that for a between-subjects study). Each subject usually participates in 1–3 **sessions**, the scheduled time when an actual experiment is conducted. Each session contains many **runs** in which the brain is repeatedly scanned and the hypothesis is tested. Within each run, functional data is accumulated and stored as **volumes**—three-dimensional constructions of brain space. A complete volume is typically acquired every 1–3 seconds. Each volume contains a complete set of **slices**, two-dimensional representations of a plane of the brain that are only 1 voxel thick. A **voxel** is the smallest functional unit of brain space that can be analyzed for changes in signal intensity over time. In fMRI, a voxel is usually 1–5 mm^3. Remember that a voxel of fMRI data is essentially four-dimensional: three dimensions are the location of the voxel in space (x, y, z) and the fourth dimension is the time in which the signal intensity was recorded. Thus, an fMRI experiment generates huge quantities of data: 10–20 subjects, each with 1–3 sessions, each with several experimental runs, each with several volumes of brain space, each with several brain slices, each with a grid of voxels that represent the data! Therefore, computers with ample storage space are required to record and store data until the results can be analyzed after the collection is complete.

Manipulating Neural Activity During an Experiment

One of the limitations of studying the human nervous system is that it is very difficult or even impossible to functionally perturb the brain during an experiment. For example, if activity in a specific part of the brain is correlated with performance on a task, it would be interesting to determine if a loss of that part of the brain caused a deficit in the ability of the subject to perform the task. While such a loss-of-function experiment could be performed on another model organism with a lesion study, this is, of course, impossible in humans. Likewise, it would also be interesting to stimulate neural activity in a particular brain

BOX 1.2 Walkthrough of a Simple fMRI Experiment

Let's consider a hypothetical situation: Suppose you work in a laboratory that uses fMRI to study cognition in humans. Your research group becomes interested in the neural basis of how a person recognizes and identifies familiar people. One way to frame this question is "Are there specific brain regions that are active in response to familiar individuals versus nonfamiliar individuals?"

How would you go about designing an fMRI study to answer this question? A good starting point would be to identify the independent and dependent variables in this experiment. The independent variable, the variable that varies from trial to trial, is a person who is familiar or unfamiliar to the subject. This immediately leads to another decision to make as an investigator: the choice of stimuli. Will the stimulus be a picture of an individual? The sound of the individual's voice? The individual's name? In this example, let's make the independent variable a person's name that is visually displayed on a screen. Therefore, in each trial, the subject will see a name on a screen. Because the independent variable is the stimulus and not the identity of the subject, this is a "within subjects" study.

The dependent variable is the change in BOLD signal intensity in the brain. We might hypothesize that a specific brain region will show a significant difference in signal intensity between the two conditions, but in this experiment such a specific hypothesis is unnecessary, as we can scan the entire brain and identify all regions that significantly differ between the two stimulus conditions.

What will be the subject's task during the experiment? If all we ask the subjects to do is stare at the screen, they may become bored, distracted, and even fall asleep. Therefore, we must give the subjects a specific task so that they fully attend to the stimuli. In this example, a good task may be to indicate whether the name on the screen is a man's or a woman's. Each subject could be presented with a handheld device with two buttons, one for man and one for woman. At the end of the imaging session, we can measure the accuracy of the responses to ensure that the subject was fully attentive throughout the experiment.

Obviously, we will need to generate a list of names for both the "familiar" and "unfamiliar" categories of stimuli. Each list should contain a balanced list of males and females, old and young, and so forth, to ensure that there are no confounding variables in the study. Familiar names could include famous celebrities and politicians, while the unfamiliar names could include combinations of first and last names that are not well-established celebrities. How can we ensure that the subject knows all of the "familiar" names and none of the "unfamiliar" names? This is impossible to confirm before the experiment begins, but after the experiment is over, each subject could complete a brief survey in which he or she indicates the level of familiarity of each name, perhaps on a scale of 1 to 10. This survey can then be utilized when analyzing the functional imaging data to ensure that each stimulus is adequately categorized.

The next major decision will be the best task paradigm to present stimuli to subjects. A blocked design will allow us to present multiple stimuli of the same category (familiar or unfamiliar names) to a subject over a 1- or 2-minute period, a condition that may allow for maximal signal intensity in active brain regions. However, we may wish to evaluate the degree of signal intensity for each name presented to the subject. In this case, we may wish to pursue an event-related design, which will allow us to correlate the degree of signal intensity with the degree of familiarity with a famous name, as indicated by the subject on a survey after the experiment is over.

Once the IRB approves the study and our proposal to use human subjects, we are ready to reserve scanner time and start collecting data! We will need 10–20 subjects for this experiment, but we should recruit at least 20–40 because some of the data will need to be discarded. For example, some subjects may fail to show up, some may fall asleep in the scanner, or some may turn out to have medical conditions or treatments that may preclude them from participation in the study. In order to recruit subjects, our lab could offer $25/hour for their time. After the imaging session ends, we will provide them with the familiarity survey and then start to analyze the data.

For each subject, we morph the structural coordinates from their brain scans to match a normalized template system, such as the MNI template. Then we can proceed with a voxelwise comparison in signal intensity between the "familiar name" trials and the "unfamiliar name" trials. Individual brains can then be pooled across all subjects for each condition. The final figure will be the difference in signal intensity between the two conditions, voxel-for-voxel, superimposed on a T1 image from the MNI template. Because the template contains coordinates for discrete brain regions, we can interpret the specific regions activated in response to familiar or unfamiliar names.

region to determine if performance improves on a task. Such gain-of-function experiments are also possible in model organisms, with stimulating microelectrodes or over-expressed genes driving activity in a specific population of neurons. Again, this is impossible in typical, noninvasive fMRI experiments.

However, an emerging technology called **transcranial magnetic stimulation (TMS)** is making it possible to perform loss- and gain-of-function experiments in humans (Figure 1.18). TMS attempts to reversibly activate or inactivate regions of the human brain. In this technique, a coil that generates magnetic field pulses is positioned near a subject's head. The changing magnetic fields induce weak electric currents at a specific focal point on the surface of the brain. Depending on the brain region and the strength of the magnetic field impulse, the electrical activity may either stimulate neural activity, or cause a hyperpolarized state in which neural activity is temporarily inactivated. Therefore, it is theoretically possible to perform a loss- or gain-of-function experiment using this technology. One of the limitations of this technology is that TMS can only induce electrical activity on the outer surface of the brain—it is currently impossible to stimulate or inactivate deep brain structures that may be of interest to an investigator. However, there are thousands of publications that have used TMS to selectively activate and inactivate activity in the human

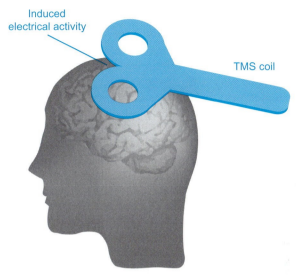

Induced electrical activity

TMS coil

FIGURE 1.18 Transcranial magnetic stimulation. A coil is held near a person's head, and a magnetic field stimulates electrical activity in a superficial brain region. Depending on the strength of the magnetic field and the brain region, this electrical activity can cause either a temporary loss-of-function or gain-of-function in that brain region.

cerebral cortex. This technology can also potentially be used for therapeutic treatment of diseases and is being heavily researched in the treatment of depression and motor diseases.

An additional strategy for investigators to perform loss-of-function experiments is to attempt to find patients with lesions (caused by an accident or stroke) to specific brain regions. Because many institutions that utilize fMRI scanners are associated with hospitals, identifying and recruiting patients with lesions can be coordinated between primary researchers and neurologists. However, these patients may present lesions that are not tightly localized to the brain area of interest, there may be compensatory mechanisms that develop over time, and it may be more difficult for these patients to perform specific tasks. All of these caveats must be considered before including these patients in experiments, as well as when evaluating these experiments in the literature.

Postexperimental Data Analysis

The end result of most fMRI experiments is a colorful figure depicting the activation or inactivation of a particular region of the brain that is correlated with a stimulus or task. These figures can seem deceptively simple to produce, yet require rigorous methods of data analysis and interpretation. The process of converting the raw data of many individual subjects to a meaningful figure is described below.

Data Analysis

What is an investigator to do with the huge quantity of data acquired from multiple research subjects? In most analyses, the data from each subject is analyzed separately first and then pooled together into a common subject pool. Because there is variability in every subject's brain structure, data must first be fit into a common three-dimensional brain template. Structural data from each individual's brain is therefore stretched and warped to fit specific anatomical landmarks in a precise mapping system. One of the most widely used coordinate systems for normalizing fMRI data is called **Talairach space**, based on the stereotaxic measurements of a single postmortem brain. This coordinate system is ubiquitous in brain imaging research as it defines brain regions and **Brodmann's areas** into stereotaxic coordinates, allowing investigators to make anatomical comparisons among different brains. Another commonly used coordinate system is called the **MNI template** (MNI stands for Montreal Neurological Institute, the institution that established this system). This template is a probabilistic mapping system based on the averages of hundreds of individual brain scans and scaled to match the landmarks within the Talairach atlas.

After an individual subject's brain is adjusted to either the Talairach or MNI templates, signal intensity is compared at different time points according to the task paradigm. There are two general strategies of analyzing signal intensity in the brain: **voxelwise analysis** or **region-of-interest (ROI)** analysis. Most fMRI studies present data using voxelwise analysis, in which each voxel is analyzed for significant differences in signal intensity in time between two experimental conditions. In ROI analysis, the brain is divided into a set of discrete regions assigned by the investigator. Rather than voxel-by-voxel comparisons, entire brain regions are compared for significant differences in signal intensity. ROI analysis can provide information about specific structures, but can initially be more time-consuming, as each voxel has to be assigned to a specific ROI. Sometimes an investigator will first analyze data using a voxelwise analysis and then analyze discrete ROIs.

Each individual's fMRI data is analyzed for significant differences, and then this data is pooled together for all the individuals within a certain subject pool. For a within subjects study, the subject pool is the same and the investigator determines if there is a significant difference in signal intensity between two different stimuli or tasks. For a between subjects study, the subject pools are different and the investigator determines if there is a significant difference in signal intensity between the two groups for a stimulus or task. Keep in mind that data analysis methods vary widely among different labs, and new analysis techniques are constantly being developed. We describe general methods here, but detailed descriptions of the methods used should be included in any publication. Following this data analysis, the statistically significant differences in the pooled subject data are presented as figures for publication.

Preparation of Figures

The most common functional imaging figures presented in the literature take the form of color-coded brain activation data superimposed on a structural image of a brain slice (Figure 1.19). As mentioned previously, the brain slice is almost always a T1-weighted image because these images present the greatest contrast between brain regions and are subjectively more aesthetically pleasing to most people. These brain slices can either be taken from one of the subjects during the experiment, or from a stock set of images from the Talairach or MNI templates. Sometimes, investigators superimpose imaging data on an "inflated brain" in which the **sulci** and **gyri** of the brain are expanded into a balloon-like shape in order to better show neural activity in the sulci (Figure 1.20).

It is important to realize that these fMRI figures are statistical comparisons in signal intensity over time for a grid of voxels, and that color-coded activation does not just "light up" during a scanning session. The color-coded voxels superimposed over the structural brain image are essentially a grid of numbers (Figure 1.19). Each voxel is assigned a statistical value that represents the difference in signal intensity for that region of brain space between two conditions. The exact color scheme is chosen by the investigator to reflect the statistical magnitude of signal intensity: usually bright colors represent high differences in intensity, while darker colors represent more subtle differences in intensity.

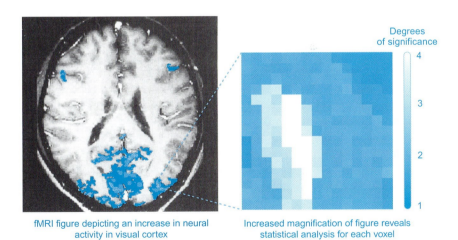

fMRI figure depicting an increase in neural activity in visual cortex

Increased magnification of figure reveals statistical analysis for each voxel

Degrees of significance

FIGURE 1.19 Preparation of an fMRI figure for publication. A typical fMRI figure contains a representation of the changes of BOLD signal intensity over time superimposed over a T1-weighted structural image. It is important to remember that the colored representation of signal intensity is essentially a grid of numbers, with each number representing a measure of statistical significance. (fMRI image reprinted with kind permission of Springer Science + Business Media and Jens Frahm from Windhorst, U. and Johansson, H. (eds.), 1999. *Modern Techniques in Neuroscience Research,* Ch. 38: Magnetic Resonance Imaging of Human Brain Function, p. 1064, Fig. 5.)

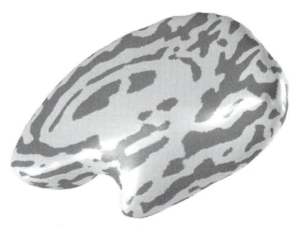

FIGURE 1.20 **An inflated representation of the right hemisphere of the human brain used to show activity within sulci.** The dark gray regions represent the exposed sulci of the inflated brain, while the light gray regions represent gyri. (Courtesy of Dr. Rory Sayres.)

CONCLUSION

The goal of this chapter was to provide a thorough introduction to the technologies that currently exist to image the structure and function of the brain, as well as the approaches an investigator may take when designing and analyzing functional imaging experiments. These technologies are more complicated than the brief descriptions found here, and readers interested in more detailed explanations should consult the Suggested Readings section at the end of this chapter. However, this survey of techniques hopefully provides an appreciation for the ingenuity of scientists and engineers in developing and improving these techniques over the past 10–20 years. Imaging neural function can be performed based on the electrical, magnetic, metabolic, and light-scattering properties of distinct brain regions. Producing so many different methods of imaging the brain is a remarkable achievement, and it will be exciting to observe, either as fellow scientists or participants, how these methods are combined and improved over the next 10–20 years.

SUGGESTED READING AND REFERENCES

Books

Frahm, J., Fransson, P. & Kruger, G. (1999). Magnetic resonance imaging of human brain function. In *Modern Techniques in Neuroscience Research*, Windhorst, U. & Johansson, H., eds, pp. 1055–1082. Springer, Berlin.

Huettel, S. A., Song, A. W., McCarthy, G. (2004). *Functional Magnetic Resonance Imaging*. Sinauer, New York.

Mori, S. (2007). *Introduction to Diffusion Tensor Imaging*. Elsevier. Amsterdam.

Schild, H. H. (1990). *MRI Made Easy. . . Well Almost*. Schering AG, Germany.

Review Articles

Friston, K. (2002). Beyond phrenology: what can neuroimaging tell us about distributed circuitry? *Annu Rev Neurosci* **25**, 221–250.

Logothetis, N. K. (2008). What we can do and what we cannot do with fMRI. *Nature* **453**, 869–878.

Miller, G. (2008). Neuroimaging. Growing pains for fMRI. *Science* **320**, 1412–1414.

Mori, S. & Zhang, J. (2006). Principles of diffusion tensor imaging and its applications to basic neuroscience research. *Neuron* **51**, 527–539.

Savoy, R. L. (2005). Experimental design in brain activation MRI: cautionary tales. *Brain Res Bull* **67**, 361–367.

Primary Research Articles—Interesting Examples from the Literature

Dagher, A., Leyton, M., Gunn, R. N., Baker, G. B., Diksic, M. & Benkelfat, C. (2006). Modeling sensitization to stimulants in humans: an [11C]raclopride/positron emission tomography study in healthy men. *Arch Gen Psychiatry* **63**, 1386–1395.

deCharms, R. C., Maeda, F., Glover, G. H., Ludlow, D., Pauly, J. M., Soneji, D., Gabrieli, J. D. & Mackey, S. C. (2005). Control over brain activation and pain learned by using real-time functional MRI. *Proc Natl Acad Sci U S A* **102**, 18626–18631.

Gil-da-Costa, R., Martin, A., Lopes, M. A., Muñoz, M., Fritz, J. B. & Braun, A. R. (2006). Species-specific calls activate homologs of Broca's and Wernicke's areas in the macaque. *Nat Neurosci* **9**, 1064–1070.

Glasser, M. F. & Rilling, J. K. (2008). DTI tractography of the human brain's language pathways. *Cereb Cortex* **18**, 2471–2482.

Gonsalves, B. D., Kahn, I., Curran, T., Norman, K. A. & Wagner, A. D. (2005). Memory strength and repetition suppression: multimodal imaging of medial temporal cortical contributions to recognition. *Neuron* **47**, 751–761.

Hariri, A. R., Mattay, V. S., Tessitore, A., Kolachana, B., Fera, F., Goldman, D., Egan, M. F. & Weinberger, D. R. (2002). Serotonin transporter genetic variation and the response of the human amygdala. *Science* **297**, 400–403.

Kuhnen, C. M. & Knutson, B. (2005). The neural basis of financial risk taking. *Neuron* **47**, 763–770.

Logothetis, N. K., Pauls, J., Augath, M., Trinath, T. & Oeltermann, A. (2001). Neurophysiological investigation of the basis of the fMRI signal. *Nature* **412**, 150–157.

McClure, S. M., Li, J., Tomlin, D., Cypert, K. S., Montague, L. M. & Montague, P. R. (2004). Neural correlates of behavioral preference for culturally familiar drinks. *Neuron* **44**, 379–387.

Pascual-Leone, A., Meyer, K., Treyer, V. & Fehr, E. (2006). Diminishing reciprocal fairness by disrupting the right prefrontal cortex. *Science* **314**, 829–832.

Whalen, P. J., Kagan, J., Cook, R. G., Davis, F. C., Kim, H., Polis, S., McLaren, D. G., Somerville, L. H., McLean, A. A., Maxwell, J. S. & Johnstone, T. (2004). Human amygdala responsivity to masked fearful eye whites. *Science* **306**, 2061.

Protocols

Ferrera, V., Grinband, J., Teichert, T., Pestilli, F., Dashnaw, S. & Hirsch, J. (2008). Functional imaging with reinforcement, eyetracking, and physiological monitoring. JoVE. 21. http://www.jove.com/index/details.stp?id=992, doi: 10.3791/992

Haacke, E. M. & Lin, W. (2007). Current protocols in magnetic resonance imaging. In *Current Protocols*, Downey, T., ed.. John Wiley & Sons, New York.

Websites

fMRI 4 Newbies: http://fmri4newbies.com/
The Basics of MRI: http://www.cis.rit.edu/htbooks/mri/

Chapter 2

Animal Behavior

After reading this chapter, you should be able to:

- Compare and contrast the advantages and disadvantages of common model organisms used to study animal behavior
- Discuss considerations for designing a behavioral assay
- Understand and describe common behavioral assays used in rodents, invertebrates, and primates

Techniques Covered

- **Rodent motor assays:** running wheel, homecage activity, open field test
- **Rodent motor coordination assays:** rotarod test, footprint pattern assay, hanging wire assay, vertical pole test
- **Rodent sensory assays:** visual cliff assay, startle response, taste assays, olfactory assays
- **Rodent nociception assays:** tail flick assay, Hargreaves assay, hot plate assay, Von Frey assay, formalin assay
- **Rodent spatial learning and memory assays:** Morris water maze, Barnes maze, radial arm maze
- **Rodent nonspatial learning and memory assays:** classical conditioning, operant conditioning, novel object recognition, delayed match to sample and nonmatch to sample assays
- **Rodent social assays:** resident-intruder assay, social approach/avoidance assay
- **Rodent anxiety assays:** open field test, elevated plus maze, defensive marble burying, Geller-Seifter conflict test
- **Rodent depression assays:** forced swim test (Porsolt test), tail suspension test, sucrose preference test
- ***Drosophila* behavioral assays:** locomotion, flight, sensory assays (vision, olfaction, taste), learning and memory assays, courtship, aggression
- ***C. elegans* behavioral assays:** locomotion, sensory assays (mechanosensation, thermosensation, chemosensation)
- **Nonhuman primate behavioral paradigms**

Humans are the only animals capable of verbally reporting what they feel, what they perceive, and what they know. To gain insight into the emotional, perceptual, and cognitive processes of other animals, a scientist can only observe their behavior. Throughout the history of neuroscience and psychology, scientists have developed and used batteries of tests that probe animals for clues about their mental status, from their ability to perceive a stimulus to whether or not they are emotionally depressed. There are now dozens of assays commonly described in the literature, with novel assays continually being developed to study a growing list of behaviors and psychiatric diseases.

The goal of the behavioral neuroscientist is not only to characterize an animal's behavior, but also to identify and describe the genetic, biochemical, and cellular correlates of that behavior. Therefore, behavioral neuroscience does not exist in isolation, but rather in combination with systems and molecular techniques. A scientist can monitor neural activity using electrophysiology or optical methods (Chapters 4 and 7) during a behavioral task to correlate electrical events with specific behavioral outputs. A scientist can also determine which genes, proteins, and/or neurons are necessary or sufficient for a behavior to occur by investigating the effects of their perturbation (Chapters 3, 4, 11, and 12). Thus, the behavioral paradigms described in this chapter are used in conjunction with many other techniques described in this book.

The goal of this chapter is to describe the general strategies scientists use to measure behavior in rodents, invertebrates, and primates. First we will discuss some of the issues a scientist must consider when choosing and performing behavioral assays. Then we spend the rest of the chapter surveying the most commonly used behavioral assays described in the literature. This will not be a comprehensive overview, but a flavor of the methods used to elucidate the neural basis of behavior.

CONSIDERATIONS FOR CHOOSING AND PERFORMING A BEHAVIORAL ASSAY

There are many factors involved in the design of a behavioral assay. These factors include choosing a model organism, choosing a behavioral paradigm, reducing variability between individuals, and validating the behavior as a useful way to understand the nervous system.

Choosing an Appropriate Model Organism

A scientist should consider two major factors when choosing an animal model to study behavior: the natural ethological capabilities of an animal and the additional experiments the scientist will use to complement behavioral experiments. **Ethology** refers to the natural capability of an animal to perform specific behaviors. Worms and flies can be good models for studying behaviors in which animals interact with the environment, such as sensory transduction, food seeking, and motor control. Flies have also been used to study addiction, learning and memory, courtship, circadian rhythms, and other behaviors. Mice and rats are able to perform more complicated cognitive tasks that investigate emotion, learning, and social behavior. Primates are capable of highly intelligent behaviors, such as complex decision making, identifying faces, and using tools. In addition to these standard model organisms, scientists can choose to study other animals with specialized adaptations that make them an optimal model for certain behavioral tasks (Box 2.1).

A scientist must also consider the behavioral assay in the context of other experiments. If the goal of the study is to perform a forward genetic screen and identify novel genes involved in a behavior, the best choice of animal model will be a genetically tractable organism, such as a worm, fly, or mouse. If the goal is to record from several neurons in the brain while the animal performs a behavior, a larger animal is necessary, such as a large rodent or primate. Rodents are often used in behavioral assays because they are amenable to genetic, molecular, and electrophysiological approaches.

It is also important to consider the ethical and moral implications of using a specific animal model in an experiment. The use of animals in research is both a necessity and a privilege that a scientist should never fail to appreciate (Box 2.2).

Choosing an Appropriate Behavioral Paradigm

Whether investigating a commonly studied behavior or developing a new behavioral paradigm from scratch, a scientist should consider many factors when

BOX 2.1 Neuroethology

Neuroethology is the study of the neural basis of an animal's natural behaviors. This term can be applied to traditional research animals, like rodents and primates. However, it is often used to describe nontraditional research animals that have evolved unique adaptations, making that animal a particularly good model organism for a particular study. For example:

- Echolocating bats have enlarged brainstem auditory structures useful for studying the auditory system.
- Various species of birds learn and practice song, allowing scientists to study learning, consolidation of learning, and memory.
- Barn owls hunt in the dark and have an amazing capacity for sound localization, as well as coordinating visual and auditory information.
- Common toads readily discriminate predators from prey in the visual scene based simply on their shapes.

These are just some of the many examples of nontraditional animal models useful to neuroscientists for studying specific aspects of neural circuitry involved in behavior.

BOX 2.2 Ethical Considerations for Animal Use

Using animal models in biological research is necessary to make conclusions relevant to a whole living system. There are common ethical considerations that must be made when deciding to use animal models or which animal model to use. This is not only an ethical issue for the individual investigator; the use of animals must also be justified in grants and to animal oversight committees that exist to maintain the welfare of research animals. Federal, state, and institutional guidelines and committees govern the use of animals in research. Each institution has an **Institutional Animal Care and Use Committee (IACUC)** that must approve protocols involving animals and ensure that the research is justified.

When using animals in experiments, distress must be minimized while maximizing benefit to humans. Animal welfare guidelines throughout the world are based on the principles proposed by Russell and Burch in *The Principles of Humane Experimental Research* (1959)—the 3 Rs:

- *Replacement:* If there is any way to do the research without using animals, it should be done. This includes using computational modeling and cell culture methods. In the case of behavioral research, modeling is impossible without actually performing the task. However, another component of replacement is the use of "less-sentient" animal species: choosing to study invertebrates rather than vertebrates, or mice rather than monkeys.
- *Reduction:* Use the minimum number of animals to obtain scientifically valid data. If possible, use the same animals for multiple experiments. Also, investigators can perform analyses to determine the minimum number of animals required to provide enough statistical power for significant results.
- *Refinement:* Procedures must minimize distress and pain and enhance animal well being. Scientists must continually monitor animals for signs of distress and ensure their overall well being. This is also important for the consistency and interpretability of experimental results; stress has numerous effects on physiology that can confound results.

Following these guidelines is standard practice for developing protocols and justifying research goals.

preparing to use an assay. First and foremost, the assay must be quantifiable, with the ability to measure discrete, easily observable variables. For example, consider an experiment in which a scientist wishes to measure drinking behavior between two groups of rodents. The scientist has many potential variables to measure: the total volume of liquid consumed, the rate of water consumption, the number of licks applied to a liquid delivery system, and so on. In the literature, data are usually presented as bar graphs or scatter plots, with different graphs representing different quantifiable aspects of each behavioral assay.

A scientist must also consider that animals, especially animals in which an investigator has manipulated gene or protein expression or neural activity, may exhibit abnormal behaviors for reasons the scientist may not expect.

For example, animals injected with a pharmacological agent may take a statistically longer time to complete a maze than control animals injected with saline. However, this effect may not be due to problems in spatial navigation or memory, but altered motor behavior that indirectly increases the amount of time spent in the maze. Therefore, when choosing a behavioral assay, scientists must ensure that there are additional experiments to test alternate explanations of any differences in animal behavior. It is also important to examine the over-all health of the animal (body weight, heart rate, etc.) to ensure that animals don't perform abnormally because they are sick.

Variability in Individuals

There are many factors that can increase the variability of a model organism's performance on a behavioral task, making it difficult to achieve consistent results. One factor is an animal's genetic background, which can have profound effects on behavior. Consider mice as an example: there are over 10 strains of mice commonly studied in research laboratories. Some strains behave very differently than others. For example, C57/BL6 mice are known to be much more aggressive than the FVB strain. An investigator performing social experiments on C57/BL6 mice might therefore observe different results than another investigator performing the same experiments using FVB mice. Strain-specific differences are also present in other model organisms, such as worms, flies, rats, and even primates. Therefore, it is important to always per-form behavioral assays in the same strain and, if possible, the same generation of animal. Comparisons between genetically modified animals and wild-type animals should be performed on individuals from the same litter.

Environmental variables can also impact behavior. These variables include animal handling, diet, circadian rhythms, seasonal changes, social interactions, and so on. For example, an investigator may observe different behaviors on a Monday compared to a Thursday if it turns out that an animal technician changes the animal cages (and thus handles the animals) each Monday. Alternatively, an individual animal may perform differently on a behavioral assay if it is housed with other animals compared to if it is housed alone. These factors require a scientist to foresee and minimize any aspects of an animal's environment that may introduce variability in behavior. Additionally, scientists may need to perform behavioral assays on large numbers of animals to produce reliable measurements and results.

Using Animal Behavior as a Model for Human Behavior

One of the primary goals of biomedical research is to learn more about humans by studying animal models. Is it possible to learn about humans by studying behaviors in other species? Can scientists extrapolate and generalize the results of an animal behavioral assay to people? Is it possible to model human neurological and psychiatric disorders, such as Alzheimer's disease and depression, using rodents?

The answers to these questions depend on the exact behavior studied, but scientists generally gauge the ability of an assay to provide information about human behavior by examining the assay's **validity**. There are three components to the validity of an animal experiment as a model of human behavior: face validity, construct validity, and predictive validity. A model has **face validity** if the model's behavior is similar to the analogous human behavior. For example, mouse models of autism spend statistically less time engaging in social behaviors than wild-type mice, just as many humans diagnosed with autism have difficulty socially interacting with others. **Construct validity** refers to a state in which the animal model and human model have the same underlying genetic or cellular mechanism that may result in a certain behavior, such as a transgenic mouse engineered to have the same mutation that causes Alzheimer's disease in humans. A model has **predictive validity** if treatments used in human patients have the same effect on the animal model. For example, some antidepressants prescribed to humans decrease behavioral measures of depression in rodent models.

It is possible for animal models to exhibit one form of validity but not others. Different types of experiments may benefit from different forms of validity, though the more forms of validity an animal model exhibits, the greater its relevance and value to humans. Note that validity is not something that can be objectively measured, but a concept judged by scientists and their colleagues within various subfields of neuroscience.

Now that we have discussed some general features of behavioral paradigms, we describe many commonly used assays in modern neuroscience research.

RODENT BEHAVIORAL PARADIGMS

Rodents are the most commonly used mammalian model organisms in behavioral neuroscience because they can also easily be used for complementary molecular, genetic, and electrophysiology experiments. The following assays represent common approaches used by investigators to measure motor function, sensation, learning and memory, social interactions, and emotion.

Locomotor Activity

Locomotion assays are used to determine the net motor activity of an animal over a given time period. These assays are useful for determining if two cohorts of animals (e.g. knockout animals compared to wild-type littermates) have the same baseline activity. If measurements are recorded continuously over days and weeks, these assays are also useful for studying **circadian rhythms**—the regular, roughly 24-hour-cycle of stereotyped biochemical, physiological, and behavioral processes.

Running Wheel

A running wheel is the simplest way to measure the locomotor activity of an animal over time. The wheel is coupled to a device that measures the total number of revolutions and the speed of each revolution. If the animal is free to enter or exit the running wheel voluntarily, then this assay is limited by the fact that an animal may be active off the wheel when activity is not measured. However, a running wheel exerts significantly more energy than simple movements a rodent may perform around its cage, and therefore running wheels can serve as reliable measurements of activity.

Homecage Activity

Locomotor activity can be measured in specially designed cages that project infrared beams from one side of the cage to the other. Each time an animal moves around the cage, it breaks the beam and a computer records the time and position (Figure 2.1). Alternatively, if cages with infrared beams are not available, a scientist can place a camera above the cage and record activity. Computer tracking programs can then be used to statistically analyze the total locomotor activity over time. Depending on the specific setup, a scientist can examine horizontal activity, vertical activity, time spent in various regions of the cage, and total distance traveled.

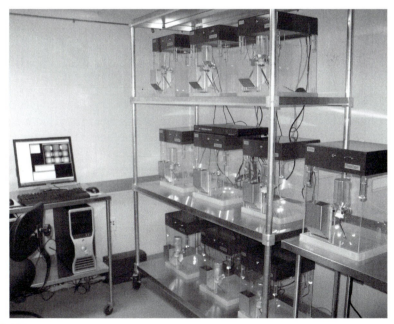

FIGURE 2.1 Locomotor activity can be monitored in a specially designed home cage. (Courtesy of the Stanford Behavioral and Functional Neuroscience Laboratory.)

FIGURE 2.2 Exploratory locomotor activity can be monitored in an open field testing chamber. (Courtesy of the Stanford Behavioral and Functional Neuroscience Laboratory.)

Open Field Locomotion Test

An **open field test** takes place in a large cubic box, usually measuring 1 meter long × 1 meter wide × 1 meter high (Figure 2.2). The top of the cube is typically left uncovered. An animal is placed in the middle of the bottom surface, and its movements are recorded over the course of minutes to hours as it moves around and explores its environment. After the experiment is completed, computer tracking programs analyze the movements of the animal over time. This assay can measure horizontal activity, time spent in various regions of the open field, and the total distance traveled. This assay can also be used to measure anxiety (described later).

Motor Coordination and Balance

Assays for motor coordination and balance are used to test the proper functioning of an animal's motor and vestibular systems. These tests can be used in isolation or to exclude motor deficits as a reason for poor performance in other behavioral experiments. For example, if pharmacological or genetic manipulation of an animal results in abnormal performance on a separate behavioral assay, such as a learning and memory assay, a scientist could ensure that this

FIGURE 2.3 **A rotarod tests balance and motor coordination**. Multiple animals can be tested at one time with a device that has individual lanes to record each animal's latency to fall. (Courtesy of the Stanford Behavioral and Functional Neuroscience Laboratory.)

abnormal behavior was due to a specific component of the nervous system rather than a secondary effect caused by general motor deficits.

Rotarod

A **rotarod** is a device consisting of a cylindrical rod suspended above a padded base (Figure 2.3). A scientist places a rodent on the rod and gradually accelerates the rotation until the rodent falls to the bottom of the chamber. Most animals are able to maintain balance for many minutes, but rodents with deficits in motor coordination or balance fall more quickly. This assay can also be used as a measure of motor learning, as most animals perform better on the rotarod assay after consecutive training days.

Footprint Pattern Assay

A scientist performs a **footprint pattern assay** by dipping an animal's paws in ink and allowing it to walk down a paper-lined tunnel. These footprints can reveal gait abnormalities and **ataxia**—abnormal muscle coordination (Figure 2.4). Dipping the forepaws and hindpaws in different colors of ink allows a scientist to measure a variety of gait parameters: distance between each stride, changes in stride length, variability around a linear axis, width between left and right hindpaws, regularity of steps, and overlap between fore- and hindpaws. Modern automated versions of this task can also track pressure and walking speed. Ataxic gait is characterized by a highly variable stride length and path.

Hanging Wire Assay

The **hanging wire assay** is used to measure neuromuscular deficits. An investigator places a rodent on a cage lid made of a wire grid, shaking the lid so the rodent grips onto the wire. The investigator then gently flips the lid so the rodent hangs upside down from the lid. To prevent falling, the rodent grasps the wire, requiring both balance and grip strength. The investigator measures the time it takes before the rodent falls to a padded surface. A normal rodent can hang upside down for several minutes, but a 60-second cutoff is usually used for experimental purposes.

FIGURE 2.4 Footprint pattern analysis can reveal gait abnormalities. Footprint patterns of a wild-type (+/+) mouse compared to a mutant (stg/stg) showing a wider stance (proximity of blue forelimb and black hindlimb footprints) and stumbling (arrow). (Reprinted from Meng, H. et al. (2007). BDNF transgene improves ataxic and motor behaviors in stargazer mice. *Brain Research* 1160: 47–57, with permission from Elsevier.)

Vertical Pole Test

In a **vertical pole test**, a rodent is placed on the center of a horizontal pole wrapped in cloth to provide traction. The pole is gradually lifted toward a vertical position, and the investigator records the amount of time and the angle at which the animal falls off. Normally, rodents can move up and down the pole without falling before the pole reaches a 45-degree angle.

Sensory Function

Abnormal sensory function is usually more difficult to observe than abnormal motor function, as a rodent must demonstrate that it cannot detect a sensory stimulus. In many sensory assays, a rodent uses motor output to report detection of a sensory stimulus; thus, these assays are sensitive to confounds due to deficits in motor function.

Visual Cliff Assay

A **visual cliff assay** evaluates the ability of a rodent to see the drop-off at the edge of a horizontal surface. The edge of a box or table can serve as a horizontal plane with a vertical drop to a lower horizontal plane, such as the floor. Each surface is coated with a black-and-white checkerboard pattern to enhance the detection of visual planes (Figure 2.5). A clear sheet of Plexiglass is placed

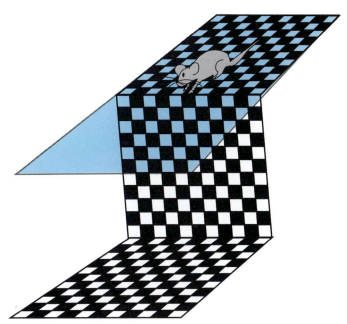

FIGURE 2.5 Visual cliff assay. A checkerboard pattern enhances the perception of a drop-off. Rodents that can see this drop-off will pause at the edge, while blind rodents will feel the Plexiglass surface and continue walking without pause.

over the drop-off, extending across the "cliff" so an animal cannot actually fall. Normally, rodents visually detect the drop-off and approach the edge cautiously, stopping at the cliff. Blind rodents approach the cliff but continue walking, unable to see the edge but able to feel the Plexiglass sheet. However, rodents with normal vision can also use their whiskers to feel the Plexiglass extension, so the whiskers are often cut prior to an experiment.

Startle Response Assay

In a **startle response assay**, a scientist exposes a rodent to an unexpected and disruptive sensory stimulus and measures the degree to which the animal responds, typically exhibited as an eye blink, body flinch, or overall muscle contraction. These stimuli can be visual (a bright light), auditory (loud and disrupting noise), or tactile (unexpected touch or air puff). Investigators can either observe and score the number of times a startle response occurs or use specialized sensors, such as an **electromyogram**, to more quantitatively measure the degree of the response. This assay can be confounded by deficits in emotional and motor components of the startle reflex circuitry. This assay can also be used to measure animal learning by measuring extinction of the startle response following the stimulus over successive trials.

Taste

Investigating an animal's ability to taste usually involves a choice assay, in which different gustatory compounds are placed in different solutions in water bottles. The scientist can then measure the volume of water in each bottle to determine the animal's taste preferences. It is possible to investigate whether an animal can detect bitter tastes by placing quinine, an extremely bitter compound, in a water solution and determining if the animal consumes less water than a solution with no quinine. Likewise, it is possible to investigate whether an animal can detect

---

BOX 2.3 Prepulse Inhibition (PPI)

If the startle response assay is performed after presenting a weaker sensory stimulus first, the startle response is reduced. This phenomenon is known as **prepulse inhibition (PPI)**, and it reflects the nervous system's ability to prepare for a strong sensory stimulus after a small warning (the prepulse). PPI is also thought to reveal changes in sensorimotor gating, the ability to filter out unnecessary sensory input; human patients with attention deficit disorder, Alzheimer's disease, and schizophrenia show deficits in PPI where the prepulse does not reduce the startle response as much. This can be a useful paradigm to test animal models of these disorders, since the startle response assay, and thus PPI, can essentially be carried out and analyzed in the same way for animals and humans. Examining deficits in PPI has been the major behavioral assay used to understand rodent models of schizophrenia and test potential antipsychotic medications that restore PPI.

sweet tastes by placing sucrose or saccharine in a water solution and determining if an animal drinks more water than a solution with no sweet compounds.

Olfaction

A scientist can test for normal olfactory processing by measuring how long it takes for an animal to find a piece of familiar, palatable food (like a cookie, cheese, or chocolate chip) hidden within the cage.

Nociception

Nociception is the ability to detect a noxious stimulus, usually perceived as pain. Nociceptive assays rely on physical indicators of discomfort, such as withdrawal reflexes, licking, and vocalizations. These assays are usually used to either investigate the neural basis of pain or the therapeutic potential of analgesic drugs. Genetically or pharmacologically manipulated animals that exhibit decreased sensitivity to pain must be monitored closely during these assays to prevent tissue damage induced by nociceptive stimuli.

Tail Flick Assay

In the **tail flick assay**, a high-intensity beam of light is aimed at a rodent's tail. In normal animals, the beam of light produces a painful heat sensation, causing a reflex that moves the tail. The investigator measures the amount of time it takes before the tail flicks to the side. Sex, age, and body weight can affect a rodent's response, so all animals used for these experiments (experimental and control animals) must be similar to avoid confounding results.

Hargreaves Assay

The **Hargreaves assay** is similar to the tail flick assay, but the high-intensity beam of light is aimed at the rodent's hindpaw rather than the tail. The investigator measures the time it takes for the rodent to withdraw its paw.

Hot Plate Assay

In the **hot plate assay**, a scientist places a rodent on a heated surface and prevents the animal from leaving by enclosing it within a tall cylinder (Figure 2.6). The surface is calibrated so that a normal animal will react within about 10 seconds of exposure (usually 52–53° C). The investigator records the latency and amount of time an animal reacts to the heat stimuli. Individual reactions vary: licking a rear paw is a reliable indicator of discomfort, though some animals will jump or vocalize.

Von Frey Assay

A **Von Frey assay** is used to examine the sensitivity to pinch and mechanical stimuli. Von Frey hairs are fine-gauge metal wires. The investigator pokes the hindpaw of an animal standing on an elevated mesh platform by inserting a

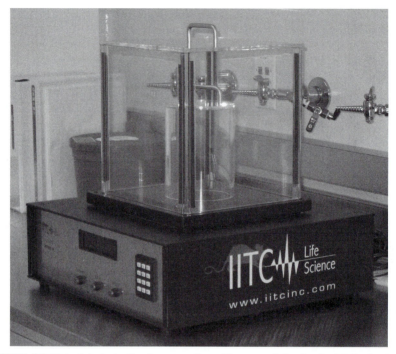

FIGURE 2.6 Hot plate test of pain perception. (Courtesy of the Stanford Behavioral and Functional Neuroscience Laboratory.)

Von Frey hair through the mesh from below. Normal rodents usually react by withdrawing or licking their paws and possibly vocalizing.

Formalin Assay

A **formalin assay** is used to examine an animal's sensitivity to noxious chemical stimuli. An investigator injects a small volume of formalin, a noxious chemical, into a rodent's hindpaw and records the total amount of time the rodent spends licking, biting, or withdrawing the affected limb. This activity is usually assessed in two phases: one phase starting immediately after the injection and lasting about 10 minutes and another phase starting 20 minutes after the injection and lasting about an hour. The second phase is a response to tissue damage caused by inflammation.

Spatial Learning and Memory

Spatial learning and memory are ethologically relevant behaviors to a mouse or rat. In the wild, these animals spend most of their lives locating good nesting spots, searching for food, and avoiding predation. Therefore, rodents are great model species to test assays for spatial navigation and memory.

FIGURE 2.7 The Morris water maze is one of the most common rodent tests for learning and memory. (Courtesy of the Stanford Behavioral and Functional Neuroscience Laboratory.)

Morris Water Maze

The **Morris water maze** is probably the most frequently used spatial learning and memory paradigm. Rodents are placed in a large circular pool of opaque water (Figure 2.7). Their natural dislike of water makes rodents highly motivated to escape from the pool. In initial trials, a visible platform is placed within the pool so that an animal can emerge from the water. A scientist places visual cues around the pool so the animal can relate the spatial environment with the location of the platform. In subsequent trials, the platform is hidden just below the surface of the water, and a scientist measures the time required for the rodent to swim to the platform to escape the water.

In the first few trials, the rodents become progressively better at swimming directly toward the platform, demonstrating their ability to learn the platform's location. To test memory at the end of training, the investigator completely removes the platform to measure the amount of time the animal spends searching in each quadrant of the pool. This experiment measures the ability of the animal to identify a spatial location based on the visual cues placed around the chamber. Because water is stressful to rodents (in mice more than rats), it is important for a scientist to account for stress-induced performance effects.

FIGURE 2.8 Barnes maze. Rodents learn which one of the holes contains a hidden drop box they can use to escape from light. (Courtesy of the Stanford Behavioral and Functional Neuroscience Laboratory.)

Barnes Maze

The **Barnes maze** consists of a circular table with holes around the circumference, placed in a room with visual cues in the periphery (Figure 2.8). Most of these holes lead to an open drop to the floor, but a single hole leads to a "drop box," a dark box in which the animal can hide. A rodent is naturally motivated to avoid open spaces and bright lights, and therefore attempts to find the drop box. In initial trials, the scientist gently leads the animal to the drop box. In subsequent trials, the animal is placed in the center of the table and must find the drop box on its own. After a few trials, rodents typically remember which hole contains the drop box and quickly proceed in a direct path toward the hole. Investigators can measure the amount of time to find the correct hole, the number of incorrect holes explored, and the length of the exploratory path. The Barnes maze is considered to be less stressful than the Morris water maze.

Radial Arm Maze

The **radial arm maze** consists of an array of "arms," usually eight or more, that radiate from a central starting point (Figure 2.9). At the end of each arm is a cup that may or may not contain a food reward. Animals are trained to recognize that only one of the arms will contain food. Investigators measure the amount of time it takes for animals to find the arm leading to food, as well as the number of times it traverses an arm it has previously visited. Exploring a previously visited arm indicates that the animal did not remember previously choosing that spatial path. This is a relatively difficult task for rodents, requiring several days or weeks to train rats and many weeks to train mice. Unlike the Morris water maze

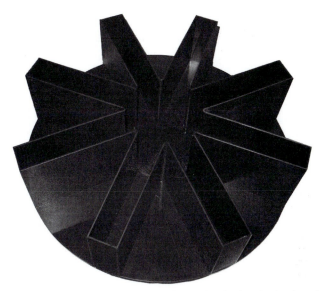

FIGURE 2.9 **Radial arm maze**. (Courtesy of the Stanford Behavioral and Functional Neuroscience Laboratory.)

and Barnes maze, the radial arm maze does not use distant visual cues to aid spatial learning.

Nonspatial Learning and Memory

The neural circuitry that mediates nonspatial learning and memory differs from that mediating spatial learning and memory. Thus, multiple assays using both nonspatial and spatial-based tasks are used to evaluate learning and memory. Nonspatial learning and memory assays include innate learning, such as motor performance improvement through practice and novel object recognition, as well as learned associations through conditioning. While classical (Box 2.4) and operant (Box 2.5) conditioning can be used directly to assay an animal's ability to learn, these paradigms are often used to train and test animals on more sophisticated tasks. For example, the delayed match to sample and nonmatch to sample assays use operant conditioning techniques to examine an animal's working memory. These techniques are described below.

Novel Object Recognition

The **novel object recognition** paradigm assesses an animal's innate ability to distinguish an old from a new object. Rodents are naturally curious and will take time to explore novel objects. When presented with both a new object and a previously explored object, normal rodents remember the old object and spend relatively more time exploring the new object. A scientist measures the amount of time spent interacting with both objects.

BOX 2.4 Classical Conditioning

In **classical conditioning** (also called **Pavlovian conditioning**), the investigator couples an initially neutral stimulus, such as a tone or light, with a salient stimulus, such as food or an electric shock. The neutral stimulus is referred to as the **conditioned stimulus (CS)**, while the salient stimulus is referred to as the **unconditioned stimulus (US)**. The unconditioned stimulus elicits a reflexive response, such as salivation or fear behaviors. By pairing the conditioned and unconditioned stimuli, the conditioned stimulus alone is eventually able to elicit the reflexive response of the unconditioned stimulus. Examples of Pavlovian conditioning are **cued fear conditioning** and **contextual fear conditioning**. Fear conditioning involves using foot-shock (US) to induce freezing behavior (response). By associating an auditory or visual cue (CS) with the foot-shock, animals will freeze in response to the sensory cue alone, without any foot-shock, demonstrating a learned association. Contextual fear conditioning measures freezing behavior when the animal is returned to the training chamber, but without foot-shock. Displaying the response without the unconditioned stimulus shows the animal has learned the association between a location and the shock. Cued conditioning should be measured in a new testing chamber with a very different context (shape of the box, visual cues, lighting, odors, textures) but with the auditory or visual cue from training to test the specific learned association between the cue and the shock.

Delayed Match to Sample/Nonmatch to Sample

Operant conditioning (Box 2.5) can be used to train an animal to make a response such as poking its nose into a hole or pressing a lever for a reward in response to a stimulus, often a light appearing above the lever. To test working memory, the investigator trains the rodent to perform an operant response under specific conditions—for example, pressing a specific lever when multiple levers are presented. In a **match to sample** task, a light appears above one lever, and the animal must choose to press the lever beneath that light. In a **nonmatch to sample** task, a light appears, and the animal must choose to press the opposite lever. After learning these tasks, the scientist adds artificial delays, such as withdrawing the lever, between the time when a cue appears and the time that an animal responds. The longer the delay, the longer the animal must remember the light cue that was presented, thus testing the animal's working memory.

Social Behaviors

Rodents display a variety of innate social behaviors. Mating, parenting, nesting, and grooming are all innate behaviors that can be observed within an individual's cage. Rodents also exhibit specific behaviors based on social recognition when multiple animals are exposed to one another. Exploratory sniffing allows animals to decide on an appropriate social response, such as

> **BOX 2.5 Operant Conditioning**
>
> In **operant conditioning**, a particular voluntary response (e.g., lever press) can be increased by positive reinforcement or decreased by punishment. Operant conditioning can be used in all model organisms to test learning and memory, though often this is the standard method to train more complex tasks that require a specific response, like pressing a lever to indicate a choice. To use reward or appetitive stimuli for conditioning, animals are usually restricted from normal amounts of food or water to motivate them to work for a food or water reward. Using punishment or aversive stimuli, such as a shock, does not usually require additional motivation, though these types of stimuli can be stressful. To train animals, the reinforcement stimulus is given when the animal makes the response so they learn the association between the reinforcement and the response. An animal will continue to provide a desired response if it is paired with a positive stimulus. Alternatively, an animal will avoid a particular response if it is paired with a negative stimulus.

defending territory or attempting copulation. Two standard assays are well represented in the literature: the resident-intruder assay and social approach/avoidance assays.

Resident-Intruder Assay

The **resident-intruder assay** measures territorial behavior in males. A scientist adds an "intruder" animal to the cage of a "resident" animal and measures specific aggressive behaviors: tail rattling, quivering or thrashing of the tail, physical attacks such as biting or clawing, wrestling, and chasing. The investigator can then compare the amount of time each animal spends investigating the other (following, sniffing, grooming) compared to the amount of time displaying aggressive behavior.

Social Approach/Avoidance

If a scientist presents a rodent with a choice between another rodent and an inanimate object, most of the time will be spent interacting with the other animal, referred to as social approach. Animals with abnormal social behaviors may spend relatively little time with the other animal, referred to as social avoidance. For example, in some mouse models of autism, there is no difference in the amount of time spent exploring the object and exploring the other mouse. These behaviors are thought to be analogous to the asocial behaviors exhibited by humans diagnosed with autism.

Anxiety

Paradigms that measure anxiety generally involve placing rodents in naturally stressful environments, such as brightly lit, open spaces. These assays are

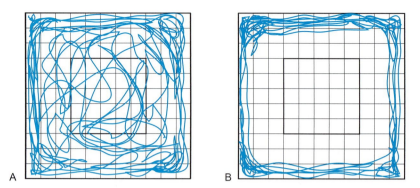

FIGURE 2.10 **The open field test as a measure of anxiety.** Observing locomotor exploration patterns can reveal differences in anxiety. **(A)** After being acclimated to the chamber, rodents will normally explore the entire area, but **(B)** a rodent model of anxiety will stay near the perimeter of the chamber.

useful for investigating brain regions thought to modulate anxiety, as well as testing novel anxiolytic drugs designed to reduce anxiety. In addition to the common assays described here, investigators can also measure physiological correlates of anxiety, such as defecation and freezing behavior.

Open-Field Test

The **open-field test** (Figure 2.2), previously described as an assay of locomotor activity, can also be used to measure anxiety. A rodent placed in a bare, open chamber will initially stay near the walls and avoid the center (Figure 2.10). Normal animals typically acclimate to the chamber and eventually explore the center area. More anxious animals spend significantly less time in the open area and more time closer to the walls. The investigator can monitor the animal's activity with a camera placed above the open field, and computer software measures time spent in the center compared with the periphery. Anxiolytic drugs increase the amount of time rodents spend in the open area.

Elevated Plus Maze

An **elevated plus maze** is a four-armed platform resembling the shape of a plus sign positioned two to three feet above the ground (Figure 2.11). Two arms of the maze have sides and two do not. A scientist places an animal in the center of the maze and allows it to move freely to any of the arms. Normal animals avoid the open arms, spending relatively more time in the protected, closed arms. Anxiolytic drugs increase the proportion of time rodents spend in the exposed arms.

Defensive Marble Burying

Rodents tend to bury objects present in their environment, such as glass marbles, within the bedding of their cages. Animals that are more anxious tend

FIGURE 2.11 **The elevated plus maze**. (Courtesy of the Stanford Behavioral and Functional Neuroscience Laboratory.)

to bury more objects in the 30–60 minutes after they are introduced. To perform a **defensive marble burying assay**, a scientist places 10–20 marbles in an animal's cage and quantifies the number of marbles buried after 30 minutes. Anxiolytic drugs tend to decrease the number of marbles buried over time.

Geller-Seifter Conflict Test

The **Geller-Seifter conflict** test is one of the oldest assays to study anxiety and remains a robust test of anxiolytic drug effects. A scientist uses operant conditioning (Box 2.5) to train an animal to press a lever for a food reward. In the beginning of an experiment, the scientist deprives the animal of food so it is motivated to press a lever to get the food reward. However, during the experiment, the investigator pairs a lever-press with an unpleasant electrical shock. Thus, the animal must decide whether to receive food while getting shocked or not receive food at all. Anxious animals tend to press the lever significantly fewer times that nonanxious animals. Anxiolytic drugs (but not other psychoactive drugs) increase the number of lever presses.

Depression

How is it possible to measure depression in a mouse or rat? Animal paradigms that investigate depression in animal models assay for the feelings of hopelessness and despair experienced by humans diagnosed with depression. Two

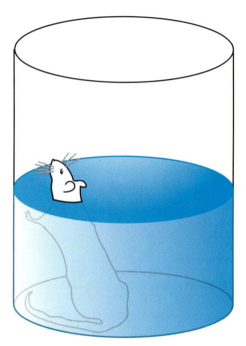

FIGURE 2.12 **Performance on the forced swim (Porsolt) test is used as a test of behavioral despair to model depression in rodents**.

methods a scientist can use to generate a rodent model of depression are creating conditions of learned helplessness and chronic mild stress. The **learned helplessness** paradigm exposes an animal to aversive stimuli at random intervals. Theoretically, this treatment creates a condition in which the animal experiences a lack of control, exhibiting symptoms of behavioral despair. Applying conditions of **chronic mild stress** (leaving the lights on all the time, tilting the animal's cage, wetting the animal's bedding, applying intermittent air puffs) over a period of days to weeks can also theoretically lead to conditions of behavioral despair and helplessness. These conditions lead to differences in behavior on the assays of depression described here. Importantly, antidepressants reverse these behaviors, demonstrating the predictive validity of these animal models.

Forced Swim Test (Porsolt Test)

In the **forced swim test** (also called the **Porsolt test**), rodents are placed in a small, confined space, such as a large graduated cylinder filled halfway with water (Figure 2.12). Initially, there is a period of vigorous activity during which the animal tries to escape. Eventually, the animal ceases vigorous activity and exhibits a characteristic immobility in which it only moves to maintain its head above water. This physical immobility is thought to be an indication of behavioral despair. Investigators measure the amount of time between when the animal is placed in the chamber and the onset of immobility. Rodent models of depression exhibit a decrease in the time spent trying to escape, but this decrease is reversed with antidepressants.

Tail Suspension Assay

In the **tail suspension assay**, either a device or the investigator holds the rodent in the air by its tail. The animal's natural reaction is to move vigorously to escape. Depressed animals give up more quickly than nondepressed animals. Antidepressants increase the amount of time the animals struggle. An advantage of this paradigm over the forced swim test is that tail suspension is less stressful to the animal because there is no exposure to water.

Sucrose Preference Test

The **sucrose preference test** is hypothesized to model anhedonia—the loss of desire to pursue an activity that was once enjoyable. Normally, rodents that are allowed to choose between drinking tap water or water with sucrose will choose the water with sucrose. Rodent models of depression have a significantly reduced preference for water with sucrose, or no preference at all, but this effect is reversed by administration of antidepressants.

DROSOPHILA BEHAVIORAL PARADIGMS

One of the powerful advantages to using *Drosophila*, the fruit fly, as a model organism is the relatively easy ability to identify and manipulate genes. Coupling amazing genetic tools with behavioral assays allows scientists to investigate the genetic basis of behavior. The lab of Seymour Benzer famously used *Drosophila* to study genes necessary for learning, memory, vision, circadian rhythms, nociception, and many other behaviors. Many other labs have used *Drosophila* to study sleep, addiction, courtship, and aggression.

As with any animal model used to study behavior, age, diet, and environmental factors must be strictly controlled when performing behavioral experiments to attain consistent results. If experiments are performed over multiple days, the scientist should perform assays at the same time each day to avoid confounding results due to circadian rhythms. Applying anesthesia or handling the flies just prior to behavioral tests can also affect the outcome.

Locomotor Behavior

Locomotor behavior in flies can be studied similarly to locomotor behavior in rodents. A scientist can project infrared beams of light through a housing chamber, allowing flies to break the beams and thus indicate their movement over time. Alternatively, a scientist can record a housing chamber with a video camera, and tracking software can record the activity of many flies over

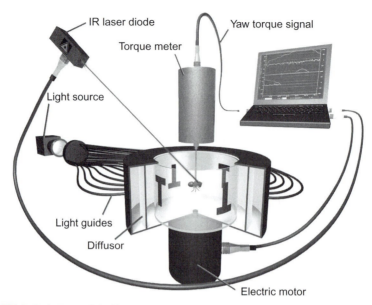

FIGURE 2.13 **A *Drosophila* flight simulator**. This device can be used to measure free-flight behaviors or flight responses to visual stimuli, as well as to train flies by using operant conditioning methods that use a laser to generate heat reinforcement. (Courtesy of Dr. Björn Brembs.)

time. Locomotor behavior is useful when studying motor function, circadian rhythms, and the effects of alcohol and other drugs.

Flight

Traditional locomotor assays measure movement in an enclosed container. Another ethologically relevant movement for a fly is flight. Therefore, scientists have developed specialized **flight simulators** that suspend a fly from a thin pin and project visual stimuli to depict the external, visual environment (Figure 2.13). These simulators contain torque meters to record turning responses to various stimuli, allowing analysis of sensory and motor reflexes. These flight simulators can also be used to study learning and memory, as well as sensorimotor processing. Investigators can deliver visual and/or olfactory stimuli to the fly in the simulator and measure flight responses to test spatial learning.

Sensory Function

Sensory modalities commonly studied in the fly include vision, olfaction, and taste. These modalities are all ethologically relevant for a fly, which spends most of its life navigating through space using visual and chemical cues in search of food or responding to chemical signals produced by other flies.

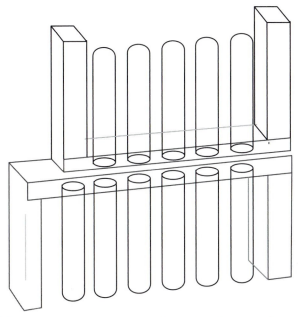

FIGURE 2.14 The countercurrent apparatus can be used to isolate a population of phototaxing flies.

Vision

Flies exhibit a natural tendency to move toward light, a phenomenon known as **phototaxis**. In order to test proper visual function, scientists can use a device called a **countercurrent apparatus**, made up of a series of test tubes (Figure 2.14). Flies are placed in one test tube, which is gently tapped so that the flies fall to the bottom. The test tube is then laid horizontally across from a second test tube. This second tube is illuminated with a fluorescent light. The scientist waits approximately 15 seconds before preventing the flies from crossing between the two tubes. The flies trapped within the second tube are considered to have demonstrated more of a phototaxic response. This process is repeated several times, with flies demonstrating a phototaxic response being used for the next trial. The end result is a fractionated population of flies that vary in their phototaxic responses.

Vision can also be assayed using a device called a **T-maze** (Figure 2.15). This apparatus consists of two arms oriented in opposite directions—a "T" shape—with an additional loading arm. Flies are placed into the loading arm and moved to the center of the other two arms of the T-maze, where they can travel to either arm. Each arm contains a different visual stimulus—for example, two different wavelengths of light, bright light versus darkness, flashing light versus constant light, and so forth. Thus, a T-maze allows scientists to

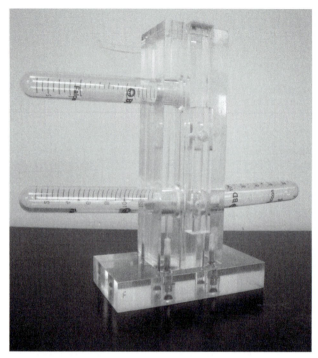

FIGURE 2.15 A *Drosophila* T-maze can be used to test fly sensory preferences or to train and test flies on conditioning training.

understand the visual preferences of a fly, as well as investigate the genes necessary and/or sufficient for these preferences to occur.

Olfaction

Olfaction assays can also take place in a T-maze. Each arm of the maze contains a different odor, and flies decide which odor to move toward. Investigators can use this assay to determine whether an odor is attractive, repulsive, or neutral.

In an olfactory trap assay, odors absorbed in a cotton swab or filter paper are placed at the bottom of a chamber with a pipette tip that allows flies to get in but not get out. By simply waiting a certain amount of time for flies to experience the baited odor, an attraction index can be calculated from the ratio of flies trapped in the odor vial compared to a water control. An attraction index of 1 indicates complete attraction to the odor, while an attraction index of 0 indicates either no preference for an odor or anosmic flies.

There are many simple avoidance assays that assess behavioral responses to olfactory cues. In the **olfactory avoidance** (also called **dipstick**) assay, a scientist inserts a piece of filter paper or cotton swab containing a specific odor into a fly vial, then measures the distance that the fly maintains from the offending odorant. A similar assay is the **chemosensory jump assay** or **olfactory jump response**, which exploits the tendency for flies to exhibit a startle response when encountering a novel odor.

Taste

Hungry flies exhibit an unconditioned reflex called the **proboscis extension response (PER)**. A scientist can elicit this response by applying taste ligands to gustatory receptor neurons on the leg or proboscis. However, the scientist must first starve the fly, as well as ensure that the fly is healthy. This reflex can be used to test specific gustatory receptor responses and sensitivity to different taste ligands.

A scientist can also assay taste using a **feeding acceptance assay**. Starved flies are provided a choice between appetitive, aversive, or neutral stimuli, each dyed a different color. The amount of each stimulus ingested in the dark can be scored by examining the color of the fly's stomach.

Learning and Memory

Learning and memory assays can be performed in *Drosophila* using a T-maze and classical conditioning. An odor is paired with an electric shock. This odor is then presented in one arm of a T-maze. Flies that avoid this arm are considered to have learned the association. This assay has been widely used in genetic screens that assay both learning and memory, as well as olfaction.

Investigators can also assay learning and memory on individual flies in the flight simulator through classical or operant conditioning. Visual or olfactory stimuli can be paired with heat to influence the turning response of a fly.

Social Behaviors

There are many social behaviors that a scientist can study using *Drosophila*. Many of these behaviors are innate (as opposed to learned), so they can be used to investigate the specific genes necessary and/or sufficient for these behaviors to occur. Perhaps the two most commonly studied behaviors in the fly are courtship and aggression.

Courtship

Flies exhibit a stereotyped courtship ritual composed of several distinct stages: (1) orientation, (2) tapping, (3) wing song, (4) licking, (5) attempted copulation, and (6) copulation. Because these behaviors are innate, scientists can use them to study the genetic basis of a hard-wired behavior. Groups of male and/or female flies can be placed into tiny chambers and recorded with a video camera. Later, an investigator scores the behavior, measuring a variety of parameters related to the timing, performance, and progression of these different stages.

Aggression

In order to study aggression, a scientist places two male flies in a chamber. The male flies naturally begin to fight, and the scientist can record many different parameters: which fly initiated the fight, the outcome, the interfight interval, the order of events for each fight, and so on.

C. ELEGANS BEHAVIORAL PARADIGMS

The relatively simple and well-understood neural circuitry of *C. elegans*, coupled with the relative ease with which to study genetic alterations, make this species a powerful animal model to study the genetic basis of behavior. Indeed, many genes in the worm are named after the behavioral phenotypes that led to their discovery: *Unc* (uncoordinated) mutants have disrupted locomotion, *Egl* (egg-laying) mutants have disrupted egg-laying, and *Mec* (mechanosensation) mutants have disrupted responses to touch. Most behavioral assays in *C. elegans* examine motor and sensory functions.

Locomotor Behavior

Locomotor behavior in the worm can now be automatically tracked using computer software called **Worm Tracker** (Figure 2.16). This software allows for a detailed analysis of the sinusoidal movement of worms, allowing scientists to determine the genetic and cellular components that make these movements possible.

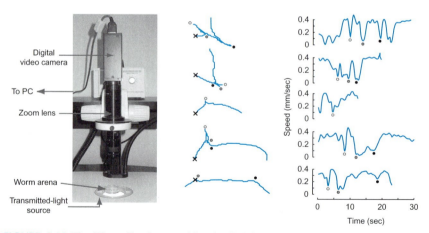

FIGURE 2.16 The Worm Tracker provides detailed information about the movements of individual worms. Representative tracks generated by the worm tracker can indicate starting position (x), path length, speed, and turning events (circles). (Reprinted from Ramot, D. et al. (2008). The parallel worm tracker: a platform for measuring average speed and drug-induced paralysis in nematodes. *PLoS ONE* 3(5): e2208, under the Creative Commons Attribution License.)

Sensory Behavior

C. elegans has proved very useful for investigating sensory transduction, as worms depend on detecting mechanical, thermal, and chemical stimuli to stay alive and find food.

Mechanosensation

A gentle touch with a fine hair to the body, head, or tail of a worm results in different behavioral responses. For example, if touched on the nose, worms move backward. If touched on the body, worms immediately stop and sometimes move away. Because the complete circuitry diagram of *C. elegans* is known, it is possible to use these behavioral assays to map functional connections between different neurons in different locations of the worm's body.

Thermosensation

Worm behavior is strongly influenced by environmental conditions—temperature, population density, and feeding status. Animals avoid temperatures at which they've been starved, and accumulate at temperatures associated with food. This requires thermosensation, as well as a memory for the associations. There are various methods for creating thermal gradients on which to test worms: a simple method is to place frozen glacial acetic acid on one end of a plate and an incubator on the other end. A more sophisticated method is to use a thermoelectric device to maintain a steady heat gradient across a plate. Investigators can then measure thermotactic migration by placing worms on the plate and identifying the temperature where migration stops.

Chemosensation

Chemosensation can be thought of as a worm's sense of taste or smell. A scientist can assay attraction or avoidance of specific compounds by dissolving these compounds in agar and placing them on opposite sides of a Petri dish. Worms are placed at the intersection of the two chemosensory cues, and the scientist determines the number of worms that migrate to each compound.

NONHUMAN PRIMATE BEHAVIORAL PARADIGMS

Nonhuman primates can be used to study the neural basis of complex motor actions, sensory perception, and cognitive processes such as decision making, attention, and complex learning. Experiments almost always couple electrophysiological recordings with behavior so that scientists can correlate the firing of neurons with distinct behavioral events. In addition to simply measuring neural activity, it is possible to manipulate neural activity using electrical and pharmacological methods.

One of the greatest strengths of primates as research models is their genetic, anatomical, and behavioral similarities to humans, allowing scientists to extend results to theories about how the human brain works. Primates are also capable of more complex behaviors and tasks than rodents and other animal models. In fact, many of the specific tasks performed by nonhuman primates are the same or adapted versions of tasks performed by humans in cognitive neuroscience or psychology studies.

The most commonly used monkeys used in research labs include rhesus macaques (*Macaca mulatta*) and cynomolgus (also known as crab-eating or long-tailed) monkeys (*Macaca fascicularis*). These animals can be very expensive to purchase and maintain. Therefore, each scientist typically only uses one to three monkeys over a 3- to 5-year period. While other behavioral paradigms described in this chapter typically require large numbers of individuals to achieve statistical significance, experiments involving nonhuman primates typically only need to be performed on a minimum of two animals. This is primarily due to differences in what the scientist measures: in rodent assays, scientists typically measure quantifiable behaviors. In primate assays, scientists typically measure response properties of neurons. Therefore, statistical power is achieved not by the number of animals studied but by the number of neurons recorded.

Primate behavioral/electrophysiological experiments usually follow a specific structure: an investigator designs a behavioral task, trains an animal to perform the task, completes a surgical procedure on the animal to allow access to the brain, conducts many experimental trials while recording from neurons in the brain, and analyzes data outside of the experiment to correlate activity with specific, behavioral events.

There are no standard behavioral tasks for primates as there are for rodents and invertebrates. Each investigator designs tasks specific to his or her research question, creating stimuli to maximize the response properties of the neurons to be studied. However, there are many common elements to primate experiments: they usually involve a single primate, working alone in a chamber for a juice reward, staring at a visual stimulus on a screen (Figure 2.17). Sometimes animals are required to make eye movements to follow moving dots on a screen; other experiments require that monkeys follow a path with their fingers; while still others require a monkey to press a lever in response to a remembered cue. Like different levels of a video game, each trial within a

| Fixate | Targets On | Motion On | Delay Period | Go Cue | Hold Target |

FIGURE 2.17 A primate behavioral task. Here, the monkey has been trained to make saccadic eye movements in the direction indicated by overall movement of a random dot stimulus. The monkey fixates at a point on the screen where the motion stimulus will appear. The fixation point disappears indicating the monkey can saccade to the correct target to receive a reward. (Courtesy of Rachel Kalmar.)

task is slightly different, with some trials more difficult than others. Once the animal learns how to "play the game," an investigator can design dozens or hundreds of individual trials with different conditions.

Training a monkey to perform a task is often the most time-consuming aspect of an experiment. Much of this time is dedicated to simply acclimating the monkey to research life: sitting in a chair, learning how to accept a juice reward, becoming comfortable working alone in a chamber, and so forth. Most training involves operant conditioning (see Box 2.5), in which successful completion of a task is reinforced with a juice reward. Depending on the task, monkeys are trained to perform at an 80–90% success rate. In order to increase success, the monkeys are often given restricted amounts of food and water before the task begins so they are motivated to work for juice.

During the actual experiments, a monkey performs many repetitions (trials), completing a task for juice rewards. The number of trials performed in a day is primarily determined by the monkey, who will only work for a certain amount of time. Like people, different monkeys exhibit different work ethics. After the completion of the daily experiments, scientists return monkeys to their cages and analyze the data. Many investigators only record from primates every other day. Some experiments, such as experiments measuring learning over time, take place over consecutive days.

CONCLUSION

This chapter has provided a framework for thinking about behaviors in commonly used animal models, surveying many specific assays frequently found in the literature. These techniques are only a fraction of the assays that investigators can use to elucidate the neural basis of behavior. Most published studies, especially studies using rodents, utilize many different behavioral assays in combination to confirm hypotheses and strengthen conclusions.

SUGGESTED READING AND REFERENCES

Books

Conn, P. M. (1993). *Paradigms for the Study of Behavior*. Academic Press, San Diego.

Conn, P. M. (2008). *Sourcebook of Models for Biomedical Research [Digital]*. Humana Press, Totowa, N.J.

Connolly, J. B. & Tully, T. (1998). Chapter 9: Behaviour, learning, and memory. In *Drosophila: A Practical Approach*, Roberts, D. B., ed., 2nd ed. IRL Press at Oxford University Press, Oxford, pp. xxiv, pp. 389.

Crawley, J. N. (2007). *What's Wrong with My Mouse?: Behavioral Phenotyping of Transgenic and Knockout Mice*, 2nd ed. Wiley-Interscience, Hoboken, N.J.

Russell, W. M. S., & Burch, R. L. (1959). *The principles of Humane Experimental Technique [Print]*. Methuen, London.

Whishaw, I. Q., & Kolb, B. (2005). *The Behavior of the Laboratory Rat: A Handbook with Tests*. Oxford University Press, New York.

Primary Research Articles—Papers First Describing Standard Behavioral Assays

Barnes, C. A. (1979). Memory deficits associated with senescence: A neurophysiological and behavioral study in the rat. *J Comp Physiol Psychol* **93**, 74–104.

Geller, I. & Seifter, J. (1960). The effect of meprobamate, barbiturates, d-amphetamine and promazine on experimentally induced conflict in the rat. *Psychopharmacologia* **1**, 491–492.

Hargreaves, K., Dubner, R., Brown, F., Flores, C. & Joris, J. (1988). A new and sensitive method for measuring thermal nociception in cutaneous hyperalgesia. *Pain* **32**, 77–88.

McKenna, M., Monte, P., Helfand, S. L., Woodard, C. & Carlson, J. (1989). A simple chemosensory response in Drosophila and the isolation of acj mutants in which it is affected. *Proc Natl Acad Sci USA* **86**, 8118–8122.

Morris, R. G. M. (1981). Spatial localisation does not depend on the presence of local cues. *Learning and Motivation* **12**, 239–260.

Olton, D. S. & Samuelson, R. J. (1976). Remembrance of places passed: spatial memory in rats. *J Exp Psyc: An Behav Proc* **2**, 97–116.

Ramot, D., Johnson, B. E., Berry, T. L., Jr., Carnell, L. & Goodman, M. B. (2008). The parallel worm tracker: A platform for measuring average speed and drug-induced paralysis in nematodes. *PLoS ONE* **3**, e2208.

Wicks, S. R., de Vries, C. J., van Luenen, H. G. & Plasterk, R. H. (2000). CHE-3, a cytosolic dynein heavy chain, is required for sensory cilia structure and function in Caenorhabditis elegans. *Dev Biol* **221**, 295–307.

Woodard, C., Huang, T., Sun, H., Helfand, S. L. & Carlson, J. (1989). Genetic analysis of olfactory behavior in Drosophila: A new screen yields the ota mutants. *Genetics* **123**, 315–326.

Primary Research Articles—Interesting Examples from the Literature

de Bono, M. & Bargmann, C. I. (1998). Natural variation in a neuropeptide Y receptor homolog modifies social behavior and food response in C. elegans. *Cell* **94**, 679–689.

Moretti, P., Bouwknecht, J. A., Teague, R., Paylor, R. & Zoghbi, H. Y. (2005). Abnormalities of social interactions and home-cage behavior in a mouse model of Rett syndrome. *Hum Mol Genet* **14**, 205–220.

Ramot, D., MacInnis, B. L., Lee, H. C. & Goodman, M. B. (2008). Thermotaxis is a robust mechanism for thermoregulation in Caenorhabditis elegans nematodes. *J Neurosci* **28**, 12546–12557.

Richter, S. H., Garner, J. P. & Wurbel, H. (2009). Environmental standardization: cure or cause of poor reproducibility in animal experiments? *Natual Meth* **6**, 257–261.

Rothenfluh, A. & Heberlein, U. (2002). Drugs, flies, and videotape: the effects of ethanol and cocaine on Drosophila locomotion. *Curr Opin Neurobiol* **12**, 639–645.

Shahbazian, M., Young, J., Yuva-Paylor, L., Spencer, C., Antalffy, B., Noebels, J., Armstrong, D., Paylor, R. & Zoghbi, H. (2002). Mice with truncated MeCP2 recapitulate many Rett syndrome features and display hyperacetylation of histone H3. *Neuron* **35**, 243–254.

Protocols

Brembs, B. (2008). Operant learning of Drosophila at the torque meter. JoVE. 16. http://www.jove.com/index/details.stp?id=731, doi:10.3791/731

Crawley, J. N. (2007). *Short Protocols in Neuroscience: Systems and Behavioral Methods: A Compendium of Methods from Current Protocols in Neuroscience.* John Wiley, Hoboken, N.J.

Current Protocols in Neuroscience. Chapter 8; Behavioral Neuroscience; 2005. John Wiley and Sons, Inc: Hoboken, N.J.

Current Protocols in Neuroscience. Chapter 9: Preclinical Models of Neurologic and Psychiatric Disorders; 2007. John Wiley and Sons, Inc: Hoboken, N.J.

Duistermars, B. J. & Frye, M. (2008). A magnetic tether system to investigate visual and olfactory mediated flight control in Drosophila. JoVE. 21. http://www.jove.com/index/details. stp?id=1063, doi: 10.3791/1063

Hart, A. C., ed. (2006), *WormBook*, ed. The *C. elegans* Research Community, WormBook, doi/10.1895/wormbook.1.87.1, http://www.wormbook.org

Komada, M., Takao, K. & Miyakawa, T. (2008). Elevated plus maze for mice. JoVE. 22. http:// www.jove.com/index/details.stp?id=1088, doi: 10.3791/1088

Mundiyanapurath, S., Certel, S. & Kravitz, E. A. (2007). Studying aggression in Drosophila (fruit flies). *J Vis Exp* (2) pii.155. doi 10.3791/155. JoVE. 2. http://www.jove.com/index/details. stp?id=155, doi: 10.3791/155

Nunez, J. (2008). Morris water maze experiment. JoVE. 19. http://www.jove.com/index/details. stp?id=897, doi: 10.3791/897

Shiraiwa, T. & Carlson, J. R. (2007). Proboscis extension response (PER) assay in Drosophila. JoVE. 3. http://www.jove.com/index/details.stp?id=193, doi: 10.3791/193

Whishaw, I. Q., Li, K., Whishaw, P. A., Gorny, B. & Metz, G. A. (2008). Use of rotorod as a method for the qualitative analysis of walking in rat. JoVE. 22. http://www.jove.com/index/ details.stp?id=1030, doi: 10.3791/1030

Witt, R. M., Galligan, M. M., Despinoy, J. R. & Segal, R. (2009). Olfactory behavioral testing in the adult mouse JoVE. 23. http://www.jove.com/index/details.stp?id=949, doi: 10.3791/949

Chapter 3

Stereotaxic Surgeries and *In Vivo* Techniques

After reading this chapter, you should be able to:

- Describe the process of performing stereotaxic surgeries in rodents and primates, from applying anesthetic agents to monitoring recovery
- Describe common implants that allow permanent access to the brain
- Discuss methods of measuring neural activity and neurochemistry *in vivo*
- Discuss methods of manipulating the nervous system *in vivo*

Techniques Covered

- **Stereotaxic surgeries in animals:** sterile fields, anesthesia, stereotaxic positioning of animals, accessing the brain, implantation, recovery
- **Implants that allow access to the brain:** sealable chambers, cannulae, cranial windows
- **Measuring activity:** electrophysiology, fluorescent activity indicators
- **Measuring neurochemistry:** microdialysis, voltammetry, amperometry
- **Invasive manipulations:** physical lesions, cooling, pharmacology, microstimulation, electrolytic lesions, viral gene delivery

An *in vivo* experiment is any procedure that takes place in a whole, living organism. This kind of experiment contrasts with an *in vitro* experiment that takes place in a controlled environment outside a living organism, such as a cell culture dish or test tube. There is a great trade-off between these two conditions: although *in vivo* experiments accurately reflect the physiology and behavior of a living animal, it can be relatively difficult to measure or manipulate activity in cells and tissues. In an *in vitro* experiment, a scientist has a high degree of control over the extracellular environment and can perform accurate measurements of physiological activity or biochemical expression; however, these results may not correspond to conditions inside a living organism. Chapter 13 discusses *in vitro* methods, including their advantages and limitations.

The purpose of this chapter is to describe the rationale and process of penetrating the brain in a living animal. This allows a scientist to measure the activity and neurochemistry of cells *in vivo*, as well as manipulate their activity and biochemistry to determine effects on behavior. In order to gain access to an animal's brain, scientists must perform a **stereotaxic surgery**, a survivable procedure that allows injection of a substance into the brain or implantation of a permanent device. After describing the process of performing these surgeries in vertebrate animals, we will survey methods of measuring and manipulating physiological activity *in vivo*. Scientists can measure the activity of neurons using electrophysiology or visualization of fluorescent probes. Alternatively, scientists can manipulate specific brain regions using physical, pharmacological, electrical, or genetic methods. These manipulations may affect specific behaviors, as discussed in the previous chapter, and/or affect neural circuits or gene and protein expression, as discussed in the next few chapters.

STEREOTAXIC SURGERIES

A stereotaxic surgery is an invasive procedure used to precisely target specific regions of the brain. The word *stereotaxic* is derived from the roots *stereos*, meaning "three-dimensional," and *taxic*, meaning "having an arrangement." Thus, a stereotaxic surgery places an animal brain within a three-dimensional coordinate system so an investigator can place neurobiological probes and reagents in discrete brain regions. These tools can be used to measure neural activity, sample bioactive substances within the extracellular environment, or manipulate neural function.

Just as sailors use maps and astronomical reference points to navigate the globe, scientists use **brain atlases** and anatomical landmarks to navigate the brain. Published atlases provide three-dimensional coordinates of brain structures for a variety of animals, including rodents, primates, and even birds and bats (Figure 3.1A). The coordinates of brain structures are defined in terms of distances to anatomical landmarks visible as seams on the skull: bregma and lambda. **Bregma** is defined as the intersection between the sagittal and coronal sutures of the skull; **lambda** is defined as the intersection between the lines of best fit through the sagittal and lambdoid sutures (Figure 3.1B). In order to properly align an animal brain so that its structures are consistent with the coordinates in a brain atlas, a scientist must correctly position the animal's head on a **stereotaxic instrument** (Figure 3.2). This specialized equipment holds an animal in place and uses a fine-scale micromanipulator to precisely measure and target distances in $1\,\mu m$ increments.

Brain atlases provide a useful three-dimensional guide for targeting discrete brain regions. However, because different animal strains and individuals vary, scientists must empirically determine the actual coordinates for a region of interest before any experiments take place. Scientists can verify the correct targeting of brain regions by examining histological sections of the brain after

A

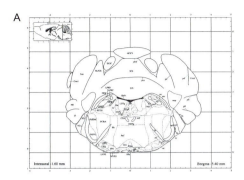

B

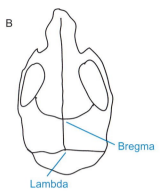

Bregma

Lambda

FIGURE 3.1 Stereotaxic atlases and anatomical landmarks. (**A**) A stereotaxic atlas presents the locations of distinct brain structures in three-dimensional coordinates. To use these coordinates, a scientist must properly align an animal onto a stereotaxic instrument and identify landmarks on the skull. (**B**) Bregma and lambda are commonly used landmarks for stereotaxic targeting of brain structures. (A: Reprinted from Paxinos, G. and Franklin, K. (2001). The Mouse Brain in Stereotaxic Coordinates, 2nd ed. with permission from Academic Press/Elsevier: New York.)

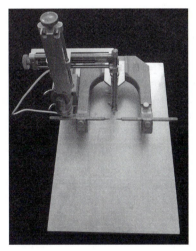

FIGURE 3.2 A rodent stereotaxic instrument. This equipment is used to precisely hold an animal's head in place and target specific brain structures according to three-dimensional coordinates.

experiments have ended. It is also possible to examine the position of brain structures and implants using MRI.

As with any procedures involving vertebrate animals, an Institutional Animal Care and Use Committee (IACUC) must approve procedures before any experiments take place. This committee evaluates survival surgery procedures to ensure that pain and distress are minimized. This is important for the well-being of the animals, as well as for the experiments themselves, as pain and distress can alter physiology and behavior.

In the following section we describe general procedures for performing stereotaxic surgeries in animals. We first describe procedures in rodents, as the majority of stereotaxic surgeries in neuroscience are performed on mice and rats. Then we describe surgeries in nonhuman primates. These procedures can theoretically be adapted for use in any mammalian organism, as well as birds and reptiles.

Stereotaxic Surgeries in Rodents

Scientists routinely perform stereotaxic surgeries on mice and rats because of the variety of useful studies that can be performed using these model organisms. Unlike smaller animals with less complex nervous systems, such as flies and worms, rodent brains are organized similar to the brains of larger mammals, including humans. Unlike larger animals that are not studied in large numbers, such as cats or primates, rodents can be used in large sample sizes to increase the statistical power of experimental results.

Creating a Sterile Environment

The first step in any invasive surgical procedure is to prepare sterile working conditions. **Sterile** doesn't simply mean clean, but also aseptic—free from microorganisms that can invade the brain and lead to subsequent infection. Some aspects of the surgical environment can be made clean, but not sterile. For example, a scientist's lab coat, scrubs, face mask, and gloves can decrease the possibility of contamination but can never be 100% sterile. Cleaning the surgical environment with alcohols and specialized disinfectant solutions can also decrease the possibility of contamination but is also not 100% sterile.

In general, there are two ways to sterilize implants and tools: the use of an autoclave or a tabletop heat sterilizer. An **autoclave** is a large appliance that increases pressure to heat water above its natural boiling point (Figure 3.3A). The high-temperature steam kills any microorganisms growing on tools and reagents placed inside the autoclave chamber. A heat **sterilizer** is a smaller device that can sterilize tools or fine instruments when an autoclave is not convenient, such as sterilization between surgeries (Figure 3.3B). These sterilizers are often filled with thousands of tiny glass beads that allow the heat to evenly surround the instruments. The only aspects of the surgical environment that must be kept completely sterile are the instruments and probes that will come in contact with the brain, as well as a dedicated surface to place these tools when they are not in use.

Just before a surgery takes place, a scientist cleans the surgical environment and sterilizes tools and probes in an autoclave or tabletop sterilizer. Some probes cannot handle the high heat of sterilization and can be soaked

A

B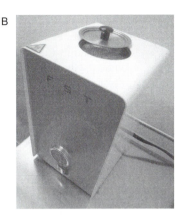

FIGURE 3.3 Equipment for sterilization. (**A**) An autoclave uses high-temperature steam to sterilize contents. (**B**) A bead sterilizer uses high temperatures to sterilize small instruments and surgical tools before and between surgeries.

in 70% ethanol or another disinfectant prior to use. It is usually necessary to create a **sterile field**, a dedicated surface that can serve as a resting place for tools when they are not in use. In order to create a sterile field, a scientist can purchase sterile cloths/papers commercially or can autoclave a large strip of gauze. If the tips of probes or instruments come in contact with gloves or other non-sterile surfaces, they must be considered nonsterile.

Anesthesia

Once the environment is clean and ready for a surgery to take place, a scientist can anesthetize a rodent with either pharmacological agents or gas anesthetics. A widely used pharmacological agent is a cocktail of ketamine/xylazine. **Ketamine** is an anesthetic that acts by inhibiting N-methyl d-aspartate (NMDA) and hyperpolarization-activated cyclic nucleotide-modulated (HCN1) ion channels. **Xylazine** is a sedative and analgesic that serves as an agonist for α_2 adrenergic receptors. A ketamine/xylazine cocktail typically anesthetizes an animal for 1–2 hours. Reinjection of the cocktail is not recommended, as long-term exposure can lead to bradycardia (abnormally slow heart rate), blindness, seizures, or death.

As an alternative to pharmacological agents, scientists can administer gas anesthetics, which are especially useful for long surgeries. A common gas anesthetic is **isoflurane**, an inhalable ether that likely works by disrupting synaptic transmission. To anesthetize the animal, the scientist places the rodent in a small, ventilated chamber. Gas is released into the chamber and the scientist waits 1–3 minutes until the rodent stops all movements. Then, the animal is quickly placed on a bite bar or provided with a nose mask that receives a steady supply of oxygen mixed with the gas anesthetic.

After an animal is anesthetized, a scientist should verify that the animal is indeed unconscious by testing for muscle responses to stimuli. For example, pinching the animal's tail produces a withdrawal of the leg in an animal that is not fully unconscious. Only after the animal shows no signs of motor responses should the scientist begin a surgical procedure.

Investigators who wish to perform stereotaxic surgeries may consult the animal facility at their institution for recommended anesthetic agents, doses, and protocols.

Positioning a Rodent on a Stereotaxic Instrument

After cleaning the surgical area and properly anesthetizing an animal, the scientist places the animal on a stereotaxic instrument. The apparatus is designed to position the animal's head in a precise orientation so that three-dimensional stereotaxic coordinates remain consistent from animal-to-animal (Figure 3.2).

First, the scientist inserts earbars into the animal's ear canals. This placement allows the head to be tilted up and down but not side to side. To restrict movement up and down, the animal's teeth are positioned on a bite bar and the nose is held firmly in a nose brace. It is crucial that the head is precisely fixed in place so that the scientist can accurately traverse the three-dimensional brain.

In order to ensure proper orientation of the brain, the scientist must place the head in the **flat skull position**, such that the top surface of the skull is flat from bregma to lambda in the rostral-caudal direction. To achieve this position, the skull is exposed by a careful incision down the midline of the skin. The skin is gently moved aside and the skull is cleaned. A micromanipulator on the stereotax is used to measure three-dimensional coordinates in the x, y, and z axes. The scientist uses the micromanipulator to ensure the top surface of the skull is flat. If the skull is not flat, the scientist can raise or lower the nose brace/bite bar until the flat skull position is achieved.

Accessing the Brain

Once the animal's head is in the flat skull position, the investigator can target a discrete brain region using three-dimensional coordinates. Starting from bregma or lambda, the scientist orients the micromanipulator over the appropriate area of the skull. A dental drill is used to perform a **craniotomy**, the process of creating a small hole in the skull to expose the brain. Sometimes bleeding occurs at this stage, and the investigator must remove excess blood with a cotton swab or thin tissue. If a scientist intends to implant a delicate probe, such as a glass electrode, it is often necessary to also remove a section of dura, the thin meningeal layer of tissue surrounding the brain.

Penetrating the Brain

Once the brain is exposed, the scientist can use the micromanipulator to deliver a bioactive substance or neurobiological probe into the brain. For example, the scientist can deliver a pharmacological or viral agent into a brain region, or implant a device such as a cannula or optical window (described below). Permanent implants can be held in place with dental acrylic and/or cement. These cements are specifically designed to adhere small objects to the surface of bone. If properly applied, implants can remain securely adhered to the skull throughout the life of the animal.

Finishing the Surgery

After the adhesive has fully hardened, the final tasks are to seal the incision, clean the affected area, and help the animal recover. Depending on the length of the incision, a scientist can use sterile **sutures** or glues designed for tissues to seal the skin around an implant. After the skin is sealed, the top of the animal's head can be lightly treated with an iodine or Betadine® solution to clean the affected area and help prevent infection.

During surgery, an animal can lose much body heat due to blood loss and low heart rate. Therefore, recovering animals should be placed in a recovery cage on a heat pad or under a warm light. As the animal recovers, a scientist should inject further analgesic and/or antibiotic agents and should always monitor the animal until it becomes conscious before returning it to an animal storage room. It will take the animal from a few days to a full week to recover from the surgery. During this time, the animal should be inspected for any signs of pain or discomfort, including little movement, hunched posture, and a lack of eating or drinking. Further injections of an analgesic and/or antibiotic may be necessary to aid in the recovery of the animal.

Stereotaxic Surgeries in Nonhuman Primates

Stereotaxic surgeries in primates are conceptually similar to surgeries in rodents: the investigator anesthetizes the animal, exposes the skull, performs a craniotomy, applies dental acrylic to implant probes, sutures the skin, and carefully monitors the animal throughout the recovery process. The major difference between primate and rodent surgeries is that primates are limited resources and individual investigators may only use one to three monkeys over the course of single studies lasting 4–6 years. Therefore, investigators have no room for error and take extra steps to ensure correct targeting of brain regions. For example, although primate brain atlases exist, scientists often produce structural MR images (Chapter 1) for each individual weeks before the actual surgery takes place so they can target specific regions more accurately.

The purpose of most primate brain surgeries is to implant devices for future electrophysiology experiments. In these experiments, the scientist trains the monkey to perform a task and then records neural activity in a specific part of the brain. In order to perform these experiments, a scientist usually implants three devices:

- **A sealable chamber.** After performing a craniotomy, a scientist implants a chamber around the hole with a screw cap (Figure 3.4A). This allows the surface of the brain to be exposed at any time after the surgery for the insertion of a recording electrode, but protected by the cap at other times.
- **A headpost.** The purpose of this post is to keep the primate's head fixed in place during experiments. It is especially important to stabilize a monkey's head for the presentation of visual stimuli when the monkey must fix his gaze upon a target and make decisions about complicated visual scenes. A headpost can be as simple as a bolt that attaches to the top of the apparatus where the monkey sits during experiments.
- **An eye coil.** The eye coil is also used to ensure that the monkey correctly fixes its gaze during the presentation of visual stimuli. A coil is basically a wire loop that is implanted around the outer circumference of the eye. This coil is usually hidden beneath the eyelids after the surgery. The wire is threaded beneath the skin from the eye to the top of the head, where it can be attached to additional cables to measure eye movements during experiments. An eye coil is not necessary for experiments in which there is no need to monitor the animal's eye movements. Many laboratories implant and adjust these devices over two or three surgical sessions so that their placement is accurate, firm, and precise.

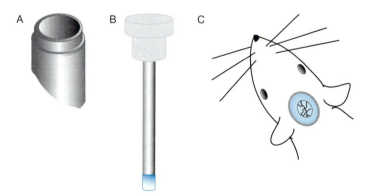

FIGURE 3.4 Implants that allow long-term access to the brain. (**A**) A sealable chamber, allowing chronic access of electrodes and other probes. (**B**) A cannula—a narrow, cylindrical tube affixed to the surface of the skull and placed directly above a brain region of interest. (**C**) An optical window, allowing visual access to superficial brain structures.

IMPLANTS FOR LONG-TERM ACCESS TO THE BRAIN

After a scientist completes a stereotaxic surgery, access to the brain is lost. However, there are a variety of implantable devices that allow long-term access to the brain for chronic injections or physiological recordings. The most common implants include sealable chambers, cannulae, and cranial windows, each described below.

Sealable Chambers

A sealable chamber is a round, hollow well with one end affixed to the skull and the other end capped with a screw-top (Figure 3.4A). This implant allows an investigator chronic access to a specific region of the brain for electrophysiology. Care must be taken to keep the chamber clean and as sterile as possible. Whenever not in use, the chamber must always be closed to prevent outside contaminants from penetrating the brain.

Cannulae

A **cannula** (Figure 3.4B) is a common device used to access deep structures within the brain. It is composed of a narrow, cylindrical tube that a scientist can insert into the brain and permanently affix to the skull. Cannulae can be made of plastic, glass, or steel of various sizes, though thinner cannulae are less damaging to neural tissue. Once implanted, a scientist can deliver multiple substances or probes: pharmacological agents, viral vectors, narrow electrodes, and even optical probes to deliver light stimulation or image deep brain structures. At the end of experiments, scientists verify the correct placement of cannulae by histological examination of brain sections.

Cranial Windows

There are many neuroscience techniques that allow imaging the structure and function of living cells (Chapters 6 and 7), such as labeling neurons with fluorescent probes. In order to visualize these probes *in vivo*, it is necessary to create an observation "window" through the skull (Figure 3.4C). To create such a window, a scientist performs a craniotomy and replaces the missing section of skull with a glass coverslip. Alternatively, scientists can thin and polish an area of skull so that it becomes translucent. These cranial windows allow imaging of the same brain region over long periods of time. It is usually necessary to implant a headpost in addition to a window so that an animal's head can be properly aligned and a scientist can locate the same cells and structures over weeks and months.

Now that we have described stereotaxic surgeries and implants allowing continued access to the brain, we survey various techniques that allow scientists to measure and manipulate neural activity *in vivo*.

MEASURING NEURAL ACTIVITY *IN VIVO*

Neural activity can be measured *in vivo* using electrophysiology or visualization of fluorescent proteins. These methods will be described in greater detail in other chapters, but we mention them here in the context of *in vivo* approaches to studying the brain after stereotaxic surgeries.

Electrophysiology

Electrophysiology techniques are used to record the electrical activity of neurons. We will discuss these techniques in much greater detail in Chapter 4. In brief, there are three fundamental types of electrophysiological probes that can measure activity *in vivo*:

- **Single electrodes**. A single electrode can be inserted through an implanted chamber on the surface of an animal brain. These electrodes are often mounted onto a microdrive, which itself can be mounted on the chamber on the skull. The microdrive precisely raises or lowers the electrodes to specific depths in the brain.
- **Multielectrode array**. A **multielectrode array** is a grid of dozens of electrodes capable of recording the activity of multiple neurons on the surface of the brain. Although extremely small, the size and mass of arrays are often too large to be inserted into deep brain structures. Therefore, they are usually implanted on the outer surface of the brain.
- **Tetrodes**. A **tetrode** is a type of multielectrode array consisting of four active electrodes. Unlike a microelectrode array, a tetrode is narrow and can be easily inserted into relatively deep brain structures.

Fluorescent Activity Indicators

Fluorescent dyes and genetically targeted activity sensors, such as calcium indicators, can be used as indirect measures of neural activity. Cranial windows are used to observe these indicators *in vivo*. We describe fluorescent indicators in much more detail in Chapter 7.

MEASURING NEUROCHEMISTRY *IN VIVO*

It is often important to know how specific neurochemicals, such as neurotransmitters, hormones, or peptides, change in concentration over time. Two techniques are commonly used to identify and measure neurochemicals *in vivo*: microdialysis and voltammetry. These techniques can also be used *in vitro*, but their use in living animals allows scientists the opportunity to correlate changes in neurochemical expression in discrete brain regions with specific behaviors or physiological functions.

Microdialysis

Microdialysis is a technique used to sample chemical substances from the extracellular fluid in a specific location in the brain. Based on the principles of diffusion, a microdialysis probe creates a concentration gradient between the extracellular fluid in the probed brain region and a continuously perfused physiological solution (Figure 3.5). The probe is made of a semipermeable membrane connected to two thin cannulae, with perfusion solution flowing into and out of the probed brain region. Because of the concentration gradient, substances present in the extracellular fluid will passively diffuse through the probe's membrane into the collected solution, the dialysate. This dialysate is collected continuously into a vial for a set amount of time or until enough fluid is collected to be analyzed using sensitive chemical methods, such as HPLC (high performance liquid chromatography), radioimmunoassay, or mass spectrometry.

Numerous factors affect the molecules that can be collected into the dialysis solution: the characteristics of the neurochemicals (size, charge, solubility), properties of the dialysis membrane itself (material, pore diameter), flow speed of perfusion, composition of perfusion solution, and the density of the brain

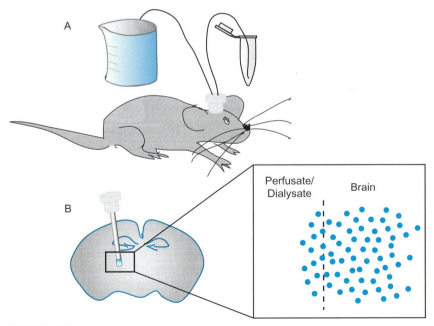

FI1GURE 3.5 *In vivo* **microdialysis. (A)** A microdialysis probe can be implanted through a cannula into an awake, behaving rodent. Scientists perfuse a physiological solution in through the probe and collect the dialysate containing neurochemicals. **(B)** Neurochemicals in the probed brain region pass through the semipermeable membrane of the microdialysis probe through diffusion.

tissue surrounding the microdialysis probe. These factors affect both the identity and the amount of substance that can be collected. Thus, the total amount of substance collected in the dialysate over a set amount of time, known as **absolute recovery**, is not the true extracellular concentration. Instead, scientists present the concentration as **relative recovery**, the relative concentration of the substance in the dialysate from the probed brain region compared to the concentration in the perfusion solution.

It is possible to determine the actual concentration of a substance using at least two different methods: (1) measuring the relative recovery for different flow rates and extrapolating the information to a zero-flow rate, where recovery should theoretically be 100%; and (2) perfusing with known concentrations of the substance to determine the equilibrium point at which the collected concentration does not change, due to the lack of a concentration gradient. Some investigators calculate extracellular concentrations by calibrating probes *in vitro*, immersing the probes in beakers with known concentrations of the substance. However, because of the differences in the diffusion properties in a simple solution compared to the complex environment of the brain, it is not possible to equate *in vitro* with *in vivo* recovery to determine the true extracellular concentration of substances.

There are a number of advantages to using microdialysis to measure neurochemicals in the brain. Microdialysis collection of extracellular fluid can be followed by sensitive chemical assays that clearly identify most types of small molecules. In addition, microdialysis can be combined with simultaneous extracellular electrophysiological recordings and stimulation, even in freely moving animals. Another use of the microdialysis probe is **reverse microdialysis**: the delivery of drugs through the perfusion fluid. This allows for greater control over the flow of the pharmacological agent, leading to a more physiological rate of introduction compared with pressure injection through a cannula.

The greatest disadvantage of microdialysis is low temporal resolution. The concentration of neurochemicals that can be collected in the dialysate is typically very low, requiring a high volume of fluid to be collected for analysis. This can lead to long sampling periods (1–10 minutes), which is adequate for long-term changes but cannot detect neurotransmitter release on a physiological time scale.

Voltammetry and Amperometry

Voltammetry is a technique used to detect neurochemicals capable of undergoing oxidation reactions. These neurochemicals include neurotransmitters such as serotonin and the catecholamines (e.g., epinephrine, norepinephrine, and dopamine). A scientist inserts a carbon fiber microelectrode into the brain and applies a specific voltage. When the chemicals encounter the surface of the electrode, they undergo an oxidation reaction, releasing electrons that produce a measurable change in current. The magnitude of the current is proportional to the number of molecules oxidized. Therefore, it is possible to detect the presence and relative concentration of these neurochemicals at a physiological timescale.

A commonly used form of voltammetry is **fast-scan cyclic voltammetry** (FCV), which has a high temporal resolution (fractions of a second). In FCV, the voltage of the electrode is shifted back and forth from a nonoxidizing potential to an oxidizing potential within milliseconds. The scientist can then plot the amount of measured current versus the applied voltage, producing a **cyclic voltammogram** (Box 3.1). Because electroactive neurochemicals are oxidized and reduced at different potentials, a cyclic voltammogram can be used as a fingerprint to identify the specific neurotransmitter detected at the electrode.

Another subtype of voltammetry is **amperometry**. Unlike in FCV, amperometry holds the electrode at a specific, constant voltage. All molecules that can be oxidized at that potential will be detected. Thus, one limitation of amperometry compared with FCV is that it is difficult to identify the specific compound detected by the electrode solely based on the shape of the current. However, in amperometry, measured currents can be averaged over longer time periods, allowing more precise measurements of the relative concentrations of neurochemicals.

Compared to microdialysis, voltammetric techniques offer a much higher spatial and temporal resolution for measuring neurochemicals. The higher spatial resolution is due to the small size of the carbon fiber microelectrode compared with the larger size of the microdialysis membrane. The temporal resolution of voltammetry (fractions of a second) is much higher than the standard sampling rate of microdialysis, which occurs every 1–10 minutes. However, voltammetric techniques can only be applied to electroactive chemicals and cannot be used to assay other bioactive molecules present in the extracellular environment.

BOX 3.1 Data Analysis: The Cyclic Voltammogram

In cyclic voltammetry, the carbon fiber microelectrode measures current as it cycles between oxidizing and non-oxidizing potentials (Figure 3.6A). A scientist can use this data to plot the current (I) versus voltage (V), an I/V curve similar to those found in some electrophysiology experiments (Figure 3.6B, Chapter 4, Box 4.2). The rapid, large changes in potential produce a high background current. However, these background currents are stable over a period of minutes, and thus, can be subtracted from the experimental I/V curve (Figure 3.6C). Subtracting this background current produces the cyclic voltammogram, the "fingerprint" that identifies an electroactive substance.

Another way scientists present voltammetry data is to show current traces over time to indicate changes in the substance's concentration, demonstrating the responses to electrical, pharmacological, or behavioral stimuli (Figure 3.6D). Three-dimensional graphs can be used to display changes in both voltage and current over time, with the current encoded in color (Figure 3.6E).

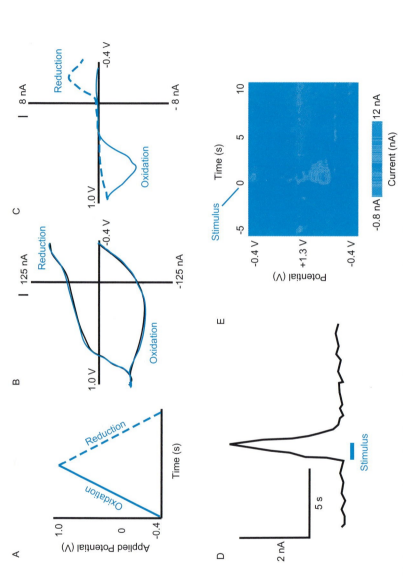

FIGURE 3.6 Analyzing and presenting cyclic voltammetry data. (A) A voltage is applied to the microelectrode that causes oxidation and reduction of nearby neurochemicals. (B) The current (I) is recorded at these oxidizing and reducing potentials. Large background currents (black) are generated from the rapid changes in potential, but they can be subtracted to produce the (C) cyclic voltammogram signature of a neurochemical. Notice the difference in scale between (B) and (C). Plotting current over time (D) can reveal changes in the neurochemical's concentration in response to a stimulus. (E) 3D plots can be used to show changes in both potential and current over time.

MANIPULATING THE BRAIN *IN VIVO*

The ability to penetrate the brain and stereotaxically target specific brain regions provides scientists with many experimental opportunities for determining the role of these regions in animal behavior. For example, in a loss-of-function experiment, a scientist can inactivate a brain region to determine if it is necessary for a behavior to occur. Alternatively, a scientist can perform a gain-of-function experiment and stimulate neural activity to determine whether activity in that region is sufficient to cause a behavior. In order to perform these loss-of-function or gain-of-function experiments, a scientist can manipulate the living brain using physical, pharmacological, electrical, and genetic methods.

Physical Manipulation

Ablation, the permanent removal or destruction of tissue, is perhaps the oldest, most widely used technique to study the necessity of a particular brain region for a biological function. A scientist can physically lesion the brain by excising or aspirating a piece of brain tissue. Despite their place in neuroscience history, permanent physical lesions are a relatively crude method for performing loss-of-function experiments in the brain. It is difficult to control or determine the precision of the lesion or the impact of removing tissue on the remaining tissue. Because ablating an area damages neuronal connections as well as the circulatory system, regions far from the lesion site can be affected. Furthermore, behavioral assays examine the functions of spared brain tissue and, because some time may be required to allow the animals to recover from surgery, the spared brain tissue may be able to recover functions normally controlled by the ablated region. All this can lead to high variability among animals. Fortunately, there are now reversible, nonphysical methods of creating lesions in the brain (described below).

An alternative method to permanent ablation is the use of cold temperatures to produce temporary lesions. Cooling a region of the brain temporarily slows and inactivates cell physiology. In contrast to permanent lesions, reversible inactivation tests the necessity of the cooled structure at the time of the lesion, rather than the ability of spared brain tissue to perform in the absence of the lesioned site. Also, because cooling is temporary, each animal can serve as its own control, making results more reliable and allowing the same animal to be used to examine the effect of the cooled brain region on different tasks. However, temperature-modifying probes are unable to access deep, internal structures without damaging overlying structures. They also tend to require restraint on the animal to maintain the probe's position, restricting the types of behavioral tasks that can be examined.

Pharmacological Manipulation

An investigator can manipulate the nervous system pharmacologically using one of hundreds of compounds that selectively activate or inactivate specific types of ion channels or receptors. An **agonist** is a compound that can bind and activate an endogenous receptor, thus mimicking an endogenous ligand. An **antagonist** can also bind to an endogenous receptor but does not cause activation. "Therefore, it prevents the endogenous ligand from binding, inhibiting its biological activity." Other pharmacological agents target biochemical events that take place within the cell, such as transcription of DNA, degradation of proteins, synthesis of new proteins, or transduction of cell signaling molecules.

Where do pharmacological agents come from? Some drugs are designed in pharmaceutical labs to target specific receptors or other proteins. However, many plants and animals naturally produce pharmacological compounds as neurotoxins for self-defense. Scientists have purified these toxins as a way to manipulate neural activity *in vivo* and *in vitro*. For example, many fish of the order Tetraodontiformes (the most famous being Pufferfish) produce a compound called tetrodotoxin (TTX). This toxin is widely used to block action potentials by binding to the pores of voltage-gated sodium channels. Another commonly used agent to inhibit action potentials is tetraethylammonium (TEA), a potassium-selective ion channel blocker.

Many commonly used pharmacological agents target GABAergic neurons and receptors. GABA receptor antagonists include bicuculline and metrazol and have a net effect of inhibiting GABAergic inhibitory neurons. GABA agonists increase the effect of inhibition, the most popular being a drug called muscimol. This drug is often used to inhibit neural circuitry during *in vivo* electrophysiology experiments.

Some drugs can be delivered to the brain via the bloodstream, allowing a scientist to perform an **intraperitoneal (i.p.) injection**. However, many drugs do not cross the blood-brain barrier. In this case, a scientist can perform an **intracerebroventricular (i.c.v.) injection** into the lateral ventricles of the brain. Substances in the cerebrospinal fluid in the ventricles can then diffuse throughout the extracellular matrix of the brain. Alternatively, a scientist can perform a small, local injection into a discrete brain region.

Injections into the brain can be performed either by pressure injection, or **microiontophoresis**, the process of using a small electrical current to drive substances out of a glass electrode. The advantage to microiontophoresis is that it allows for careful delivery to a small, local area of the brain. For chronic application of a pharmacological agent, a scientist can use reverse microdialysis or osmotic pumps that infuse the agent through an implanted cannula.

Electrical Manipulation

The most commonly used method to stimulate action potentials in neurons, especially *in vivo*, is **microstimulation**. A microelectrode is placed near a neuron of interest and current is applied at a fixed frequency and time. The electrode elicits action potentials by changing the extracellular environment such that voltage-gated ion channels open, depolarizing the neuron. However, if too much current is applied to the electrode, the scientist can kill nearby cells, producing an **electrolytic lesion**. Thus, electrodes can be used both to stimulate neural activity and create small lesions in the brain.

Genetic Manipulation

During a stereotaxic surgery, a scientist can deliver a gene into a specific population of cells using a viral vector. Viruses can be thought of as tiny machines that attach themselves to cells and deliver DNA cargo. These genes can encode toxins, ion channels, or other transgenes that can be used to help determine the role of the virally transduced cells. We describe viral vectors in much greater detail in Chapter 10.

CONCLUSION

Stereotaxic surgeries provide the ability to target specific brain regions for observation or manipulation. Following a surgery, it is possible to measure the neural activity and neurochemistry of a small population of cells, as well as manipulate activity to determine the role of these cells in physiology and behavior. The techniques described in this chapter greatly aid in the study of behavior (Chapter 2), as well as the study of individual neurons and neural circuits, as will be described in the next few chapters. These techniques are also complemented by *in vitro* assays, described in Chapter 13.

SUGGESTED READING AND REFERENCES

Books

Boulton, A. A., Baker, G. B., Bateson, A. N. (1998). *In Vivo neuromethods*. Humana Press, Totowa, N.J.

Institute of Laboratory Animal Resources (U.S.). (1996). *Guide for the Care and Use of Laboratory Animals*, 7th ed. National Academy Press, Washington, D.C.

Paxinos, G., Watson, C. (2009). *The Rat Brain in Stereotaxic Coordinates [Print]*, Compact 6th ed. Elsevier/Academic Press, Amsterdam; Burlington, Mass.

Review Articles

Fillenz, M. (2005). *In vivo* neurochemical monitoring and the study of behaviour. *Neurosci Biobehav Rev* **29**, 949–962.

Hutchinson, P. J., O'Connell, M. T., Kirkpatrick, P. J. & Pickard, J. D. (2002). How can we measure substrate, metabolite and neurotransmitter concentrations in the human brain? *Physiol Meas* **23**, R75–R109.

Wotjak, C. T., Landgraf, R. & Engelmann, M. (2008). Listening to neuropeptides by microdialysis: echoes and new sounds? *Pharmacol Biochem Behav* **90**, 125–134.

Primary Research Articles—Interesting Examples from the Literature

Day, J. J., Roitman, M. F., Wightman, R. M. & Carelli, R. M. (2007). Associative learning mediates dynamic shifts in dopamine signaling in the nucleus accumbens. *Nat Neurosci* **10**, 1020–1028.

Grutzendler, J., Kasthuri, N. & Gan, W. B. (2002). Long-term dendritic spine stability in the adult cortex. *Nature* **420**, 812–816.

LaLumiere, R. T. & Kalivas, P. W. (2008). Glutamate release in the nucleus accumbens core is necessary for heroin seeking. *J Neurosci* **28**, 3170–3177.

Lomber, S. G. & Malhotra, S. (2008). Double dissociation of "what" and "where" processing in auditory cortex. *Nat Neurosci* **11**, 609–616.

Roitman, M. F., Wheeler, R. A., Wightman, R. M. & Carelli, R. M. (2008). Real-time chemical responses in the nucleus accumbens differentiate rewarding and aversive stimuli. *Nat Neurosci* **11**, 1376–1377.

Shou, M., Ferrario, C. R., Schultz, K. N., Robinson, T. E. & Kennedy, R. T. (2006). Monitoring dopamine *in vivo* by microdialysis sampling and on-line CE-laser-induced fluorescence. *Anal Chem* **78**, 6717–6725.

Trachtenberg, J. T., Chen, B. E., Knott, G. W., Feng, G., Sanes, J. R., Welker, E. & Svoboda, K. (2002). Long-term *in vivo* imaging of experience-dependent synaptic plasticity in adult cortex. *Nature* **420**, 788–794.

Ungerstedt, U. & Hallstrom, A. (1987). *In vivo* microdialysis—a new approach to the analysis of neurotransmitters in the brain. *Life Sci* **41**, 861–864.

Xu, H. T., Pan, F., Yang, G. & Gan, W. B. (2007). Choice of cranial window type for *in vivo* imaging affects dendritic spine turnover in the cortex. *Nat Neurosci* **10**, 549–551.

Zhou, F. M., Liang, Y., Salas, R., Zhang, L., De Biasi, M. & Dani, J. A. (2005). Corelease of dopamine and serotonin from striatal dopamine terminals. *Neuron* **46**, 65–74.

Protocols

Athos, J. & Storm, D. R. (2001). High precision stereotaxic surgery in mice. *Curr Protoc Neurosci*, Appendix 4, Appendix 4A.

Chefer, V. I., Thompson, A. C., Zapata, A. & Shippenberg, T. S. (2009). Overview of brain microdialysis. *Curr Protoc Neurosci*, Chapter 7, Unit 7 1.

Cunningham, M. G., Ames, H. M., Donalds, R. A. & Benes, F. M. (2008). Construction and implantation of a microinfusion system for sustained delivery of neuroactive agents. *J Neurosci Methods* **167**, 213–220.

Geiger, B. M., Frank, L. E., Caldera-Siu, A. D. & Pothos, E. N. (2008). Survivable stereotaxic surgery in rodents. JoVE. 20. http://www.jove.com/index/details.stp?id=880, doi: 10.3791/880.

Mostany, R. & Portera-Cailliau, C. (2008). A craniotomy surgery procedure for chronic brain imaging. JoVE. 12. http://www.jove.com/index/details.stp?id=680, doi: 10.3791/680.

Mundroff, M. L. & Wightman, R. M. (2002). Amperometry and cyclic voltammetry with carbon fiber microelectrodes at single cells. *Curr Protoc Neurosci*, Chapter 6, Unit 6 14.

Saunders, R. C., Kolachana, B. S. & Weinberger, D. R. (2001). Microdialysis in nonhuman primates. *Curr Protoc Neurosci*, Chapter 7, Unit 7 3.

Zapata, A., Chefer, V. I. & Shippenberg, T. S. (2009). Microdialysis in rodents. *Curr Protoc Neurosci*, Chapter 7, Unit 7 2.

Chapter 4

Electrophysiology

After reading this chapter, you should be able to:

- Recall the basic electrical properties of neurons and the ionic basis of membrane potentials
- Describe the basic components of an electrophysiology rig
- Compare different categories of electrophysiological recording techniques, including which questions each technique can address
- Compare different types of tissue preparations used in electrophysiology
- Describe common methods of manipulating neural activity during electrophysiological recordings

Techniques Covered

- **Standard electrophysiological methods:** extracellular, intracellular, patch clamp techniques
- **Standard electrophysiological preparations:** heterologous expression systems, primary cultures, slice cultures, anesthetized or awake animals

Electrophysiology is the branch of neuroscience that explores the electrical activity of living neurons and investigates the molecular and cellular processes that govern their signaling. Neurons communicate using electrical and chemical signals. Electrophysiology techniques listen in on these signals by measuring electrical activity, allowing scientists to decode intercellular and intracellular messages. These techniques can answer systems-level questions, such as the role of a neuron in a neural circuit or behavior. Alternatively, they can be used to investigate the specific ion channels, membrane potentials, and molecules that give each neuron its physiological characteristics.

There are a variety of ways to categorize electrophysiology experiments. In this chapter, we will categorize experiments by technique and by tissue preparation. The three main types of electrophysiology techniques are defined by

Guide to Research Techniques in Neuroscience

91

where the recording instrument, the electrode, is placed in the neural specimen. In an **extracellular recording** experiment, the electrode is placed just outside the neuron of interest. In an **intracellular recording** experiment, the electrode is inserted inside the neuron of interest. Finally, using **patch clamp** techniques, the electrode is closely apposed to the neuronal membrane, forming a tight seal with a patch of the membrane. These different recording techniques are used to examine the electrical properties of neurons both *in vitro* and *in vivo*. *In vitro* cell cultures and brain slices allow for detailed investigations of the molecules responsible for electrical signals, while *in vivo* preparations can demonstrate the role of these electrical signals in animal behavior.

The purpose of this chapter is to differentiate between the major categories of electrophysiological techniques and preparations, comparing the relative advantages, disadvantages, and common uses of each. To explain how these techniques are performed, we will survey the instruments that make up an electrophysiology setup ("the rig"). We will also examine some common methods of data analysis and presentation used in the literature. Finally, we will describe methods of manipulating neural activity during electrophysiology experiments. However, before describing electrophysiological techniques, it is necessary to have an understanding of what electrophysiology experiments attempt to measure. Therefore, we will start with a brief review of the physical principles that provide neurons their electrical characteristics.

A BRIEF REVIEW OF THE ELECTRICAL PROPERTIES OF NEURONS

The electrical activity of a neuron is based on the relative concentration gradients and electrostatic gradients of ions within the cell and in the extracellular fluid, as well as the types of ion channels present within the neuron. The difference in charge between the intracellular and extracellular sides of the membrane creates an electrical **potential**, measured in units of volts (V). A neuron membrane at **resting potential** is about $-70\,mV$. This is due to differences in the permeability of various inorganic ions, particularly sodium (Na^+), potassium (K^+), and chloride (Cl^-), as well as to the active contributions of a sodium-potassium pump (Figure 4.1). Ions moving across the membrane generate a measurable **current** (I), the movement of charge over time. The movement of ions across the membrane is limited by the membrane **resistance** (R). This resistance is generated by properties of the membrane, such as how many channels are open or closed. The relationship among the membrane potential, the current flow, and the membrane resistance is described by **Ohm's law**: $V = I \times R$. This relationship is the fundamental basis of electrophysiological techniques.

Neurons communicate by causing changes in membrane potential in other neurons. For example, a neurotransmitter binding to a ligand-gated ion channel opens the channel to allow more ions to flow through the membrane. Relative

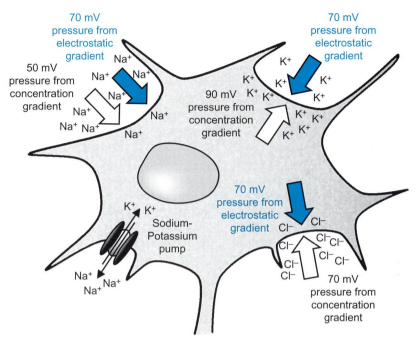

FIGURE 4.1 The ionic basis of the resting potential. The resting potential of neurons—about −70 mV—is caused by the permeability of various inorganic ions. These ions experience pressure to move in or out of the cell based on concentration gradients (differences in concentration of the ion per unit distance in the local environment) and electrostatic gradients (differences in electrical charge per distance in the local environment). In addition to passive diffusion, a sodium-potassium pump continually pumps sodium ions out of the cell and potassium ions into the cell.

to the membrane's resting potential, this current flow can make the membrane potential more positive, an effect called **depolarization**. Alternatively, current flow can make the membrane potential more negative, an effect called **hyperpolarization**. Whether a depolarization or hyperpolarization effect occurs depends on the charge of the flowing ions. This local voltage change is called a **graded potential** or **localized potential**, and its magnitude is proportional to the strength of the stimulus. A local voltage change that makes the membrane potential more positive is called an **excitatory postsynaptic potential (EPSP)**, while a local voltage change that makes the membrane potential more negative is called an **inhibitory postsynaptic potential (IPSP)**. Different EPSP and IPSP events combine to form an overall signal in the postsynaptic neuron. These localized potentials can add up in space (called "spatial summation") and time (called "temporal summation"). If enough localized potentials sum to depolarize the membrane to a threshold point, usually around −55 mV, an action potential will occur.

An **action potential**, also referred to as a **spike**, is an all-or-none, rapid, transient depolarization of the neuron's membrane. A local depolarization to the threshold potential opens voltage-gated sodium channels, and the rapid influx of sodium ions brings the membrane potential to a positive value (Figure 4.2). The membrane potential is restored to its normal resting value by the delayed opening of voltage-gated potassium channels and by the closing of the sodium channels. A refractory period follows an action potential, corresponding to the period when the voltage-gated sodium channels are inactivated. The all-or-none generation of an action potential initiates a wave of depolarization that preserves the amplitude of the voltage change as it propagates down the axon's membrane.

In a chemical synapse, depolarization stimulates the fusion of synaptic vesicles with the presynaptic membrane and the release of neurotransmitter molecules into the synaptic cleft. The neurotransmitters bind to receptor proteins associated with particular ion channels on the postsynaptic membrane. These ion channels then generate EPSP and IPSP events in the postsynaptic neuron, which can, in turn, sum to generate an action potential in that neuron.

Many more details on the electrical properties of neurons can be found elsewhere. What is important in the context of discussing electrophysiology methods is to understand that scientists can study these properties at many different levels of investigation. For example, an investigator may want to know the

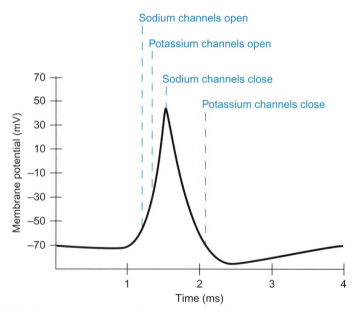

FIGURE 4.2 The ionic basis of an action potential. Localized potentials within the neuron sum to bring the membrane voltage to its threshold potential—around $-55\,mV$. This causes voltage-gated sodium channels to open, further depolarizing the membrane. Potassium channels open as the membrane potential becomes more positive. At about $25\,mV$, sodium channels close, and the membrane potential decreases until it becomes hyperpolarized. Finally, potassium channels close, and the membrane potential returns to a resting state.

frequency of action potentials in a specific neuron over time to decipher how a neuron encodes a particular stimulus or action. This kind of experiment could be performed using an extracellular recording, either *in vitro*, or in an awake, behaving animal. Alternatively, an investigator may want to know how the presence of a drug in the extracellular fluid affects the ability of a specific ion channel to pass current. This experiment could be performed using a patch clamp technique in a heterologous expression system. Whether in the context of circuit analysis or the molecular basis of the membrane potential, nearly any aspect of neuronal physiology can be investigated with the electrophysiology methods described in detail later. Before we describe these methods, let's examine the tools and equipment necessary to perform an electrophysiological recording.

THE ELECTROPHYSIOLOGY RIG

Each electrophysiology lab setup is different, reflecting the questions being addressed, the requirements of the experiment, and the personal preferences of the investigators. There is, however, a standard set of equipment necessary and desirable to record electrical signals from neurons (Figure 4.3). In general, a signal is detected by a **microelectrode**, which transmits that signal to

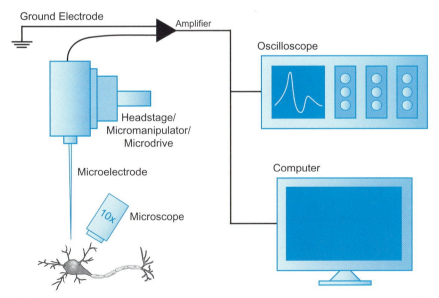

FIGURE 4.3 Some basic components of an electrophysiology rig. In general, an electrical signal is recorded by a microelectrode and passed along to an amplifier. The amplifier compares the recording to a ground electrode, and then passes along the signal to an oscilloscope and/or computer. Various other types of equipment are necessary and desirable, depending on the nature of the experiment.

an **amplifier**, an **oscilloscope**, and a **computer**. The oscilloscope presents a visual display of the membrane potential over time, which can also be heard using a **loudspeaker system**. The signal is also sent to a computer that records data and graphically represents the results of an experiment as it is occurring. An electrophysiology setup includes equipment to stabilize the microelectrode and correctly position it to record from a neuron of interest. A **microscope** is used to ensure proper placement of the electrode for *in vitro* recordings, while stereotaxic equipment (Chapter 3) is used to position electrodes for *in vivo* recordings. Box 4.1 details various parts of a standard electrophysiology rig.

BOX 4.1 Meet the Rig

What follows is a brief description of some of the equipment that may or may not be part of a scientist's electrophysiology rig. Some of this equipment is mandatory, such as a microelectrode and amplifier. Some equipment is not, such as a loud-speaker system.

Microelectrode

There are two main types of electrodes: (1) glass micropipettes filled with an electrolyte solution (2 or 3 M sodium chloride or potassium chloride); and (2) metal electrodes (usually tungsten, steel, or platinum-iridium). An important characteristic of both kinds of electrodes is their electrical resistance, which is related to the exposed tip size. Smaller tips have higher resistances, and they restrict the area from which potentials can be recorded, thus permitting the isolation of the activity of either a fiber or a cell. Large tips and low resistances pick up the activity from a number of neurons and are of limited use in efforts to identify the functional properties of single cells. Tips with very high resistances are also of little use, since they cannot record the neural activity unless they are very close to the cell membrane or actually inside a cell. Electrodes often also have significant capacitances at the metal/fluid interface. Usually, the impedance (a measure of electrical resistance plus capacitance) of the most successful electrodes is in the range of 5 to 20 MOhms when measured by an AC current at 50 Hz. Electrodes with impedances of less than 3 or 4 MOhms tend to record from more than one cell simultaneously. The spikes are smaller, and usually only a multiunit recording can be made.

Glass micropipettes are necessary for patch clamp recordings because of the way the pipette must make a tight seal with the cell membrane (as discussed later in the chapter). For intracellular or extracellular electrophysiology experiments, metal electrodes are more commonly used. These metal electrodes not only provide more stable isolation of single units than micropipettes, but they tend to sample from a larger morphological variety of cells and also help in better localization of electrode tracks to identify where recordings took place in whole brains. The main advantage of using a glass electrode in extracellular or intracellular experiments is that the pipette can be filled with a dye or other materials that can subsequently be injected into the cell or cellular environment for staining and/or pharmacological experiments.

A pipette puller is used to create glass electrodes. A glass capillary is loaded into the machine, which heats the glass and pulls it apart to end in a fine tip. For glass micropipettes that will be used in patch clamp recordings, the micropipette is also fire-polished, rounding and smoothing the electrode tip. Glass capillaries and pipette pullers come in many varieties and are commercially available. Metal electrodes are generally purchased ready-for-use.

It is important to note that *two* electrodes are necessary for any electrophysiological recording: the recording electrode itself and a reference electrode (also called a "ground" electrode) placed outside the cell of interest. A reference electrode is necessary because an electrophysiological measurement is a comparison—for example, a comparison of the potential difference across the membrane of a neuron. In extracellular electrophysiology, both electrodes are located outside the neuron but are placed in different locations in the extracellular environment.

Headstage

The headstage is the central hub that connects the electronic equipment to the tissue preparation. It contains an electrode holder that stabilizes the microelectrode during recordings and also directly connects the microelectrode to the first stage of amplifier electronics needed to detect the electrical signals. The headstage passes the signal to the main amplifier for the main signal processing. The headstage is carefully positioned by the micromanipulator and is also attached to the microdrive.

Micromanipulator

The micromanipulator is a device that allows fine movements in the X, Y, and Z axes, permitting precise positioning of the microelectrode in tissue. Good micromanipulators have fine-scale units of measurement (usually µm) and can be used to stereotaxically place the microelectrode in specific regions of brain or tissue.

Microdrive

A microdrive is used to lower or raise the microelectrode to a specific depth in tissue in very fine steps. It is usually preferable to use remote-controlled microdrive systems to eliminate hand vibration. Thus, the headstage (and consequently the electrode) can be set into place by hand using the micromanipulator and then finely adjusted in and out of tissue using a microdrive for the final approach to the cell.

Amplifier

The signal is passed from the microelectrode on the headstage to the main amplifier where amplification of the signal takes place (100–1000×). A scientist needs an amplifier to enhance the relatively weak electrical signal derived from a microelectrode, just as a radio listener must amplify the signal from an FM antenna so it can drive the speakers on a radio. The amplifier also receives signal from the reference electrode, and it is here that the signals from the two electrodes are compared. The amplifier then transmits the signal to an oscilloscope.

Oscilloscope

An oscilloscope receives the electrical signal from the amplifier and displays the membrane voltage over time. This is the major source of data output in electrophysiology experiments. While many computer programs now contain virtual oscilloscopes that also receive the signal from the amplifier, a physical oscilloscope is still a standard part of a rig to detect subtle dynamics that can get accidentally filtered out by a computer.

Loudspeaker System
Changes in voltage over time are visualized using an oscilloscope, but they can also be heard by connecting the output of the amplifier to a loudspeaker. Action potentials make a distinctive popping sound, so recording the activity of an active neuron can sound like popcorn popping. Loudspeakers can be helpful when trying to locate a neuron of interest, because different types of neurons have distinctive firing patterns. An investigator lowering an electrode into neural tissue using a microdrive can be alerted to a specific type of cell or group of nuclei by the characteristic sounds of neurons near the electrode tip.

Microscope
A microscope is almost always necessary for all kinds of physiological recordings. For extracellular recordings, a low-power dissecting microscope is usually adequate to see laminae or gross morphological features of the tissue and brain. For intracellular physiology or patch-clamp techniques, a microscope with high enough magnification power to see individual cells (300–400×) is necessary. These microscopes are usually equipped with optical contrast enhancement to view cells in unstained preparations. An inverted microscope is usually preferable for two reasons: (1) it allows easier electrode access to the sample, as the objective lens is below the chamber; and (2) it provides a larger, more solid platform upon which to bolt the micromanipulator.

Computer
Computers have greatly aided electrophysiological studies by automating stimulus delivery and electrical signal recording. Computer software makes it easy to write programs that can reproducibly introduce sensory or electrical stimuli to animals or tissue preparations and record neural responses. Computers can easily manipulate many parameters during recordings, such as the recording thresholds and stimulus delivery timing. Computers also allow simple real-time data analysis, displaying the results of an experiment, even while the experiment is occurring.

Vibration Isolation System
A vibration isolation system, usually an air table, is used to absorb tiny changes in vibration that can disturb the placement of the microelectrode. Antivibration tables are usually composed of a heavy surface on compressed-air supports.

Faraday Cage
A Faraday cage can be a simple enclosure made from conductive material that blocks external electrical interference. This is needed to eliminate noise from sensitive electrical recordings that can detect electrical activity from outside sources.

Other equipment may be necessary, depending on the exact nature of the experiment. For example, in experiments that study auditory physiology, a sound booth is necessary to block environmental sounds that could influence recordings. Pharmacological injection equipment is often included in the rig to inject or perfuse drugs and other substances into neural tissue. Special stimulus delivery systems may be necessary, depending on the stimuli needed for the recordings. Each scientist's rig is often a highly specific setup, with all of the equipment designed and calibrated for the particular needs of the experiments being performed.

Now that we have reviewed the electrical properties of neurons and surveyed the equipment needed to study these properties, we will address the various electrophysiology techniques and how they are used.

TYPES OF ELECTROPHYSIOLOGY RECORDINGS

Electrophysiological recordings can be categorized into three main types based on the placement of the electrode in relation to the cell: (1) extracellular recordings; (2) intracellular recordings; and (3) patch clamp techniques (Figure 4.4). Each technique can be used to address specific questions concerning the electrical properties of neurons. For example, questions regarding signals from neurons *in vivo* are most easily addressed using extracellular methods. Questions regarding the "open" and "closed" states of a specific ion channel in the presence of neuropeptide activators are best addressed using patch clamp techniques. Table 4.1 compares some of the questions that can be addressed using various types of electrophysiology recordings.

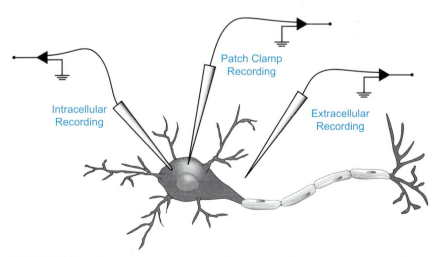

FIGURE 4.4 **The three types of electrophysiological recordings.** Each type of recording is defined by where the scientist places the recording electrode: outside the neuron (extracellular recording), inside the neuron (intracellular recording), or adjacent to the membrane (patch clamp recording).

TABLE 4.1 Questions that Can Be Addressed by Different
Electrophysiological Techniques

Extracellular	Intracellular	Patch Clamp
How does a neuron encode information in action potentials?	How does the activity (or inactivity) of one neuron affect the local potentials and action potentials of another neuron?	How do an ion channel's open and closed times depend on the membrane potential?
What is the role of a neuron in a given sensory, motor, or cognitive behavior?	How do pharmacological agents, neurotransmitters, and neuromodulators affect the local potentials and action potentials of a neuron?	How do the concentration of ions, pharmacological agents, neurotransmitters, and neuromodulators affect the current flowing into an ion channel or cell?
How does the activity (or inactivity) of one neuron affect the activity of another neuron?		How much current does a single ion channel carry?
How do pharmacological agents, neurotransmitters, and neuromodulators affect the firing of a neuron?		What contributions does a single channel provide to an entire neuron?
How is the spiking activity of a group of neurons coordinated?		

Electrophysiologists using intracellular or patch clamp methods use two additional techniques to illuminate the relationship between membrane potential and current flow: voltage clamp and current clamp. Using a **voltage clamp**, a scientist can measure currents generated by ions moving across the membrane by holding the membrane potential at a set voltage, the **holding potential**. Deviations between the membrane potential and the holding potential are corrected through a feedback system that injects current to maintain the membrane voltage at the holding potential. In this way, membrane potential is "clamped" to a set voltage. Most patch clamp experiments use a voltage clamp. Data collected using a voltage clamp can be plotted in I/V curves (Box 4.2) to examine the relationship between current and voltage or to measure fluctuations in current at a particular voltage over time (Box 4.3). In **current clamp** mode, membrane potential is free to vary, and the investigator records whatever voltage the cell generates on its own or as a result of stimulation.

BOX 4.2 Data Analysis: I/V Curves

The I/V curve (current/voltage curve) is one of the most commonly used methods of data analysis in electrophysiology. It is simply a plot of the voltage across a neuronal membrane and the associated ionic current flow through channels in the membrane. Voltage (V) is measured in units of volts, and current (I) is measured in units of amperes. I/V curves can be produced for an entire neuron (with current typically in the range of nanoamps, nA) or for an individual ion channel or class of channels within the neuron (with current typically in the range of picoamps, pA).

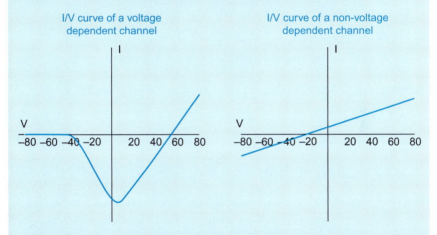

To produce an I/V curve, an investigator uses the voltage clamp technique to hold the voltage of a neuron at a specific value. Current can be recorded using one of the patch clamp methods. The I/V curve is the curve of best-fit for each data point of current for a given voltage. By convention, currents with a negative value are referred to as "inward current," while those with a positive value are known as "outward current." An inward current is the result of positively charged ions crossing a cell membrane from the outside to the inside, or a negatively charged ion crossing from inside to outside. An outward current is the result of positively charged ions crossing a cell membrane from the inside to the outside, or a negatively charged ion crossing from the outside to the inside.

The relationship between current and voltage is described by Ohm's Law:

$$V = IR \qquad V = \text{voltage}$$

$$I = \text{current}$$

$$R = \text{resistance}$$

$$1/R = G = \text{conductance}$$

This means that the slope of the I/V curve is the conductance of the membrane/channel for all ions that can pass through. In many studies where the investigator is interested in the conductance of a specific ion, pharmacological agents are used to isolate specific channels. The conductance depends on an ion channel's permeability for a specific ion (how easily the channel allows the ion to pass through), as well as the concentration of the ion in the extracellular/intracellular solution.

In the preceding example, the I/V curves could represent the current/voltage relationships for particular channels within a membrane. Note that the I/V curve for the channel on the right is linear, meaning it acts like a simple resistor and the conductance of the channel is independent of voltage. An electrophysiologist would say there is no voltage-dependent gating of this ion channel. By contrast, the I/V curve on the left is not linear. At a value of about −20 mV, there is a strong inward current. If the ion that passes through this channel is positively charged, it will flow into the cell when the membrane potential is −40 to +60 mV. This is an indication that the channel is voltage gated. In order to determine the type of ion that can flow through this channel, the ionic concentrations in the bath solution can be adjusted so that the investigator can determine which ionic species are required for current to flow through the channel.

BOX 4.3 Data Analysis: Current Over Time

Many questions in neurophysiology involve knowing how much current flows through a particular ion channel for a particular voltage or other environmental conditions. An I/V curve displays the relationship between the current that passes through a channel or membrane and the voltage across the membrane. For questions involving the ability of a cell or channel to pass current under conditions other than changes in voltage (such as the presence of a compound in the bath or the temperature of the bath), data are often presented as a plot of changes in current over time. These measurements are typically taken under voltage clamp to eliminate contributions to current dynamics due to changes in voltage.

The size of detected current changes depends on the patch clamp technique used to collect the data. For example, in whole-cell mode, the glass pipette is continuous with the cytoplasm, and a large current trace can be obtained. Current data in whole-cell mode is typically presented in nanoamps and can be applied to timescales ranging from 1 ms to even minutes in length. In the following whole-cell recording, an agonist for a ligand-gated ion channel is applied to the bath, as indicated by the blue bar.

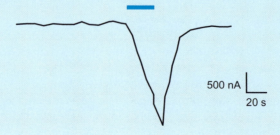

500 nA

20 s

Current data obtained from other patch clamp modes are typically smaller in value and presented in picoamps. This is because the current flowing through single channels rather than an entire cell generates a smaller signal. The timescale of these current traces is also much smaller, in the range of milliseconds instead of seconds, due to the rapid and transient events of a single channel opening and closing.

In these patch clamp modes, data are often presented as a long, continuous trace. For example, in the following experiment, data are obtained in a cell-attached mode from a single channel before and after application of a pharmacological agent:

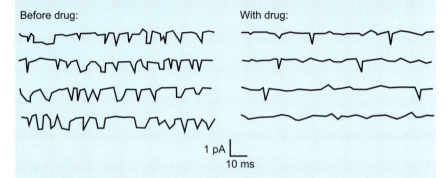

Before drug:

With drug:

1 pA
10 ms

Notice how the scale is different from the whole-cell recording. This experiment shows that the presence of the pharmacological agent reduces the probability that the channel will be open and conduct current.

Extracellular Recording

The dramatic changes in membrane voltage that occur during an action potential generate local, temporary differences in potential on the outer surface of an active neuron. Thus, action potentials can be detected in the extracellular space near the membrane of an active neuron by measuring the potential difference between the tip of a recording electrode and a ground electrode placed in a distant extracellular position. In the absence of neural activity, there is no difference in potential between the extracellular recording electrode and the ground electrode. However, when an action potential arrives at the recording position, positive charges flow away from the recording electrode into the neuron. Then, as the action potential passes by, positive charges flow out across the membrane toward the recording electrode. Thus, the extracellular action potential is characterized by a brief, alternating voltage difference between the recording and ground electrodes (Figure 4.5). The results of an extracellular experiment can be displayed and analyzed in a variety of ways (Box 4.4).

A major advantage to performing extracellular recordings is their relative ease and simplicity. They do not require the precise and delicate electrode positioning that is necessary in intracellular or patch clamp recordings. A disadvantage

Extracellular Electrode   Intracellular Electrode

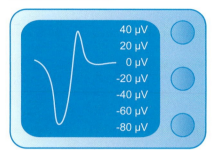
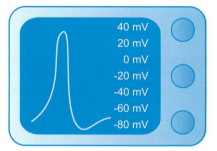

FIGURE 4.5 Recording action potentials from two separate locations. Each action potential exhibits a different waveform due to its placement relative to a neuron. Note the differences in the units of measurement for an extracellular versus an intracellular recording.

to extracellular recordings is that it is impossible to measure localized potentials. Therefore, extracellular recordings cannot report detailed information about subthreshold potential changes, such as EPSPs or IPSPs. However, the summed synaptic activity of neurons in the general vicinity of the extracellular recording electrode produces measurable **local field potentials**, the sum of all postsynaptic activity within a volume of neural tissue.

BOX 4.4 Data Analysis: Spikes

The goal of many electrophysiology experiments is to quantify the number of action potentials, or spikes, that a neuron fires in response to a specific stimulus. Usually, these experiments use extracellular recording electrodes, though intracellular electrodes can also detect action potentials. Data displaying changes in the number of spikes that occur over time following the presentation of a stimulus can be presented in several ways.

For example, consider the data collected during an extracellular electrophysiology experiment in which an electrode is placed in the auditory cortex of a mouse. The investigator presents tones to the mouse and records from a neuron close to the tip of the electrode to determine whether this neuron responds to tones of a particular frequency. The simplest way to present data from this experiment is to show a plot of **voltage over time:**

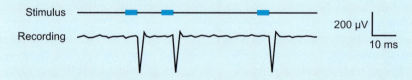

Note that changes in voltage in this plot are in the magnitude of microvolts; an intracellular electrode would record changes in voltage on the scale of millivolts. This plot clearly shows that the stimulus leads to specific changes in voltage over time. However, a simple voltage over time plot does not present the effects of multiple types of stimuli—in this case multiple frequencies of sound. To show the effects of a continuous distribution of stimuli on spike counts, there are other, more efficient methods of data analysis that can combine the results of multiple experiments with different stimuli. For example, the results of multiple frequencies can be presented as a **raster plot**.

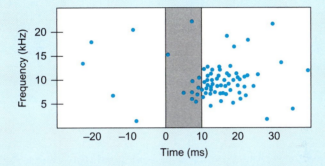

In this type of plot, each dot represents a spike as recorded by the electrode. The Y-axis shows the stimulus frequency, while the X-axis shows the time when the spike was recorded. Stimulus onset and duration are indicated by the vertical gray bar. The data are arranged so that the spikes resulting from a stimulus are aligned with the onset of the stimulus. A raster plot provides a strong visual depiction of the kinds of stimuli that give rise to the greatest number of spikes, as well as an indication of the background noise during an experiment.

Data in a raster plot can also be presented as a **peri-stimulus time histogram** (PSTH) (also referred to as a poststimulus time histogram). In this type of plot, spikes are quantified and sorted into bins of a defined size. In the following plot, the number of spikes (X-axis) is presented for bins that are each 2 kHz in size, with frequency plotted on the Y-axis. A PSTH provides an easy visualization of the stimuli that cause the greatest proportion of firing events. Different mathematical analysis tools, such as a Gaussian distribution curve, can be used to characterize the most effective stimuli and determine the tuning properties of an individual neuron.

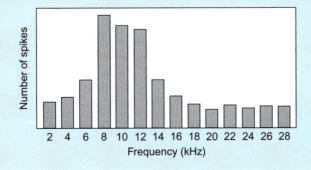

Finally, it is possible to analyze the effects of multiple stimulus variables using a three-dimensional plot. For example, if the investigator presented tones of different frequency as well as different intensity, the data could be presented using color to indicate the number of spikes.

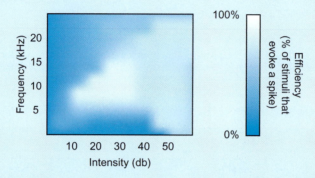

This plot shows that the neuron being investigated responds to a specific frequency of tones that produce a maximum firing rate but that the neuron can be made to produce more spikes if the intensity of the stimulus is increased.

Remember that the raw number of spikes that the investigator includes as data is determined by the settings on the electrophysiology rig and computer! For example, changing the threshold of voltage change that determines what is considered to be a spike will alter the number of spikes included in the data analysis.

Standard extracellular recordings using single microelectrodes typically measure the activity of individual cells (referred to as "units"). By using multiple electrodes, the activity of multiple cells can be recorded at the same time. A popular configuration is the use of **tetrodes**, a bundle of four individual electrodes carefully arranged into a single implant. (Figure 4.6). Grids of individual electrodes or groups of tetrodes can be arranged into a single device called a **multielectrode array** (MEA) (Figure 4.6). This allows for stimulation and extracellular recording from several neighboring sites at once. The number of microelectrodes in an array varies, but can be anywhere from four (a single tetrode) to over one hundred.

Using multiple electrodes, both single-unit activity from the individual electrodes in an array and local field potentials can be recorded. Because spiking activity can be monitored from each electrode, multiple electrodes also help isolate single unit activity. A single unit spike will depend on the cell's unique shape, size, and distance from the recording electrode. Signals are processed through triangulation methods based on the position of the electrode and the strength and characteristic shape of a recording. This results in specific and reproducible waveforms for each individual neuron recorded, enabling

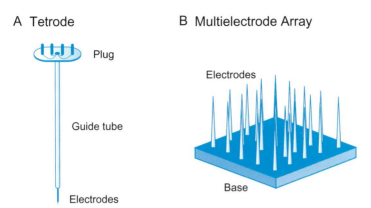

A Tetrode

Plug

Guide tube

Electrodes

B Multielectrode Array

Electrodes

Base

FIGURE 4.6 Two specialized types of electrodes for recording from more than one neuron. (A) A tetrode is composed of four microelectrode wires rolled into a single device. A scientist implants a tetrode into the brain of an animal and connects the top plug to a cable attached to the amplifier. (B) A microelectrode array is composed of 25 or more electrodes (sometimes over 100) and is used to record from neurons on the surface of the brain. Multielectrode arrays can also be used to record from slices *in vitro*.

single unit data to be obtained by recording multiple cells at once. Identifying individual cells based on their spiking activity and assigning the waveforms to particular cells is known as **spike sorting**.

By investigating the activity of dozens of neurons at a time, it is possible to answer questions regarding connectivity and timing within a neural network. A scientist can also manipulate specific neurons within the network and monitor the effect on other neurons recorded by the array. Ultimately, using MEAs to study the simultaneous response of neural circuits provides pivotal information on neuronal interactions, as well as spatiotemporal information about neural networks.

MEAs can be used to record from multiple neurons *in vitro* or *in vivo*. *In vitro*, a brain slice can be placed over a grid of many microelectrodes for extracellular recordings. *In vivo* tetrodes or MEAs can be implanted in the intact brains of live animals for multiunit recordings. Multielectrode technology has also been used in non-human primates for long-term recording. For example, network activity recorded using MEAs in the premotor cortex of monkeys while they perform reaching tasks is being studied to develop neural prosthetics that would allow animals (including humans) to move a prosthetic device using conscious thought.

Intracellular Recording

While extracellular recordings detect action potentials, intracellular recordings detect the small, graded changes in local membrane potential caused by synaptic events. Intracellular recording requires impaling a neuron or axon with a microelectrode to measure the potential difference between the tip of the intracellular electrode and the reference electrode positioned outside the cell.

Note that the measurement of an action potential is different for extracellular and intracellular recordings (Figure 4.5). From the point of view of an extracellular electrode, the difference in potential between an electrode and ground initially decreases, then increases, and then returns to baseline. From the point of view of an intracellular electrode, the potential difference initially increases and then returns to baseline. Also notice that the units of measurement are different between the two techniques. Intracellular electrodes measure potential differences in units of millivolts (mV). Extracellular electrodes measure much smaller potential differences and are often expressed in units of microvolts (μV).

In the early days of electrophysiology, intracellular recordings were used to examine the ionic basis of membrane potential. In fact, Alan Hodgkin and Andrew Huxley won the Nobel Prize in 1963 for using intracellular methods to determine the ionic basis of action potentials. However, most experiments that once used intracellular electrodes are now usually performed using patch clamp techniques due to higher signal-to-noise ratios and the ability to ask questions about the nature of single ion channels. Indeed, after developing patch clamp techniques and using them to study the function of single ion channels, Erwin Neher and Bert Sakmann were awarded the Nobel Prize in 1991.

Patch Clamp Techniques

In the patch clamp methods, a glass micropipette is used to make tight contact with a tiny area, or patch, of neuronal membrane. After applying a small amount of suction to the back of the pipette, the seal between pipette and membrane becomes so tight that ions can no longer flow between the pipette and the membrane. Thus, all the ions that flow when a single ion channel opens in the patched membrane must flow into the pipette. The resulting electrical current, though small, can be measured with a sensitive amplifier. This arrangement is usually referred to as the **cell-attached mode**. Resistance of the seal to flow between the pipette and the cell membrane must be very high to ensure that all current to be recorded flows through the pipette and does not leak out through other regions of the membrane. In practice, seal resistances in excess of 1 GigaOhm can be obtained by a cell-attached patch, often referred to as a "gigaseal."

The cell-attached method allows the investigator to study single ion channels. Minor technical modifications provide other recording possibilities. There are three other variations of the patch-clamp technique: whole-cell, inside-out, and outside-out recordings (Figure 4.7).

If the membrane patch within the pipette is disrupted by briefly applying strong suction, the interior of the pipette becomes continuous with the cytoplasm of the cell. This arrangement allows measurements of electrical potentials and currents from the entire cell and is therefore called the **whole-cell recording** method. The whole-cell configuration also allows diffusional

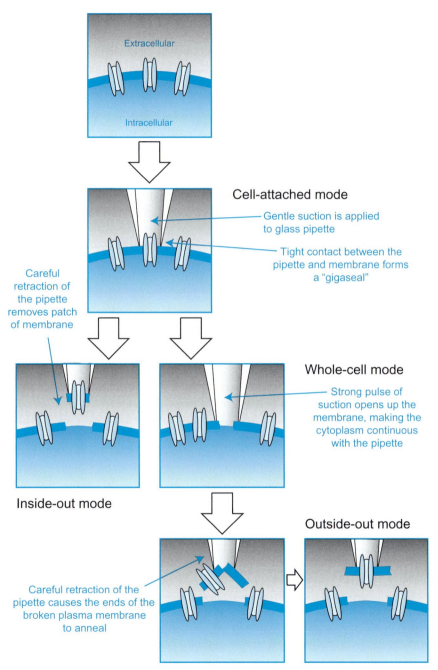

Cell-attached mode

Gentle suction is applied to glass pipette

Tight contact between the pipette and membrane forms a "gigaseal"

Careful retraction of the pipette removes patch of membrane

Whole-cell mode

Strong pulse of suction opens up the membrane, making the cytoplasm continuous with the pipette

Inside-out mode

Outside-out mode

Careful retraction of the pipette causes the ends of the broken plasma membrane to anneal

Extracellular

Intracellular

FIGURE 4.7 **The patch clamp techniques.** The four patch clamp techniques include the cell-attached mode, whole-cell mode, inside-out mode, and outside-out mode.

exchange between the pipette and the cytoplasm, producing a convenient way to inject substances into the interior of a patched cell.

Two other variants of the patch clamp technique originate from the finding that once a tight seal has formed between the membrane and the glass pipette, small pieces of membrane can be pulled away from the cell without disrupting the seal. Simply retracting a pipette that is in the cell-attached configuration causes a small vesicle of membrane to remain attached to the pipette. By exposing the tip of the pipette to air, the intracellular surface of a small patch of membrane is exposed. This arrangement, called the **inside-out recording** configuration, allows the measurement of single-channel currents with the added benefit of being able to change the medium exposed to the intracellular surface of the membrane. Thus, the inside-out configuration is particularly valuable when studying the influences of intracellular (cytoplasmic) molecules on ion channel function.

Alternatively, if the pipette is retracted while it is in the whole-cell configuration, a membrane patch is produced that reseals to expose its extracellular surface to the bath solution. This arrangement, called the **outside-out recording** configuration, is optimal for studying how channel activity is influenced by extracellular chemical signals, such as neurotransmitters, because the bath solution can be easily manipulated during recordings.

Table 4.2 compares the relative advantages and disadvantages of the four variations of the patch-clamp techniques. Sometimes electrophysiologists use a fifth method: the **perforated patch**. The purpose of this technique is to make the glass pipette continuous with the cell, as in the whole-cell mode, but without the disadvantage of cytoplasmic contents potentially leaking into the pipette. Instead of applying suction in the cell-attached mode to transition into the whole-cell mode, a chemical (often an antibiotic, such as amphotericin-B or nystatin) is applied from within the glass pipette, causing small perforations in the membrane. Thus, the perforated patch technique can be thought of as an intermediate between the cell-attached mode and whole-cell mode; this prevents the leaking of cytoplasmic contents but generates a smaller signal than whole-cell mode.

ELECTROPHYSIOLOGY TISSUE PREPARATIONS

Electrophysiology experiments can be categorized by the type of tissue preparation in addition to the type of recording. Physiological experiments *in vitro* usually fall into one of three main categories: recording from heterologous expression systems, primary cultured cells, or brain slices. While *in vitro* preparations allow probing of cell physiology in a controlled environment, conditions in the cell's natural environment *in vivo* may be different. It is possible to perform experiments *in vivo* using extracellular techniques (and more rarely, intracellular and patch clamp techniques), either in an anesthetized animal or in an awake, behaving animal.

TABLE 4.2 Relative Advantages and Disadvantages of Different Patch Clamp Configurations

	Advantages	Disadvantages
Cell-Attached	Single ion channels can be recorded and channel properties are not changed.	Exact potential of the patch is unknown.
	Physiological because cell is intact.	Not easy to control environment of single ion channels.
	Easiest patch clamp configuration to obtain.	
Whole Cell	Ability to manipulate the cell cytoplasm pharmacologically. The relatively large size of the current makes it easier to measure.	The cell is perforated so cell contents may be diluted or leak out.
Excised Patch	Recordings can be taken from individual ion channels. Excellent environmental control over both the intracellular and extracellular sides of single ion channels.	There is a risk that channel properties are changed because the cellular environment is severely altered.

In Vitro Recordings

In vitro culture preparations provide unparalleled physical and visual access to individual cells, allowing detailed studies of the molecules and proteins that affect neuronal physiology. Access to individual ion channels, functional subcellular regions, and small circuits can be gained using various *in vitro* preparations: heterologous expression systems, isolated primary cells, and brain slices.

Control over the cellular environment through the culture or bathing media aids the ability to finely dissect cellular and subcellular physiological events. For *in vitro* electrophysiology experiments, the neural specimen is incubated in a well-regulated bath solution. The composition of the solutions used during the experiments is critical because the basis of neuronal electrical properties depends on the concentration gradient of ions inside and outside the cell. When deciding what ingredients should make up the bath solution, the guiding principle is to maintain an environment that allows the observation of physiologically relevant electrical activity and channel function. In patch clamp techniques, the solution inside the microelectrode should mimic the intracellular milieu. This solution is typically composed of salts, ionic buffering agents, and a pH buffer that can be maintained at biologically relevant temperatures in order to provide cells and tissue with a simulated *in vivo* environment. To isolate the contribution of specific channels or receptors to the electrical signal, investigators frequently add to the bath solution pharmacological agents that block certain receptors, such as CNQX (to block AMPA receptors) or D-AP5 (to block NMDA receptors). One of the greatest advantages of *in vitro* recording is the ability to manipulate cells through the bath solution.

Heterologous Expression Systems

To study the physiological properties of individual ion channels or investigate the role of structure on a channel's function, scientists often use **heterologous expression systems**. A heterologous expression system is a type of cell culture system that can be easily transfected with a foreign gene, such as a gene that encodes a specific ion channel. The gene product (e.g., an ion channel of interest) is efficiently expressed, and its function can be studied with methods such as the patch clamp technique. By expressing mutant forms of a channel, scientists can observe the effect of molecular changes on a channel's electrical properties, such as the role of a specific amino acid on a gating mechanism. Heterologous expression systems are also useful for isolating the role of individual channels because of the absence of endogenous channels that may otherwise alter overall cell physiology.

Common expression systems are *Xenopus laevis* oocytes, Chinese hamster ovary (CHO) cells, and human embryonic kidney (HEK 293T) cells. These cells are usually easy to culture, readily available, have few endogenous channels, and can take up and express foreign DNA easily. The ability to combine molecular and physiological methods in a single cell system has made heterologous expression systems a powerful experimental tool for understanding the structure-function relationships of ion channels (Box 4.5).

Primary Cultures

Primary cultures of cells isolated from a region of interest in the nervous system can reveal properties close to those found *in vivo* while still providing excellent access to and control of individual channels. For example, if an investigator is interested in the physiological properties of spinal cord neurons, he or she can culture these neurons for electrophysiology experiments. Such cultures are useful for comparing the electrophysiological properties of different populations of neurons, as well as the effects of different bath solutions. For example, an investigator may want to know what happens to neuronal physiology when the extracellular concentration of a particular ion or neuropeptide increases or decreases. An investigator can also compare the physiological properties of neurons from different populations of animals—for instance, to examine changes in neuronal physiology in a genetic knockout animal.

BOX 4.5 Walkthrough of an *In Vitro* Electrophysiology Experiment

Let's say that you are interested in examining the properties of a newly discovered ion channel. Your research group has cloned the channel and created an expression construct (Chapter 9) and introduced this channel into *Xenopus laevis* oocytes (Chapter 10). We can now use this heterologous expression system, the large (~1 mm) *Xenopus* oocytes expressing our channel of interest, to answer questions about the electrical properties of the channel. You might be interested in a number of properties: What ions pass through the channel? How is the channel gated, and what stimuli open the channel to let ions pass? You might first look at the similarity of the amino acids to known channels. This can provide hypotheses about the properties of the channel, such as whether the channel is selective for specific ions or what types of compounds may affect gating properties. To examine single channel properties, patch clamp recordings will be the most appropriate technique.

Assuming the lab has established rigs on which to work, the first step in running the experiment will typically be to make the appropriate solutions and prepare fresh glass micropipettes. The composition of both the bath solution and the solution that will fill the micropipette are important because ionic concentration gradients affect the electrical properties of the membrane. The solution's osmolarity and pH will also affect pressure on the cell membrane, which can cause a cell to shrink or swell and burst if not properly regulated. In a whole-cell recording, the pipette-filling solution will be continuous with the inside of the cell, so it should closely match the cell cytoplasm. The bath solution can contain pharmacological agents that block the activity of known channels so the signal from the channel under investigation is clearer. The currents of endogenous *Xenopus* ion channels, however, are typically much smaller than those of heterologously expressed channels; thus, it is not always necessary to add blocking agents to the bath solution. For cell-attached recordings, activity of the predominant endogenous Ca^{2+}-activated Cl^- channels can be reduced by keeping the Ca^{2+} concentration low in the bath solution, by adding chelators such as EGTA and using impermeable ions instead of Cl^-, or by adding a pharmacological blocker.

Patch pipettes are generally used immediately after they are made to ensure that the tip of the micropipette is not contaminated, which can cause a poor seal to form between the membrane and the pipette. A pulled glass pipette should be fire-polished to round and smooth the tip. Now it is ready to be filled with solution and positioned in the electrode holder. Using the microscope and a micromanipulator, bring the electrode tip in a targeted cell into the field of view. Applying gentle, positive pressure to the electrode through a tube to prevent dirtying the electrode tip, insert the tip into the bath solution. By applying a test voltage pulse, the patch electrode will generate a current, and you can calculate the electrode's resistance. Observing the current response to the test pulse can help guide the electrode position. As the tip nears the cell, resistance should increase.

Gently touch the electrode tip against the surface of the cell, checking the computer output of data to see that resistance has increased, and then release the positive pressure. This can be seen as a small dimple against the cell surface as the tip is pressed against the cell. Apply gentle suction to the patch pipette, using resistance and microscopic appearance as evidence that contact has been made between the cell surface and the patch electrode tip. If you have a gigaseal, you should see the resistance rise to at least 1 GΩ. Establishing the gigaseal is one of the trickiest parts of patch-clamping—perhaps the step that makes this technique an art.

At this point, you are in cell-attached mode and can switch the amplifier to voltage clamp. Cell-attached mode can provide single-channel resolution, and you may be able to see characteristic steps in the current that demonstrate transitions between the opened and closed state of a channel. Now you can test whether changes in voltage or hypothesized ligands can activate the patched channel (or channels) by applying voltage steps or adding various agonists or antagonists to the solution filling the micropipette (as this is the extracellular environment of the patched channel).

Monitoring the effect of experimental manipulations on current will reveal open/closed times and current amplitude. Current amplitude can be analyzed to determine the conductivity and ion specificity of the channel. I/V curves (Box 4.2) reveal characteristics of voltage-activated conductance, including the reversal potential, the voltage at which there is no overall flow of ions across the membrane. The reversal potential can be used to identify the ion species that pass through the channel.

Analyzing current flow through individual channels in this heterologous expression system exposes many of the channel's properties. Further studies can use mutant versions of the channel to dissect the requirements of specific aspects of the protein's structure on its function. This knowledge can then be used to investigate the contributions of this channel of interest in intact or natural preparations: isolated neurons, brain slices, and whole animals.

Slice Cultures

The controlled *in vitro* environment in cell culture conditions can be substantially different from the *in vivo* environment. Brain slices, which preserve some endogenous connections while still providing the level of access available through cell culture conditions, can more closely mimic the *in vivo* environment. While it is possible to culture tissue slices for extended periods of time (Chapter 13), to capture conditions closest to those *in vivo*, most electrophysiological recordings from slices are from acute preparations cut the same day. At the beginning of the experiment, the brain is removed and sliced into 300–500 μm thick sections. Many neurons remain healthy despite the mechanical shock and damage of slicing, though physiological responses may be slightly altered. The brain slice is placed in a chamber that is flooded with a solution containing the proper proportion of inorganic ions, nutrients, and gases to allow the neurons to survive.

There are several compelling advantages for recording from a brain slice rather than an intact brain. First, it is much easier to make intracellular recordings from neurons if the tissue is not subject to the periodic pulsing of blood caused by a beating heart. Second, it is also easier to study neurons from a particular region of the brain if an electrode does not have to penetrate several millimeters of cells before it can get to that region. Brain slices provide much better and easier access to internal brain structures. Third, it is possible to study the pharmacology of known synapses because specific drugs or other pharmacological agents can easily be applied to a brain slice. In addition, the specific role of individual neurons in circuits can be studied because it is possible to record from both pre- and postsynaptic neurons in a known synaptic circuit.

In Vivo Recordings

Though *in vitro* recordings allow control of the environment and access to the brain, they may not accurately reflect neural activity in an intact organism. Electrophysiology experiments can be achieved *in vivo* using extracellular methods. It is possible to perform patch-clamp experiments *in vivo*, but the investigator must patch blindly (without a microscope or visual cues) and the cells must be relatively near the brain surface. Indeed, almost all studies carried out *in vivo* submerge an electrode into the brain and record extracellularly due to the fact that it is impossible to visualize the cellular environment and extremely difficult to correctly position an electrode for intracellular or patch clamp recordings.

Electrophysiology experiments can be performed using a variety of animal models, such as rodents (often for learning and memory experiments) and birds (often for song-learning experiments). Many electrophysiology experiments are carried out in nonhuman primates, since they can be trained to perform complex behavioral tasks and have similar physiology to humans.

Acute versus Chronic Recordings

While most *in vitro* recordings are performed during a single session, *in vivo* recordings can be performed both acutely and chronically. An acute recording is an experiment that uses an animal only once. For example, some animal preparations require highly invasive procedures, such as puncturing the lungs so breathing does not affect the recordings (oxygen is supplied by external pumps). As described in Chapter 3, cannulae or electrodes can be surgically implanted into an animal's brain, permitting chronic access to a recording site. Chronic recordings from implanted multielectrode arrays permit the monitoring of neural circuit activity during complex behavioral tasks over long periods of time or during a variety of different conditions and multiple tasks. Sealable chambers implanted on the skull allow an investigator to place new electrodes within the brain over periods of months in rodents, or years in nonhuman primates.

Anesthetized versus Awake Animals

In vivo electrophysiology experiments often involve animals stabilized under anesthesia. Anesthetized animals are immobile, allowing scientists to record neuronal responses to passive stimuli and characterize tuning properties of single neurons. Anesthesia affects the excitability and neurotransmission of normal neural activity, with varied effects depending on the anesthetic agent or neuron population of focus. Thus, it is important for an investigator to know the specific effects of the anesthetic used in order to be able to properly interpret the data. For some experiments, the animals are also paralyzed to eliminate motion from breathing—in these instances, artificial, controlled ventilation is provided.

Techniques used to study neurons *in vivo* have also made it possible to connect specific behaviors to the activity of individual cells through electrophysiological recordings of awake, behaving animals. An animal is first trained on one or more specific behavioral tasks. After a certain level of performance has been reached, a surgical operation is performed to create a small opening in the skull over the brain area of interest. This area is then sealed by implanting a chamber with a screwable cap (Chapter 3). After the animal has recovered from the operation, electrophysiological recordings are performed for a few hours each day with an electrode mounted on a lightweight microdrive. Single or multiunit recording while an animal performs a trained task allows scientists to examine the neural basis of various cognitive and behavioral phenomena. This is one of the few techniques available for investigating complex brain processes at a functional level with great spatial and temporal precision.

METHODS OF MANIPULATING NEURONS DURING ELECTROPHYSIOLOGY EXPERIMENTS

During neurophysiological recordings, an investigator may wish to inhibit or stimulate certain neurons for a "loss of function" or "gain of function" experiment. For example, if an investigator hypothesizes that Neuron A stimulates Neuron B, a way to test this hypothesis would be to stimulate Neuron A while recording from Neuron B. Alternatively, an investigator may want to know the contribution of a certain type of neurons (e.g., GABAergic, cholinergic, glycinergic) to a physiological process; in this instance, the investigator could inhibit only those neurons while examining the physiological and behavioral properties of a neuron of interest.

Several ways to perturb neural activity during electrophysiology experiments are described in more detail in other chapters. Physical, pharmacological, and electrical manipulations that affect neural activity and physiology are described in Chapter 3. Here we learned that lesions produced using permanent ablation or temporary cooling techniques functionally remove a region, allowing an investigator to examine the effects on the activity of the remaining tissue. Pharmacological agonists and antagonists are frequently used in electrophysiological recordings for a variety of purposes, such as isolating the electrical contribution of individual channels, investigating the effect of medication on neurotransmitter release, or determining the role of a biochemical pathway on activity. Microstimulation is another common neural manipulation that allows a scientist to use an electrode to pass tiny amounts of current into biological tissue. Electrical stimulation can generate action potentials or depolarize a cell using subthreshold changes in potential to examine more subtle contributions of neural activity. Chapter 7 describes optical methods for manipulating activity, such as using light to uncage a biologically active molecule or optogenetic techniques that use light-activated transgenic ion channels.

CONCLUSION

Electrophysiology remains the technique of choice for analyzing neural activity and the physiological properties that give rise to this activity. A wide range of techniques and tissue preparations make it possible to record the activity of neurons in a dish, a slice, or an awake, behaving animal. Many scientists, even nonelectrophysiologists, consider electrophysiology techniques to be the backbone of neuroscience research—they are the only methods that can precisely investigate the activity of neurons that produce cognition and behavior, the ultimate output of the nervous system.

SUGGESTED READING AND REFERENCES

Books

Boulton, A. A., Baker, G. B., Vanderwolf, C. H. (1990). *Neurophysiological Techniques: Applications to Neural Systems*. Humana Press, Clifton, NJ.

Hille, B. (2001). *Ion Channels of Excitable Membranes*, 3rd ed. Sinauer Associates, Sunderland, MA.

Molleman, A. (2003). *Patch Clamping: An Introductory Guide to Patch Clamp Electrophysiology*. Wiley, NY.

Nicholls, J. G., Wallace, B. G., Martin, A. R., Fuchs, P. A. (2001). *From Neuron to Brain*, 4th ed. Sinauer Associates, Sunderland, MA.

Nicolelis, M. A. L. (2008). *Methods for Neural Ensemble Recordings*, 2nd ed. CRC Press, Boca Raton, FL.

Review Articles

The Axon Guide: A Guide to Electrophysiology & Biophysics Laboratory Techniques, 3rd ed. (2008). Molecular Devices/MDS Analytical Technologies, Sunnyvale, CA.

Buzsaki, G. (2004). Large-scale recording of neuronal ensembles. *Nat Neurosci* **7**, 446–4451.

Miller, E. K. & Wilson, M. A. (2008). All my circuits: using multiple electrodes to understand functioning neural networks. *Neuron* **60**, 483–488.

Quian Quiroga, R. & Panzeri, S. (2009). Extracting information from neuronal populations: information theory and decoding approaches. *Nat Rev Neurosci* **10**, 173–185.

Super, H. & Roelfsema, P. R. (2005). Chronic multiunit recordings in behaving animals: advantages and limitations. *Prog Brain Res* **147**, 263–282.

Windels, F. (2006). Neuronal activity: from in vitro preparation to behaving animals. *Mol Neurobiol* **34**, 1–26.

Primary Research Articles—Interesting Examples from the Literature

Bliss, T. V. & Lomo, T. (1973). Long-lasting potentiation of synaptic transmission in the dentate area of the anaesthetized rabbit following stimulation of the perforant path. *J Physiol* **232**, 331–356.

Evarts, E. V. (1960). Effects of sleep and waking on spontaneous and evoked discharge of single units in visual cortex. *Fed Proc* **19**, 828–837.

Evarts, E. V. (1968). A technique for recording activity of subcortical neurons in moving animals. *Electroencephalogr Clin Neurophysiol* **24**, 83–86.

Foster, D. J. & Wilson, M. A. (2006). Reverse replay of behavioural sequences in hippocampal place cells during the awake state. *Nature* **440**, 603–680.

Salzman, C. D., Britten, K. H. & Newsome, W. T. (1990). Cortical microstimulation influences perceptual judgements of motion direction. *Nature* **346**, 174–177.

Santhanam, G., Ryu, S. I., Yu, B. M., Afshar, A. & Shenoy, K. V. (2006). A high-performance brain-computer interface. *Nature* **442**, 195–198.

Stowers, L., Holy, T. E., Meister, M., Dulac, C. & Koentges, G. (2002). Loss of sex discrimination and male-male aggression in mice deficient for TRP2. *Science* **295**, 1493–1500.

Sugrue, L. P., Corrado, G. S. & Newsome, W. T. (2004). Matching behavior and the representation of value in the parietal cortex. *Science* **304**, 1782–1787.

Wilson, M. A. & McNaughton, B. L. (1993). Dynamics of the hippocampal ensemble code for space. *Science* **261**, 1055–1058.

Protocols

Brown, A. L., Johnson, B. E. & Goodman, M. B. (2008). Patch clamp recording of ion channels expressed in Xenopus oocytes. JoVE. 20. http://www.jove.com/index/details.stp?id=936, doi: 10.3791/936.

Current Protocols in Neuroscience, Chapter 6: Neurophysiology. (2007) John Wiley & Sons, Inc.

Nicolelis, M. A., Dimitrov, D., Carmena, J. M., Crist, R., Lehew, G., Kralik, J. D. & Wise, S. P. (2003). Chronic, multisite, multielectrode recordings in macaque monkeys. *Proc Natl Acad Sci U S A* **100**, 11041–11046.

Perkins, K. L. (2006). Cell-attached voltage-clamp and current-clamp recording and stimulation techniques in brain slices. *J Neurosci Methods* **154**, 1–18.

Tammaro, P., Shimomura, K. & Proks, P. (2008). Xenopus oocytes as a heterologous expression system for studying ion channels with the patch-clamp technique. *Methods Mol Biol* **491**, 127–139.

Chapter 5

Microscopy

After reading this chapter, you should be able to:

- Define magnification and resolving power, as well as the parts of a microscope that affect these parameters
- Describe the fundamental parts of a standard light microscope and how they manipulate light to magnify an image
- Compare the relative strengths and limitations of common forms of microscopy
- Discuss issues related to image preparation and analysis

Techniques Covered

- **Light microscopy**: brightfield, phase-contrast, darkfield, differential interference contrast (DIC/Nomarski)
- **Fluorescence microscopy**: epifluorescence, confocal, two-photon laser scanning microscopy, total internal reflection fluorescence (TIRF) microscopy
- **Electron microscopy**: transmission electron microscopy (TEM), scanning electron microscopy (SEM), electron tomography (ET)
- **Microscopy data**: image processing and interpretation

The first microscopes were used in the seventeenth century to expose the microscopic world of cells and single-celled organisms for the first time. Scientific pioneers such as Robert Hooke and Anton von Leeuwenhoek, often called the "fathers of microscopy," used homemade microscopes to study cell types from a variety of living organisms. In the late nineteenth and early twentieth centuries, Santiago Ramón y Cajal used a microscope in combination with histology to produce highly detailed, seminal studies of nervous system structure.

The microscope is now an indispensable tool in neuroscience. Because organelles, glial cells, neurons, and even populations of neurons cannot be seen by the naked eye, a microscope is essential to examine the nervous system at the cellular level. Light microscopes can enlarge the image of an object up to 1000 times greater than normal, providing access to the structure of a cell and its local environment. Fluorescent microscopes provide an even greater ability to highlight individual subcellular structures. Electron microscopes can theoretically enlarge an image 1 million times greater than normal, providing unparalleled insight into the smallest neuronal structures including synapses, surface receptors, and even individual proteins.

The goal of this chapter is to provide a basic description of common forms of microscopy. First, we define the fundamental parameters and parts of a light microscope. Then, we survey the different forms of microscopy and why an investigator might choose one form over another. Finally, we examine issues related to processing and interpreting microscopy data. The information in this chapter will complement information in many other chapters, especially Chapters 6 and 7.

ESSENTIAL PRINCIPLES OF MICROSCOPY

While modern microscopes are undoubtedly more advanced than those used a few centuries ago, the fundamental parts of a microscope and the theory behind how they work together are essentially the same. Before surveying different forms of microscopy, we review the fundamental parameters and parts of standard light microscopes.

Fundamental Parameters in Microscopy

There are two important values to consider in microscopy: magnification and resolution. **Magnification** refers to how much larger the sample appears compared to its actual size. Figure 5.1 provides a sense of the relative size of nervous system components, from an entire brain down to individual atoms. The naked eye is able to perceive objects about 0.2 mm in size and larger. Therefore, it is possible to view entire brains and large neural structures, but not individual neurons or axons without additional magnification.

Resolution (or **resolving power**) refers to the minimum distance by which two points can be separated yet still be distinguished as two separate points. For example, imagine you are looking at the side of a building with a mosaic image made up of thousands of colored tiles. If you were standing a block away from the building, you would see the entire image but would be unlikely to make out the individual tiles themselves because the resolution would be too low to see that level of detail. If you move closer, the resolution becomes better and you can see the distinct tiles making up the image. Likewise, in microscopy, it can be difficult to perceive two structures as separate from one another – only by increasing the resolving power is it possible to make out distinct points in a specimen. Figure 5.1 characterizes the resolution of commonly studied neural structures, demonstrating the range of objects that can be resolved by the naked eye and the increased resolving power provided by light and electron microscopy.

Though it may seem that the resolving power of a microscope is based on the ability of the microscope to magnify an object, this is not the case. Magnifying an object does not necessarily improve the clarity or resolution of the final image. The resolution of a magnified object depends on two factors: numerical aperture and the wavelength of the light source. The **numerical aperture** (NA) is a

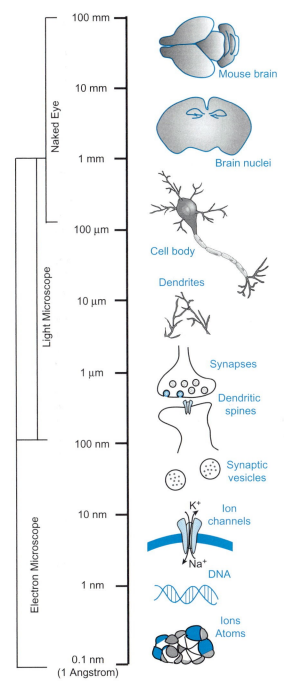

FIGURE 5.1 Magnification and resolving power in microscopy. A magnification scale from a small animal brain down to individual atoms.

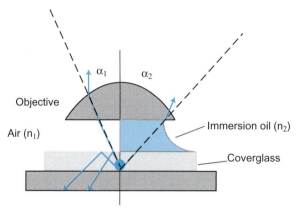

FIGURE 5.2 Numerical aperture and the angle of light entering the objective. An objective with a higher NA collects more light rays, leading to a higher resolving power. The NA depends on the angle that light enters the objective, α, as well as the refractive index of the medium, n. When light travels through material with drastically different n, such as between glass and air (n_1), it refracts more and yields a smaller α. Immersion, such as water or oil (n_2), reduces the difference and allows for a greater α.

measure of the light-collecting ability of the microscope objective, the lens that gathers and focuses light from the specimen. An objective with a higher NA will collect more light rays, leading to better resolving power. The NA of an objective depends on the angle that light enters into the objective and the index of refraction of the medium in which the objective is working (Figure 5.2). The **index of refraction** (or **refractive index**) indicates changes in the speed of light traveling through a particular medium (1.0 for air, 1.33 for water, and up to 1.55 for oils). The greater the index of refraction, the greater the NA. As light travels from one medium (e.g., air) to a medium with a higher index of refraction (e.g., oil), the angle of the light changes, allowing an objective to collect more light. This is the same principle that causes a straw in a glass of water to appear bent at the surface of the water. An objective lens immersed in oil has a higher NA than an objective in air, and thus the oil-immersed objective has a greater resolving power.

The other factor that determines the resolution of a magnified object is the wavelength of light either illuminating or emanating from the specimen. In light microscopy, it is the wavelength of light illuminating the specimen; in fluorescent microscopy, it is the wavelength of light emanating from the specimen. Wavelength is the distance between two repeating units of a light wave (Figure 5.3A). Light rays with different wavelengths appear as different colors to the human eye (Figure 5.3B). The wavelength of light affects how broad a single point of light appears. Shorter wavelengths of light (UV through green) appear sharper than longer wavelengths (red through infrared), which tend to appear more spread out, decreasing resolving power.

Therefore, the smallest object a microscope can display is a function of the magnification and resolving power used to image a specimen. Interestingly, while

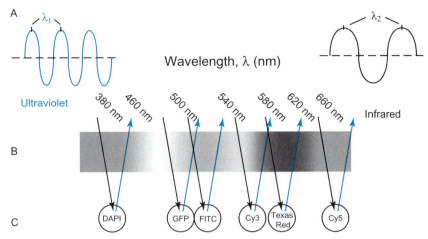

FIGURE 5.3 Light wavelength and the visible spectrum. (A) The wavelength of light, λ, is the distance between two repeating units of a propagating light wave. **(B)** Light rays of different wavelengths appear as different colors to the human eye. Shorter wavelengths (λ_1) appear violet-blue, while longer wavelengths (λ_2) appear more red. **(C)** Approximate peak excitation (black) and emission (blue) wavelengths of commonly used fluorophores.

the ability of a microscope to magnify an object is essentially unlimited, the resolution is finite. A light microscope cannot resolve details of a specimen finer than $0.2\,\mu m$—about the size of a neural synapse—no matter what the magnification. This limitation is due to the minimum wavelength of visible light; wavelengths shorter than the UV to green portion of the spectrum are invisible to the human eye. This means the resolving power of any light microscope has a maximal value (Box 5.1). The unaided human eye is limited to seeing details about $0.2\,mm$ in size. A light microscope can only resolve details of about $0.2\,\mu m$, so magnifying those details more than $1000\times$ ($1000 \times 0.2\,\mu m = 200\,\mu m = 0.2\,mm$) would not be useful. Increasing the magnification higher than $1000\times$ is like using the zoom feature on a computer to enlarge a digital picture of a fixed size: you won't get any finer details with a digital zoom-you simply make the pixels bigger. However, technologies that do not use visible light to image a specimen, such as electron microscopy, can have much higher resolving powers and can image much smaller structures than light microscopes.

Now that we have discussed magnification and resolution, let's examine the fundamental parts and design of two categories of microscopes: the compound microscope and the stereomicroscope.

The Design of a Compound Microscope

The simplest form of microscopy, one that has been used for centuries, is the magnifying glass. A magnifying glass is a convex lens that enlarges nearby objects (Figure 5.4). It works by refracting the light scattered by an object to a

BOX 5.1 The Diffraction Limit of Resolution and Ways Around It

Why does the wavelength of visible light ultimately set the limit of resolution for light microscopes? It is because of a phenomenon known as diffraction. Diffraction can be thought of as the tendency for a wave to spread out and bend around obstacles as it propagates. Light behaves like a wave, so it too can bend around obstacles, just like water ripples moving past a boulder in a river. When a single point of light passes through the opening of a microscope objective, it doesn't appear as an infinitely small single point of light; it has a finite size that looks more like a bright point with concentric rings of decreasing brightness, like the rings of ripples you see in a puddle when a raindrop hits. This ring of light ripples is known as an Airy disk. Because each point of light is really a ring of light ripples, the distance between two points needs to be large enough so the ripples don't interfere with each other. Diffraction is what causes the points of light to create these ripples. Because the ripples are the factor that limit how close together two points of light can be and still be seen as two distinct points of light, this is known as the diffraction limit of resolution. The minimum wavelength for light visible to the human eye (~400 nm) sets this diffraction limit.

New forms of superresolution microscopy attempt to get beyond the diffraction limit of resolution to image nano-scale molecular interactions. Stimulated emission depletion (STED), ground-state depletion (GSD), and saturated structured-illumination microscopy (SSIM) break the diffraction limit by controlling the spatial and/or temporal transition between a fluorophore's molecular states (e.g., dark vs. bright state). This is done by modifying the excitation light pattern to effectively shrink the size of the Airy disk. Displaying its importance in cellular and molecular neuroscience, STED was used to produce video-rate images of synaptic vesicles in live hippocampal neurons at a resolution of 60 nm. Photoactivated localization microscopy (PALM), fluorescence photoactivation localization microscopy (FPALM), and stochastic optical reconstruction microscopy (STORM) break the diffraction limit by detecting the positions of single molecules. By using photo-switchable fluorescent probes, overlapping individual fluorophores can be separated in time to construct a superresolution image.

focal point on the other side of the glass. What appears is a bigger image of the object behind the lens.

A single lens has limited magnification ability, but multiple lenses can be arranged one behind the other to multiply the magnification of each lens. This is the basis of the **compound microscope**. The term *compound* refers to the two or more lenses used in concert to magnify an object.

Most light microscopes use at least two lenses. The first lens is called the **objective lens** and is placed adjacent to the specimen (Box 5.2). Investigators can choose from multiple objective lenses to use a desired degree of magnification, usually 4×, 10×, 20×, 40×, or 100×. The second lens is located in the eyepiece and is referred to as the **ocular lens**. This lens is often set at 10×. The total magnification of the microscope is multiplicative, so if the objective lens is set at 40× and the ocular lens is set at 10×, the total magnification achieved is 400×.

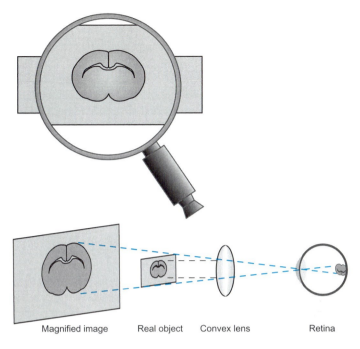

Magnified image   Real object   Convex lens   Retina

FIGURE 5.4 **The magnifying glass.** A magnifying glass is a convex lens that refracts light to make an object appear larger. (Courtesy of Dr. Dino Leone)

BOX 5.2 Meet the Microscope

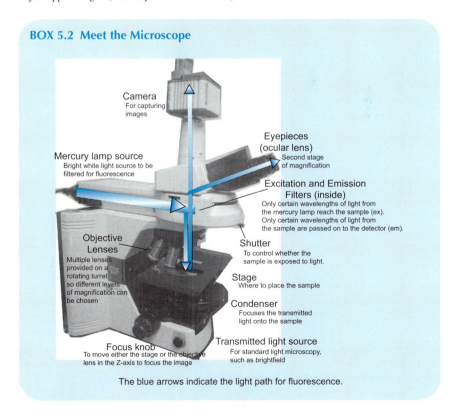

Camera
For capturing images

Mercury lamp source
Bright white light source to be filtered for fluorescence

Eyepieces (ocular lens)
Second stage of magnification

Excitation and Emission Filters (inside)
Only certain wavelengths of light from the mercury lamp reach the sample (ex). Only certain wavelengths of light from the sample are passed on to the detector (em).

Objective Lenses
Multiple lenses provided on a rotating turret so different levels of magnification can be chosen

Shutter
To control whether the sample is exposed to light.

Stage
Where to place the sample

Condenser
Focuses the transmitted light onto the sample

Focus knob
To move either the stage or the objective lens in the Z-axis to focus the image

Transmitted light source
For standard light microscopy, such as brightfield

The blue arrows indicate the light path for fluorescence.

Although every compound light microscope is unique, they manipulate light in similar ways to magnify a specimen (Figure 5.5). A **condenser** focuses light from a light source onto a specimen. Light transmitted through the

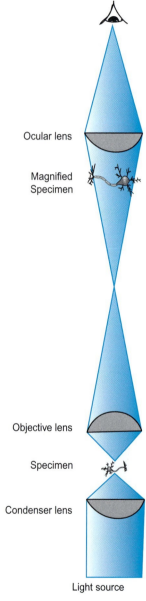

Ocular lens

Magnified
Specimen

Objective lens

Specimen

Condenser lens

Light source

FIGURE 5.5 Compound microscopes use multiple lenses to magnify a specimen. A condenser lens focuses light onto a specimen, which then transmits (or reflects) light through the objective and ocular lenses that magnify the image that reaches the eye.

specimen is then collected and magnified by the objective lens. This magnified image is focused onto the ocular lens, which essentially acts like a magnifying glass to further enlarge the image from the objective lens. Although each manufacturer designs light microscopes using unique strategies, the conceptual basis of the compound microscope remains the same.

Compound microscopes are either upright or inverted. In an **upright microscope**, the specimen is placed just below the objective (Figure 5.6A). This type of microscope is ideal for examining a specimen mounted on a glass slide. However, there is little space between the objective and specimen, making it impractical for manipulating brain sections for electrophysiology or for viewing live cells in a thick cell culture dish. In such situations, an investigator is more likely to use an **inverted microscope**, which positions the objective below the specimen with the light source and condenser located relatively far above the sample (Figure 5.6B).

The other parts of a basic compound microscope allow the investigator to adjust the magnification, focus, and light level of the microscope, as well as the precise placement of the specimen Box 5.2.

The Design of a Stereomicroscope

A **stereomicroscope**, also called a **dissecting microscope**, serves a different purpose than a compound microscope and works in a different way (Figure 5.7). The main purpose of a stereomicroscope in neuroscience is to examine the surface of brains, tissue slices, or large neural structures. It is especially useful for the fine manipulation needed during dissections, surgeries, or the

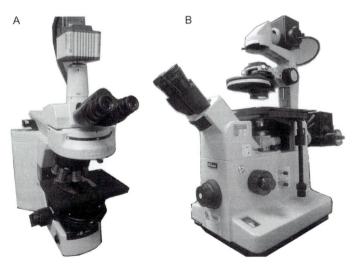

FIGURE 5.6 Upright versus inverted microscopes. (A) In an upright microscope, the specimen is placed just below the objective. **(B)** In an inverted microscope, the specimen is placed just above the objective, allowing for larger cell culture plates and access from above.

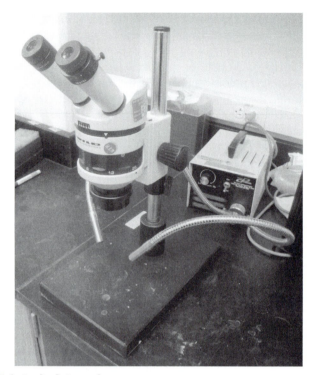

FIGURE 5.7 A standard stereomicroscope.

fabrication of small tools such as electrodes or implants. Rather than passing through a single objective and ocular lens system, light in a stereomicroscope goes through two separate lens systems. While a compound microscope directs light to both eyes from a single lightpath, a stereomicroscope directs light to each eye from two independent lightpaths. Because light from a single point on the specimen travels independently through two different paths to reach each eye, the specimen appears three-dimensional.

In addition to the appearance of the specimen, there are at least two other major differences between a stereomicroscope and compound microscope. First, stereomicroscopes typically use light that is reflected off a specimen, while compound microscopes often use light transmitted through a specimen. This makes stereomicroscopes useful for looking at specimens too thick for light to pass through. Thus, a stereomicroscope is often used to view an animal during a surgical procedure or while dissecting an animal brain into distinct regions in a dish. Second, the magnification power of a stereomicroscope is often less than a compound microscope. The ocular lenses of a stereomicroscope are usually fixed at $10\times$ and the objective lenses typically range from 0.1 to $8\times$. Therefore, a stereomicroscope is useful for examining gross neural structures, while a compound microscope is better for examining microscopic structures such as single neurons or fiber tracts.

Now that we have examined the basic concepts of microscopy, we survey the various categories of light microscopy and why each may be particularly useful for a given experiment.

LIGHT MICROSCOPY

A light microscope is any microscope that uses visible light to illuminate and image a specimen. This includes white light composed of all wavelengths, as well as the light of a specific wavelength used in fluorescent microscopy. Usually when people refer to light microscopy, they refer to nonfluorescent microscopy, even though fluorescent microscopy does, of course, use light.

The most common and general form of light microscopy is **brightfield microscopy**, in which light passes directly through or is reflected off a specimen. Most cells and tissues are transparent due to their high water content. Unless naturally pigmented or artificially stained, distinct structures are difficult to differentiate using brightfield microscopy. Thus, a variety of histological procedures have been developed to preserve and stain specimens to enhance contrast among different microscopic structures (Chapter 6). However, most of these procedures result in the death of cells to preserve the specimen.

For many experiments, working with preserved cells is not a problem. However, when an investigator wishes to magnify living, unpreserved cells or tissues, as in cell culture experiments, it is necessary to provide contrast without killing the specimen. Thus, methods of manipulating light to enhance contrast have been developed to make resolvable details stand out to the eye without requiring the application of special dyes.

Variations in density within cells cause tiny differences in the way different regions scatter light: different cellular structures have different indices of refraction. **Phase-contrast microscopy** takes advantage of these slight differences, amplifying them into larger intensity differences with high contrast that can be easily seen. Because this does not require special treatment of the specimen in order to see many details, phase-contrast is often used to examine cultured cells. **Darkfield** (or **darkground**) microscopy uses oblique illuminating rays of light directed from the side so that only scattered light enters the objective lens. Most regions of the specimen (e.g., cytoplasm) do not scatter much light so they appear dark, while highly refractive organelles that scatter large amounts of light will appear bright. **Differential interference contrast (DIC) microscopy**, also known as **Nomarski microscopy**, uses optical modifications to exaggerate changes in the light-scattering properties of cellular structures. This creates highlights and shadows around edges that create a three-dimensional textured look. Figure 5.8 demonstrates the different appearances of a single cell when viewed using these various forms of light microscopy.

FLUORESCENCE MICROSCOPY

Fluorescence microscopy takes advantage of specialized molecules called **fluorophores** that have the property of absorbing light at a specific wavelength and

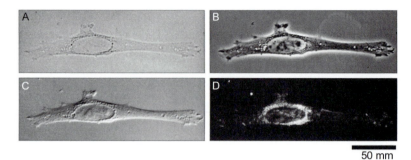

50 mm

FIGURE 5.8 Four types of light microscopy used to analyze the same cell. Various methods of light microscopy are used to enhance contrast and visualize details in unstained tissue preparations. **(A)** Brightfield microscopy, **(B)** phase-contrast microscopy, **(C)** differential interference contrast (DIC/Nomarski) microscopy, and **(D)** darkfield microscopy. (Reprinted from *Molecular Biology of the Cell,* 4th edition (2002), B. Alberts et al., Fig. 9-8, with permission from Garland Science: New York, NY.)

then emitting light at a different (typically longer) wavelength. Fluorophores that absorb blue light typically emit green light; fluorophores that absorb green light emit red; fluorophores that absorb red emit infrared, and so on. There are dozens of commercially available and commonly used fluorophore molecules each with their own characteristic absorption (excitation) and emission wavelengths (Figure 5.3C). These molecules can be linked to antibodies or other molecular probes to signify the presence of specific proteins or organelles and mark particular structures within a cell.

The principal use of fluorescence microscopy is to examine specimens labeled with these fluorescent reagents or expressing genetically encoded fluorescent proteins (Chapter 6). Light from an extremely bright source passes through an **excitation filter** that allows only light in a specific range of wavelengths through the objective to illuminate the specimen (Figure 5.9A). That light is absorbed by and excites the fluorophores in the specimen, which react by emitting a longer wavelength of light. The emitted light is collected back through the objective and passed through a second filter, an **emission filter**. The emission filter blocks extraneous wavelengths of light, including the light used to illuminate the specimen, but allows emitted light wavelengths to pass through to a detector (i.e., your eye or a camera). The optical filters make it appear as if structures tagged with the fluorophores light up against a dark background. Because the background is dark, even a tiny amount of the glowing fluorescent reagent is visible, making fluorescence microscopy a very sensitive technique.

A major appeal of fluorescent microscopy over conventional light microscopy is that structures and proteins within a cell can be labeled with a wide range of fluorophores. Because each fluorophore has its own characteristic excitation and emission spectra (the wavelengths of light they absorb and emit), it is possible to label different structures in the same sample and investigate their relative locations. For example, an investigator can use a fluorophore that emits light in the green end of the spectrum to label one protein, and a fluorophore that emits

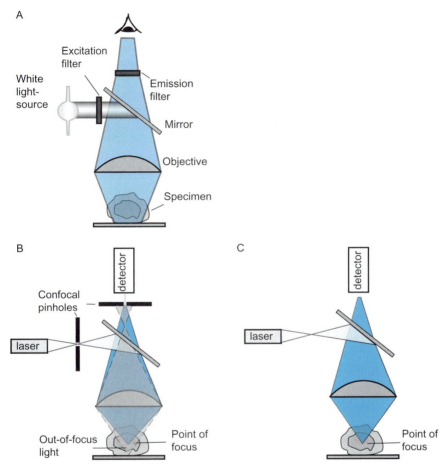

FIGURE 5.9 Comparison of different forms of fluorescence microscopy: epifluorescence, confocal, and two-photon microscopy. (A) Basic strategy of a fluorescent microscope. Light from a bright lamp passes through a filter that only allows light of the excitation wavelength through to the specimen. Light emanates from the specimen and passes through a second filter that is opaque to the excitation wavelength of light but that transmits the longer wavelength of the emitted light. **(B)** Basic strategy of a confocal microscope. Light is focused from a pinhole aperture onto a single point on the specimen. Fluorescent light emanating from the specimen is transmitted through a second pinhole aperture before hitting a detector. The two apertures are *confocal* with each other, so out-of-focus fluorescence is avoided by the detector pinhole. **(C)** Basic strategy of a two-photon microscope. Pulsed laser light illuminates the sample with light about twice as long as the wavelength required for excitation. When focused on a specific plane of section, two photons can arrive near-simultaneously at one fluorophore in that plane and deliver enough energy to excite the fluorophore. Above and below the plane in focus, light is too low-energy to excite fluorescence, eliminating out-of-focus fluorescence.

light in the red end of the spectrum to label another. Because the excitation and emission spectra of the green and red fluorophores do not overlap, they distinctly mark the locations of these different proteins within the same sample. This technique can even be used in living specimens to observe the dynamic interactions of multiple fluorescently labeled proteins and molecules.

Despite the advantages of fluorescence microscopy, there are some disadvantages. One major limitation is that fluorescent reagents cannot be illuminated indefinitely. The intensity of the light emitted from a fluorophore will decrease over time as it is continuously exposed to light, a process called **photobleaching**. This makes it necessary to limit the light exposure of fluorescently tagged specimens and capture images of them before fluorescence becomes too dim. A limitation in live cell imaging is **phototoxicity**, in which illumination leads to the death of cells expressing the fluorophore due to free-radical generation. Finally, while fluorescently labeled structures *can* be detected easily, in practice, background noise can mask the actual signal of interest. Background noise is nonspecific fluorescence that does not indicate true specific signal. One form of background can be **autofluorescence**, in which certain structures, chemicals, and organisms naturally fluoresce without the addition of a fluorophore label. If this fluorescence has the same emission wavelength as the fluorescent label, it can be difficult to determine which signal is which.

Fluorophores have advanced every area of neuroscience. Fluorescent molecules have made breakthroughs possible in studying neural structure and function, identifying the spatial relationships of proteins within a cell, examining the fine branching of neural structures, and detailing the time course of a cell's functional response to chemical or electrical stimulation. Various forms of microscopy attempt to maximize the advantages and minimize the disadvantages of fluorescent reagents. We now survey the main categories of fluorescence microscopy.

Epifluorescent Microscopy

The most elementary form of fluorescent microscopy is **epifluorescent microscopy**, also known as **wide-field fluorescent microscopy**. Epifluorescent microscopes work as just described: specimens labeled with fluorescent probes are illuminated by light of the excitation wavelength. The specimen is then viewed using a second filter that is opaque to the excitation wavelength but transmits the longer wavelength of the emitted light. Thus, the only light transmitted through the eyepiece is the light emitted by the specimen.

Epifluorescence is an excellent tool, but has one major disadvantage: light excites fluorophores throughout the entire depth of the specimen. Thus, fluorescence signals are collected not only from the plane of focus but also from areas above and below this plane. Such background fluorescence can lead to hazy, out-of-focus images that appear blurry and lack contrast. Thus, the value of epifluorescence in relatively thick specimens (>15–$30\,\mu m$) is limited. Alternative forms of fluorescent microscopy attempt to limit out-of-focus fluorescence to produce sharper, clearer fluorescent images.

Confocal Microscopy

Confocal microscopes produce clear images of structures within relatively thick specimens by selectively collecting light from thin ($<1\,\mu m$) regions of the specimen. It is the tool of choice for examining fluorescently stained neurons in brain slices or small, intact organisms such as *Drosophila* or embryonic zebrafish.

With epifluorescence, the entire specimen is illuminated at once so all the fluorophores emit light at the same time, making it difficult to see signals from a specific focal plane within a thick specimen. In a confocal system, light from a laser passes through a pinhole aperture that focuses the light at a specific depth in the specimen (Figure 5.9B). Although this pinhole aperture restricts focused illumination to one plane, other regions receive dispersed, less intense illumination and emit out-of-focus fluorescence. However, all the emitted light passes through a second pinhole aperture before reaching a detector. The key aspect of confocal microscopy is that the pinhole aperture in front of the detector is at a position that is *confocal* with the illuminating pinhole—that is, the only light to reach the detector comes from the same plane in the specimen where the illuminating light comes to a focus, minimizing background fluorescence and maintaining sharp focus on a single plane. The detected light is digitized and sent to a computer for display, storage, and subsequent manipulation.

To generate a two-dimensional image, data from each point in the plane of focus is collected sequentially by scanning across the field. Confocal microscopes can focus the laser beam at precise intervals throughout the thickness of the specimen, producing optical sections at multiple focal planes. Computer image analysis can then be used to "stack" these sections to reconstruct a three-dimensional volume with optimal contrast and resolution.

A major advantage of confocal microscopy to standard epifluorescence is the ability to produce sharp images of cells and cellular structures without background fluorescence (Figure 5.10). However, there are still limitations. Confocal lasers scan specimens point by point to form a complete image of the specimen. Unfortunately, this means that confocal microscopes expose samples to intense light for a longer time period than epifluorescent microscopes, making photobleaching and phototoxicity more problematic. The longer time to capture an image also makes it less desirable for live cell imaging of extremely fast events. In addition, though it is a great advantage to be able to create three-dimensional reconstructions by "stacking" multiple optical sections, the lateral X-Y resolution is finer than resolution in the Z (depth) dimension, so three-dimensional reconstructions may appear warped. Despite these disadvantages, confocal microscopy is the gold standard for imaging, especially for examining thick brain sections or small invertebrate organisms.

Two-Photon Microscopy

Two-photon microscopy, also referred to as **two-photon laser scanning microscopy** (TPLSM), is a further refinement of precision fluorescence microscopy. In epifluorescent or confocal microscopy, a single photon of light at a specific wavelength excites a fluorophore to emit light. In two-photon microscopy, a fluorophore can be excited by effectively absorbing two photons simultaneously. In this case, each photon by itself has a wavelength that is twice the usual excitation

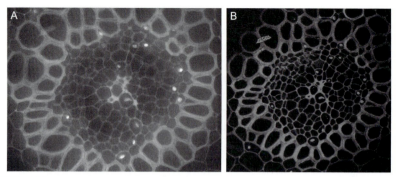

FIGURE 5.10 **Comparison of images from epifluorescence and confocal microscopy. (A)** A standard epifluorescent microscope captures fluorescence emitted from all planes, not just those in focus, leading to a hazy background, while **(B)** a confocal microscope image of the same sample is sharp and clear.

wavelength, and thus has half the energy level per photon. When a fluorophore absorbs the two lower-energy photons at the same time, each contributes half the needed excitation energy to interact with the fluorophore as if a single photon with more energy and a shorter wavelength had struck it (Figure 5.9C).

The main advantage of this method is that the near-simultaneous arrival of two photons is an extremely rare event, so fluorescence excitation is restricted to a narrow plane of focus within the specimen. This provides clearer images than traditional confocal microscopy, even in cases where relatively few fluorophores are present, because photons are only emitted from the excited fluorophores located at the focal plane. Thus, all emitted light can be collected. There is not enough energy to excite fluorophores located above and below the focal plane to produce any background fluorescence.

Two-photon microscopy can penetrate deeper into tissue ($500\,\mu m$–1 mm) than confocal or epifluorescence microscopy because the excitation wavelength is about twice as long as the usual wavelengths used. Light with longer wavelengths scatters less, allowing deeper penetration into tissue. Because excitation is restricted to a focal point, it also causes less photobleaching and phototoxicity, making two-photon microscopy particularly useful for extended observation of cells in undissected, living brain tissue in whole animals, or fine-scale neural structures in brain slices. The primary limitation of this method is that a two-photon microscope set-up requires expensive, specialized lasers and equipment. Figure 5.9 shows a comparison of the principles of epifluorescence, confocal, and two-photon microscopy.

Total Internal Reflection Fluorescence (TIRF) Microscopy

Total internal reflection fluorescent (TIRF) microscopy is another method used to eliminate hazy images with out-of-focus fluorescence by restricting the excitation and detection of fluorophores to a thin section of the specimen. In TIRF microscopy, the thin section of detection is fixed at the interface between the specimen and the surface the specimen is mounted on, usually a glass coverslip (Figure 5.11). This method is useful for investigating molecules and biochemical events that occur at

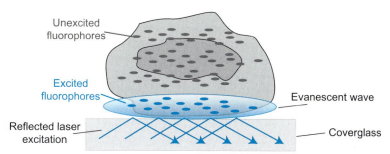

Unexcited
fluorophores

Excited
fluorophores

Evanescent wave

Reflected laser
excitation

Coverglass

FIGURE 5.11 **The basic strategy of a TIRF microscope.** Reflected laser light causes an evanescent wave of excitation at the interface between a specimen and a glass coverslip. The light
penetrates only 100 nm, ensuring that all fluorescent excitation and emission occurs only at the
surface of the specimen, usually the plasma membrane.

the cell surface. For example, a fluorophore-tagged molecule of interest may be
located on the plasma membrane as well as in the cytoplasm of the cell. Using
an epifluorescent microscope, fluorescence emitted from the membrane can be
overwhelmed by the fluorescence coming from the much larger population of molecules in the cytoplasm. A TIRF microscope allows an investigator to selectively
excite and image the membrane-bound population of this molecule.

TIRF microscopes use an evanescent wave of light, which decays exponentially with distance, to excite fluorophores exclusively at the interface between
the specimen and its adjacent surface. The evanescent wave decays at a depth
of about 100 nm into the sample. Therefore, it only effectively excites the
plasma membrane (~7.5 nm thick) and the immediately adjacent cytoplasmic
zone. TIRF microscopy is often used to selectively visualize processes that
occur at the plasma membrane in living cells with high resolution.

ELECTRON MICROSCOPY

Although there is no theoretical limit to the magnification power of a light
microscope, resolving power has a maximal value of about 200 nm due to the
fixed minimal wavelength of visible light. However, an electron has a much
shorter wavelength than a photon. By focusing a beam of electrons rather
than a beam of photons to image a specimen, **electron microscopy (EM)** can
increase the resolving power as much as 1000-fold to 0.2 nm, about the radius
of a glutamate molecule. Electron microscopy is essential for examining small
neural structures such as synapses, synaptic vesicles, and ion channels.

The major limitation to using EM is that specimens must be fixed, dehy-
drated, and chemically treated. Therefore, EM cannot be applied to living samples.
Furthermore, artifacts could appear that do not exist in the living cell. To reduce
these artifacts, investigators can rapidly freeze specimens so that water and other
components in cells do not have time to rearrange themselves or even crystallize
into ice.

There are two major categories of electron microscopy: transmission electron microscopy (TEM) and scanning electron microscopy (SEM).

Transmission Electron Microscopy (TEM)

In **transmission electron microscopy (TEM)**, a beam of electrons is aimed through a thin section of a specimen that has been chemically treated to enhance contrast. Extremely thin (<100 nm) sections of preserved tissue are stained with atoms of heavy metals, which are preferentially attracted to certain cellular components. These heavy metal stains are electron-dense, so when the electron beam hits the heavy metal atoms in the specimen, the electrons are absorbed or scattered, making electron-dense areas appear dark in an image. Electromagnets focus and magnify the image by bending charged electron trajectories. Since we cannot *see* electrons, the variations in electron intensity are converted into photons by projecting the electrons into a special detector or onto a screen that fluoresces at intensities relative to the amount of beam electrons hitting it. TEM produces two-dimensional images of thin tissue sections that are useful for studying the internal structure of cells (Figure 5.12B). TEM is frequently used to define synapses, as the postsynaptic density appears clearly as an electron-dense region at close contact with a presynaptic membrane containing neurotransmitter vesicles.

Scanning Electron Microscopy (SEM)

Scanning electron microscopy (SEM) is useful for detailed study of a specimen's surface. A high-energy electron beam scans across the surface of a specimen, usually coated with a thin film of gold or platinum to improve contrast and the

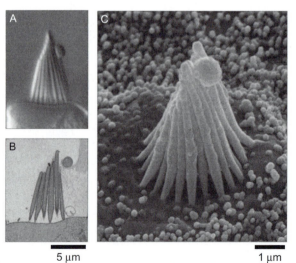

5 µm   1 µm

FIGURE 5.12 Comparison of images taken by different electron microscope techniques. **(A)** A DIC image of stereocilia projecting from a hair cell in the inner ear. Compare the light microscope image with images taken using **(B)** a transmission electron microscope or **(C)** a scanning electron microscope. (Courtesy of Dr. A. J. Hudspeth and Dr. R. Jacobs.)

signal-to-noise ratio. As the beam scans across the sample's surface, interactions between the sample and the electron beam result in different types of electron signals emitted at or near the specimen surface. These electronic signals are collected, processed, and eventually translated as pixels on a monitor to form an image of the specimen's surface topography that appears three-dimensional (Figure 5.12C). Low-energy secondary electrons excited on the sample's surface are the most common signal detected. High-energy backscattered electrons and X-rays are emitted from below the specimen surface, providing information on specimen composition.

Electron Tomography (ET)

Electron tomography (ET), also known as **electron microscope tomography**, is similar to computerized tomography mentioned in Chapter 1. In a CT scan, the imaging equipment is moved around a patient to generate different images of a slice of the brain. In ET, the specimen is tilted within the electron microscope to produce TEM images from many different perspectives to reconstruct a three-dimensional view. With a resolution of 2–20 nm, ET is now used for determining both molecular structures and the three-dimensional ultrastructure of organelles. It has been especially informative for elucidating the organization of presynaptic vesicles in an axon terminal (Figure 5.13).

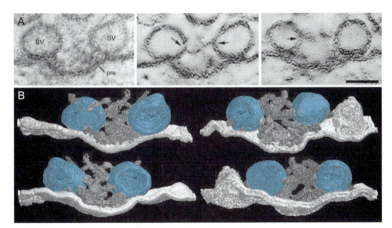

FIGURE 5.13 Example of an electron tomography image: synaptic vesicles docking at release sites in the neuromuscular junction. (A) Two-dimensional TEM images used in the ET volume reconstruction of an active zone at a frog neuromuscular junction. AZM, active zone material, denoted by arrow; sv, docked synaptic vesicles; pre, presynaptic membrane. Scale bar, 50 nm **(B)** Three-dimensional surface-rendered AZM, docked vesicles and presynaptic membrane that allow the shape, size, and associations of different components to be examined. (Reprinted by permission from Macmillan Publishers Ltd: Nature, M.L. Harlow, et al. The architecture of active zone material at the frog's neuromuscular junction, © 2001.)

TABLE 5.1 Comparison of Different Forms of Microscopy

	Description	Advantages	Disadvantages	Common Use in Neuroscience
Brightfield	Light transmitted through specimens. Contrast generated by natural pigmentation or added dyes.	Simple and inexpensive. Different-colored dyes can be used.	Most cells and tissues are transparent, so they can be difficult to see. Adding dyes to visualize structures usually requires fixation and sectioning.	Fixed samples processed using detection reagents that produce colored by-products (see Chapter 6).
Phase-Contrast	Contrast generated by changes in index of refraction because of variation in organelle densities.	Can be used on live cells. Doesn't require added chemicals or tissue processing.	Out-of-focus signals.	Tissue culture cells.
Darkfield	Specimen illuminated from the side at an oblique angle. Only scattered light is imaged.	Can be used on live cells. Sensitive. High signal-to-noise ratio. No added chemicals necessary.	Cannot see structures that do not scatter light.	Samples processed for radioactive in situ hybridization (see Chapter 6).
DIC/Nomarski	Optical methods used to exaggerate changes in index of refraction, enhancing contrast at the edges of objects. Has a 3D appearance.	Can be used on live cells. Can create thin optical sections. Can create 3D reconstructions from thin optical sections.	Can only image single, thin focal plane at a time.	Tissue culture cells and tissues.

	Description	Advantages	Disadvantages	Uses
Epifluorescent	Specimen is illuminated by excitation wavelengths and emits light from excited fluorophores throughout entire thickness.	(Advantages of all forms of fluorescent microscopy) Can be used to detect specifically labeled fluorescent molecules. Multiple fluorophores can be imaged in same sample. High signal-to-noise ratio. Sensitive. (Advantage of epifluorescence over other forms) Simpler and faster image collection methods than other forms of fluorescent microscopy (confocal, two-photon).	Out-of-focus fluorescence can cause blurry pictures that make structures difficult to resolve.	Thin physical sections of fluorescently stained tissue or cells. Time-lapse fluorescence imaging in cells.
Confocal	Out-of-focus illumination is blocked by the use of pinhole apertures, so only in-focus light is collected.	Optical sectioning ability creates sharp, in-focus images.	Intense laser illumination can cause photobleaching and phototoxicity. Long scan time.	Detecting fluorescence in thick tissues or small organisms (e.g., *Drosophila*). Time-lapse fluorescence imaging in thicker tissues (e.g., slice cultures) or for small structures (e.g., synaptic vesicles).
Two-Photon	Fluorophores in a thin focal plane are selectively excited by effectively absorbing the combined energy of two photons that cannot excite fluorophores on their own.	Longer wavelengths of laser illumination allow deeper penetration of fluorescence excitation. Reduced photobleaching and phototoxicity.	Expensive equipment.	*In vivo* imaging of intact organisms. Long-term fluorescence imaging.

(Continued)

TABLE 5.1 (Continued)

	Description	Advantages	Disadvantages	Common Utility in Neuroscience
TIRF (Total Internal Reflection Fluorescence)	Evanescent wave of illumination rapidly decays after passing the interface between a cell membrane and the surface it is on.	Produces thin optical section. Eliminates out-of-focus fluorescence from regions more than 100 nm away from surface. High resolution.	Only fluorophores at the interface between the membrane surface and the glass coverslip are illuminated; cannot image an arbitrary region within the cell.	Imaging protein dynamics at the plasma membrane in live cells.
TEM (Transmission Electron Microscopy)	Electron beams transmitted through ultrathin sections.	Nanometer resolution.	Cannot be used on live cells. Harsh processing conditions can cause artifacts. Requires specialized equipment.	Ultrastructure of cells. Synapse structure.
SEM (Scanning Electron Microscopy)	Detect secondary electrons scattered off surface of sample.	Provides 3D topological information.	Cannot be used on live cells. Harsh processing conditions can cause artifacts. Requires specialized equipment.	Topography of cells and tissues.
ET (Electron Tomography)	Rotate specimen to take TEM images from multiple perspectives and create 3D reconstruction.	Provides 3D organization information.	Cannot be used on live cells. Requires intensive computation to reconstruct TEM views.	3D cellular ultrastructure and organization.

PREPARING AND INTERPRETING MICROSCOPY DATA

Microscopy data are probably the most frequently presented form of data in neuroscience, and possibly all life sciences. They may also be the most subjective form of data to interpret. Therefore, it is important to understand standard practices used to prepare and display data to know what to look for in published results.

Image Processing

The images produced directly by the microscope are generally not the exact same images that appear in publications. Raw images are usually processed by computer software accompanying the microscope or by secondary programs such as Adobe Photoshop or NIH ImageJ. While many may feel that image processing is tampering with evidence, it is appropriate and sometimes necessary to extract meaningful information: to facilitate quantitative measurements and accurate analysis, to highlight meaningful aspects of an image, and to improve aesthetic appearance when it does not interfere with the results. The key is to faithfully represent reproducible results and not alter the inherent information contained in the image.

The most common form of processing is creating pseudo-color representations of fluorescent microscopy images taken in different channels (different emission wavelengths) to overlay them in a single image. Most microscope cameras capture light intensity in grayscale rather than color. If multiple fluorophores are imaged in a single specimen, different channels can be selected by changing the excitation and emission filters on the microscope. This allows all the different fluorophores to be imaged independently but they will all appear in grayscale. The investigator can then assign a different color to each fluorophore and merge the separate images of each fluorophore into a single image with multiple colors. This helps visualize relative intensity differences and spatial relationships for each fluorescently labeled molecule (Figure 5.14).

Other common forms of image processing include digitally enhancing contrast or removing background fluorescence. Although this processing can aid in elucidating useful information from an image, it must not introduce false elements into the image to provide support for what an investigator *wishes* to see. Many journals provide guidelines for the types of image processing that are allowed and require that all processing be reported in the Materials and Methods section.

In general, there are two important guidelines to follow when processing microscopy images for analysis and publication: (1) Any processing applied to one image of a set should be applied to *all* images within the set. For example, when removing background fluorescence from one image, it is important that all the remaining images receive the exact same digital enhancements. This is especially important when comparing images between groups. (2) An investigator should process and analyze images blind to the identity of the samples between experimental and control groups. Even with the best of intentions, scientists can subconsciously make mistakes in analysis or introduce subconscious bias. When

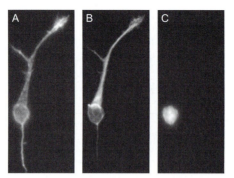

FIGURE 5.14 Merging images of different fluorophores into a single image. A sample is viewed for fluorescence in the (**A**) red (rhodamine), (**B**) green (FITC), and (**C**) UV channels (DAPI). Because these spectra do not overlap, it is possible to overlay these images, assigning each channel a distinct color, to compare the spatial relationship of fluorescent signals.

working with microscopy data, as with most data, blind processing and scoring is critical to prevent potential bias. It is important to properly represent reproducible features contained in the data itself rather than in interpretations of the data.

Interpreting Images

Some general guidelines are useful when evaluating your own or published microscope images. Judge your confidence in an image by asking some critical questions:

- *Does it look right?* Familiarity with other work in the field can provide an intuition and expectation for standard images. If the image is supposed to be of neurons, do the structures look like neurons? If the image is supposed to represent a specific brain region, can you identify that region based on the image presented? Is the image taken at an appropriate scale to make a judgment? If the author claims that expression of a gene, protein, or lesion is localized to a specific structure, does the image show that structure in the context of neighboring regions?
- *Does the illumination look uniform?* Improper microscope alignment can cause differences in signal intensity that might lead to inappropriate conclusions. If the entire field of view is not evenly illuminated, important information and resolution can be lost.
- *Does the signal-to-noise ratio allow for proper examination of the data?* Nonspecific background staining can potentially mask a specific signal. Is the contrast between the background and the signal substantial enough to produce a consistent, observable result?
- *Is the exposure and brightness optimal?* An overexposed image (Figure 5.15A) can mask details, enhance nonspecific background, or improperly suggest a high concentration of the measured signal. An underexposed

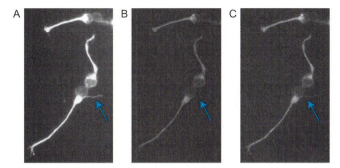

FIGURE 5.15 Over- and underexposure of an image. (A) An overexposed image of neurons.
(B) The same cells underexposed and **(C)** with the proper level of exposure. The blue arrow points
to a process that is lost when underexposed and very difficult to see under a properly balanced
exposure level. It can be difficult to find the proper exposure to see both intensely and weakly
fluorescent structures at the same time.

image (Figure 5.15B) can make it appear as if the measured signal is not
present when it is. An investigator can affect the proper exposure of an
image during both the capture of an image and the secondary image processing. When evaluating microscopy data, ensure that the conclusions are
not drawn from images with inappropriate exposure conditions or inappropriate image processing.

- *If this is the most convincing example, what do the other samples look like?*
  For obvious reasons, authors tend to publish images that best represent and
  support their conclusions. Usually, these images represent one of many individual data points. Therefore, a critical review of the data should assume that
  the other data points are of an equal or *lesser* quality than the images presented
  in a publication. If the image is suboptimal or difficult to interpret, the validity
  of conclusions reached from unpublished images should be questioned.
- *Is there quantitative analysis of the images?* Though a picture is worth a
  thousand words, subjective interpretation of data is not as strong as objective, quantifiable results. Therefore, the strongest evidence to support a
  conclusion presents images alongside complementary quantification of
  the data that demonstrates the reproducibility of the results from image to
  image. Quantification between two or more experimental sets of data is
  necessary to demonstrate any meaningful results. Common types of quantitative information calculated from images include relative counts, size and
  shape distributions, relative area occupied, variation in fluorescence intensity to indicate relative concentrations of the measured signal, and colocalization of independently labeled proteins.
- *Could fluorescent images represent poor quality control issues, such as
  photobleaching or autofluorescence?* Photobleaching could lead an investigator to believe there is no signal in a sample when the signal has actually
  faded. Alternatively, autofluorescent substances in a sample might lead an

investigator to believe there is signal where there is none. Investigators can avoid inappropriate conclusions caused by photobleaching or autofluorescence by processing all samples in the same way and by performing the appropriate positive and negative control experiments.

- *Could fluorescent images represent bleed-through of different emission spectra?* Investigators usually assume that signals from fluorophores of different wavelengths are independent of each other. However, some fluorophores have overlapping excitation or emission spectra and may be detected when not expected. **Bleed-through**, also known as **cross-talk**, occurs when the signal for one fluorophore appears when attempting to look at the signal for a different fluorophore: the signal of one fluorophore has "bled-through" the filters that are set for the other fluorophore. To control for this situation, specimens can be labeled with a single fluorophore and the signal recorded in each channel. Bleed-through of fluorophores becomes especially pertinent when an investigator claims that two signals overlap, as when two proteins colocalize. It is rare that the signal from two different fluorophores will be identical, no matter how strong the colocalization may be. Therefore, if two different channels appear exactly the same, despite use of different detection agents, the image could indicate an artifact.

CONCLUSION

Whether using photons or electrons, microscopes provide the ability to explore worlds much smaller than can be viewed by the naked eye. Although microscopes are one of the most often used tools in neuroscience, many people do not understand the essential concepts necessary to produce quality, scientifically meaningful images. This chapter has served as an introduction to microscopy and a survey of various techniques. Most microscope manufacturers (Nikon, Leica, Zeiss) provide detailed instructions in the proper use and theory behind their products. These guides can provide further information about the specific microscopes used in your laboratory. While optical methods can be used to provide enhanced contrast to make out individual structures within cells, histological preparations are utilized to examine specific processes and structures. The next two chapters present methods on preparing samples to visualize neural structure and function that depend on good microscopes and, just as important, good microscopy skills.

SUGGESTED READING AND REFERENCES

Books

Bradbury, S., Bracegirdle, B. & Royal Microscopical Society (Great Britain). (1998). *Introduction to Light Microscopy*. Bios Scientific Publishers, Springer, Oxford.

Chandler, D. E., Roberson, R. W. (2009). *Bioimaging: Current Concepts in Light and Electron Microscopy*. Jones and Bartlett Publishers, Sudbury, MA.

Herman, B. (1998). *Fluorescence Microscopy*, 2nd ed. Taylor & Francis in association with the Royal Microscopical Society, Abingdon, UK; New York.

Pawley, J. B. (2006). *Handbook of Biological Confocal Microscopy*, 3rd ed. Springer, NY.

Review Articles

Brown, C. M. (2007). Fluorescence microscopy—avoiding the pitfalls. *J Cell Sci* **120**, 1703–1705.

Kapitza, H. G. (1994). *Microscopy from the Very Beginning*, 2nd revised ed. Carl Zeiss, Oberkochen, Germany.

McEwen, B. F. & Marko, M. (2001). The emergence of electron tomography as an important tool for investigating cellular ultrastructure. *J Histochem Cytochem* **49**, 553–564.

Pearson, H. (2007). The good, the bad and the ugly. *Nature* **447**, 138–140.

Saibil, H. R. (2007). How to read papers on three-dimensional structure determination by electron microscopy. In *Evaluating Techniques in Biochemical Research*, Zuk, D., ed. Cell Press, Cambridge, MA, http://www.cellpress.com/misc/page?page=ETBR.

Schneckenburger, H. (2005). Total internal reflection fluorescence microscopy: technical innovations and novel applications. *Curr Opin Biotechnol* **16**, 13–18.

Waters, J. C. & Swedlow, J. R. (2007). Interpreting fluorescence microscopy images and measurements. In *Evaluating Techniques in Biochemical Research*, Zuk, D., ed. Cell Press, Cambridge, MA, http://www.cellpress.com/misc/page?page=ETBR.

Primary Research Articles—Interesting Examples from the Literature

Harlow, M. L., Ress, D., Stoschek, A., Marshall, R. M. & McMahan, U. J. (2001). The architecture of active zone material at the frog's neuromuscular junction. *Nature* **409**, 479–484.

Jung, J. C., Mehta, A. D., Aksay, E., Stepnoski, R. & Schnitzer, M. J. (2004). In vivo mammalian brain imaging using one- and two-photon fluorescence microendoscopy. *J Neurophysiol* **92**, 3121–3133.

Westphal, V., Rizzoli, S. O., Lauterbach, M. A., Kamin, D., Jahn, R. & Hell, S. W. (2008). Video-rate far-field optical nanoscopy dissects synaptic vesicle movement. *Science* **320**, 246–249.

Protocols

Coling, D. & Kachar, B. (2001). Theory and application of fluorescence microscopy. *Curr Protoc Neurosci*, Chapter 2, Unit 2.1.

Smith, C. L. (2008). Basic confocal microscopy. *Curr Protoc Mol Biol*, Chapter 14, Unit 14.11.

Spector, D. L., Goldman, R. D. (2006). *Basic Methods in Microscopy: Protocols and Concepts from Cells: A Laboratory Manual*. Cold Spring Harbor Laboratory Press, Cold Spring Harbor, NY.

Websites

Davidson, M.W., Molecular Expressions: http://micro.magnet.fsu.edu/.

Nikon Microscopy U: http://www.microscopyu.com.

Olympus Microscopy Resource Center: http://www.olympusmicro.com/.

Zeiss MicroImaging: http://www.zeiss.com/us/micro/home.nsf.

Chapter 6

Visualizing Nervous System Structure

After reading this chapter, you should be able to:

- Explain the process of preparing neural tissue for histological procedures
- Describe methods for visualizing gross cellular morphology
- Describe methods for visualizing gene and protein expression
- Describe methods for visualizing neural circuitry and connections between different brain regions

Techniques Covered

- **Tissue preparation:** fixation, embedding, sectioning
- **Visualizing structure:** basic histological methods (nuclear stains, fiber stains), Golgi stain, intracellular/juxtacellular labeling
- **Visualizing gene and protein expression:** *in situ* hybridization, immunohistochemistry, array tomography, enzymatic histochemistry, reporter genes
- **Visualizing circuitry:** anterograde tracers, retrograde tracers, transsynaptic tracers

A person can learn a lot about a complicated system simply by examining its structure. For example, imagine a satellite image of a city taken from space. How might you learn about what it would be like to live in the city just by examining the city from that vantage point? You could start by separating the image into component parts—for example, distinguishing buildings from roads. Larger buildings clustered together could reveal popular hubs of activity, while smaller, scattered buildings might indicate residential communities. You would learn even more by identifying the different categories of buildings, classifying different structures as schools, grocery stores, gas stations, shopping centers, and so forth. Major highways and freeways would indicate routes for long-distance travel; narrower roads would show local areas of transit. By simply examining the structure of the city, one could form educated guesses about how the city functions.

Similarly, one can learn a lot about the nervous system simply by examining its structure. Just like distinguishing between buildings and roads in a city, a neuroscientist can begin to examine the brain by discriminating between cells and fiber tracts. Cells, like buildings, can be classified into different structural and functional groups, with each cell type expressing a unique combination of genes and proteins that specify its role in the brain. Thick white matter tracts show long-distance sites of neural communication, while smaller fiber tracts show local communication networks between populations of neurons. Learning about the structure of distinct brain regions can lead to educated hypotheses about the function of different neurons and how they influence neural circuits and behavior. Indeed, in the early twentieth century, Santiago Ramón y Cajal formed many hypotheses (many of which turned out to be correct) about the function of neural systems simply by examining their structures.

The goal of this chapter is to survey techniques used to investigate the structure and connectivity of the nervous system. Using these techniques, it is possible to classify cells based on location, morphology, gene/protein expression profiles, and connections with other cells. These methods can be used in combination with techniques mentioned in Chapter 7 to investigate neural function.

TISSUE PREPARATION

Neural tissue is soft, delicate, and degradable. Therefore, in order to accurately study neural structure, an investigator's first goal is to keep the tissue as close to its living state as possible. Investigators process tissue by (1) fixing, (2) embedding, and (3) sectioning before proceeding with visualization methods. Fixing and embedding stabilize tissue to capture its current state, while sectioning makes the tissue thin enough for light to pass through so internal structures can be examined under a microscope.

Fixation

Fixation is the process of using chemical methods to preserve, stabilize, and strengthen a biological specimen for subsequent histological procedures and microscopic analysis. This process preserves cells and tissues by strengthening molecular interactions, disabling endogenous proteolytic enzymes, and killing microorganisms that might otherwise degrade the specimen. Thus, fixation terminates any ongoing biochemical reactions by "fixing" proteins into place, a process that kills cells that aren't already dead.

There are two different categories of chemical fixatives: cross-linking fixatives and dehydrating fixatives. Which fixative an investigator chooses to use depends on the type of subsequent histology to be performed. **Cross-linking** fixatives create covalent chemical bonds between proteins in tissue and include organic compounds with aldehyde groups, such as formaldehyde, paraformaldehyde, and glutaraldehyde. These fixatives are good at preserving structure and are often used when processing cells and tissues for light microscopy. Another cross-linking fixative, osmium tetroxide, causes molecules to oxidize and is often used as a secondary fixative for electron microscopy. Dehydrating fixatives disrupt lipids and reduce the solubility of protein molecules, precipitating them out of the cytoplasmic and extracellular solutions. Common dehydrating fixatives include methanol and acetone.

A scientist can fix cells in culture simply by adding a chemical fixative to the culture chamber. Whole brains can be fixed either by immersion or perfusion. Immersion refers to placing small brains, or even entire animals, in fixative solutions. The amount of time necessary for the fixative to penetrate the entire tissue is variable, depending on the size of the brain: a fly brain may need to be fixed for only a few minutes, while a mouse brain may require days. Perfusion is the process of delivering a fixative through an animal's cardiovascular system. To perfuse an animal, a scientist anesthetizes the animal, carefully opens the rib cage to expose the heart, and then penetrates the left ventricle with a needle connected to a perfusion pump. The left ventricle pumps blood throughout the entire body and can therefore deliver a fixative to the nervous system. First, the scientist pumps a buffered saline solution to rinse the blood out of the veins and arteries, and then pumps a cross-linking reagent to fix all the tissues in the body. Perfusion fixation is thorough, quick (can be performed in 5–15 minutes, depending on the animal), and adequately fixes all cells in the brain if properly performed.

Some histological procedures do not require fixation or even require that a brain *not* be fixed. For example, the use of radioactively labeled ligands to localize neurotransmitter receptor-binding sites depends on those sites remaining intact, and fixation could disrupt their structure. Each histological method determines whether fixation is necessary, which fixatives to use, and when the fixation process should occur.

Embedding

Embedding is the process of surrounding a brain or tissue section with a substance that infiltrates and forms a hard shell around the tissue. This stabilizes the tissue's structure and makes it easier to cut. Standard embedding materials include gelatin, paraffin wax, and plastic.

Sectioning

Sectioning is the process of cutting a brain into thin slices for subsequent histological procedures or microscopic examination. This process is almost always necessary to examine the structure of neurons within the brains of vertebrate model

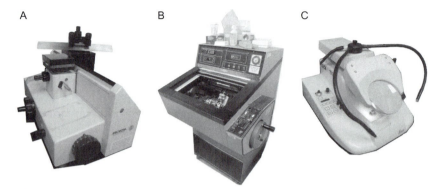

FIGURE 6.1　**Equipment used to section tissue.** (**A**) Microtome (**B**) Cryostat (**C**) Vibratome.

organisms. Sectioning the brain allows histological reagents access to cells within the slice, as well as allows a scientist to examine brain structures with a microscope. There are three common machines used to section tissue (Figure 6.1, Table 6.1):

- A **microtome** (*micro* = "small," *tome* = "cut") is an instrument used to section tissue using steel, glass, or diamond knives. A specialized microtome called a **freezing microtome** is commonly used to section frozen tissue. Before cutting, the tissue must be cryoprotected by soaking in a sucrose solution. This minimizes artifacts caused by freezing. Using a freezing microtome, a scientist can collect 25–100 μm sections off the blade for subsequent histochemical processing. These sections are usually stored in a buffered saline solution until use. Other types of microtomes are used to section paraffin-embedded tissue or to create the ultrathin sections required for electron microscopy.
- A **cryostat** is basically a microtome housed in a freezing chamber that allows sectioning at temperatures of –20° to –30° C. Maintaining the tissue and blade at low temperatures allows a scientist to section nonfixed, frozen tissue. Cryostats are able to cut thin, 10–50 μm sections. A scientist can mount sections onto slides within the cryostat or store sections in a buffered saline solution until use.
- A **vibratome** cuts tissue with a vibrating knife, something similar to an electric vibrating toothbrush. The main function of a vibratome is to section tissues that are not frozen. The knife vibrates side-to-side so quickly that a soft brain can be cut into 100–400 μm sections. Thus, vibratomes are useful for avoiding artifacts, changes in morphology, or interruption of biochemical activities caused by freezing. A vibratome is also necessary for experiments in which a slice of brain tissue must be kept alive, as in tissue culture or electrophysiology studies.

Once a brain is sectioned, a scientist can either mount the sections directly onto slides or can keep the sections floating in buffered saline for subsequent histological analysis.

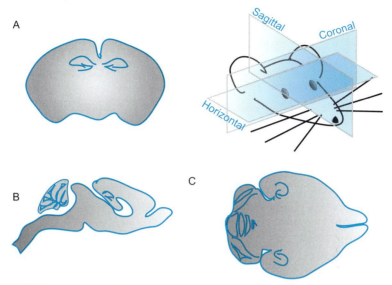

FIGURE 6.2 **Planes of sections.** Standard orientations for sectioning brain tissue. **(A)** Coronal **(B)** Sagittal **(C)** Horizontal/transverse.

Brains are typically cut in one of three standard orientations: coronal, sagittal, or horizontal (Figure 6.2). The **coronal plane** divides the brain in the anterior-to-posterior direction, revealing a section of the brain that is complete from top-to-bottom and left-to-right (ear to ear). The **sagittal plane** shows the complete top-to-bottom of the brain, cutting it into left and right portions. A **midsagittal** cut is a section that perfectly divides the left and right hemispheres of the brain. The **horizontal plane** divides the brain into superior (top) and inferior (bottom) sections, such that an area of the brain is complete from left-to-right and anterior-to-posterior.

There are some preparations for which brain sectioning is not necessary. For example, scientists may wish to examine **whole-mount preparations** of intact brains or animals. Also, the brains of *Drosophila* and other invertebrates are usually small enough so that confocal or two-photon microscopy (Chapter 5) can provide good images of individual neurons within neural structures.

Now that we have discussed some common methods of preparing neural tissue for histological analysis, we describe fundamental methods of visualizing the gross morphology of the brain.

VISUALIZING MORPHOLOGY

Although lipids, proteins, carbohydrates, and other organic molecules provide the nervous system with its structure and durability, over three-quarters of the mass of the brain is water. This composition imparts significant consequences

TABLE 6.1 Comparison of Different Sectioning Methods

	Essential Features	Standard Section Thickness
Microtome	Brains are frozen Preserves structures better than cryostat	Medium: 25–100 µm
Cryostat	Brains are frozen Sections are cut in a cold chamber	Thin: 10–50 µm
Vibratome	Brains do not need to be frozen Vibrating knife Can be used to cut tissue that will be kept alive	Thick: 100–400 µm

for any scientist interested in studying the structure of the brain, most notably that unprocessed brain sections are almost completely transparent. Unless treated by histochemical stains, the structure of the nervous system will remain invisible and appear as a wafer-thin slice of gelatin mounted on a glass slide. Treating brain sections with various dyes enhances the contrast between different brain structures and allows an investigator to visualize cells, fibers, or other features of neural systems. The methods described here each provide contrast to specific structures in the brain so that they stand out from other aqueous structures in a neural specimen.

Cell Body Stains

A **basophilic stain** is used to visualize cell bodies. The term *basophilic* is derived from *baso-,* meaning "base" and *-philic,* meaning "loving" and conveys the fact that these stains are basic (pH >8) and good for staining acidic molecules. As DNA and RNA molecules are acidic, basophilic dyes label structures enriched in nucleic acids, such as cell nuclei and ribosomes. A scientist can use these stains to examine the cytoarchitecture of individual neurons with high magnification or to visualize the macroscopic features of distinct brain regions, such as layers of cerebral cortex or subdivisions of hypothalamic nuclei (Figure 6.3A).

Common basophilic stains used in brightfield microscopy include cresyl violet, hematoxylin, and thionine. A special category of basophilic stains, referred to as **Nissl stains**, specifically label RNA within cells, providing contrast to ribosomes and rough endoplasmic reticulum, which are enriched in neurons. Cresyl violet is one of the most commonly used Nissl stains. Redistribution of Nissl-stained structures in injured or regenerating neurons allow Nissl stains to reveal the physiological state of some neurons.

In fluorescent microscopy, common markers of cell nuclei include DAPI, Hoechst (bis-benzamide), and propidium iodide (PI). These dyes intercalate

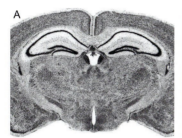 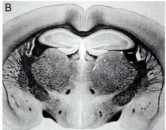

FIGURE 6.3 Comparison of basophilic and fiber stains. (A) Cresyl violet Nissl stain of a coronal mouse brain section. **(B)** Coronal section stained for myelin (A, Reprinted with permission from Rosen G.D. et al. (2000) The Mouse Brain Library @ www.mbl.org. Int Mouse Genome Conference 14: 166. www.mbl.org; B. Courtesy of Dr. Pushkar Joshi).

within the helical spirals of DNA in the cell nucleus. Both DAPI and Hoechst emit blue light when excited by UV light, while PI emits red light when excited by green light. PI is useful in assays of cell death because it is membrane impermeant; therefore, it labels dead but not healthy cells.

Fiber Stains

Fiber stains label white matter tracts by staining myelin, the fatty substance that insulates and provides electrical resistance to axons. These stains are useful for visualizing the major fiber tracts that project throughout the brain (Figure 6.3B). However, the high density of fibers in most brain regions makes it impossible to trace individual axons from cell body to synapse. When used in combination with basophilic stains in adjacent tissues sections, a scientist can identify more anatomical structures than if either stain is used alone. Table 6.3 lists some commonly used cell and fiber stains.

There are various protocols useful for labeling myelin, such as the Weigert or Weil methods that stain myelin dark blue or black. A scientist can also stain myelin using lipophilic fluorescent stains.

Golgi Stain

The Golgi stain is a classic technique used to completely label individual neurons and their processes (Box 6.1). There are two major advantages to using a Golgi stain: (1) neurons are labeled in their entirety, including cell bodies, processes, and even microscopic structures such as dendritic spines; (2) only 5–10% of the total number of cells are labeled. Therefore, individual neurons stand out in a background of numerous unlabeled cells (Figure 6.4). A scientist cannot control which neurons in a tissue slice become labeled, but newer techniques allow a scientist to duplicate the effects of a Golgi stain in a genetically defined population of neurons (Chapter 11). However, this method is still used to define and trace the elaborate morphology of neurons, especially in investigations of neurodegeneration.

BOX 6.1 Historical Use of the Golgi Stain

The Golgi stain has an important place in the history of neuroscience. In the late nineteenth and early twentieth centuries, Santiago Ramón y Cajal used this technique for his seminal studies of the cellular morphology of the nervous system. Among his many achievements was the discovery that neurons are distinct, independent cells that communicate via synapses. His work heavily contributed to the acceptance of the **neuron doctrine**: the hypothesis that neurons are the basic structural and functional units of the nervous system. Cajal also used the Golgi stain to reveal the morphological diversity of neurons, as well as to trace neural circuits in the retina, hippocampus, cerebellum, and other brain regions. In 1906, Cajal shared the Nobel Prize with Camillo Golgi, the scientist who developed the Golgi stain, for elucidating the structure of the nervous system.

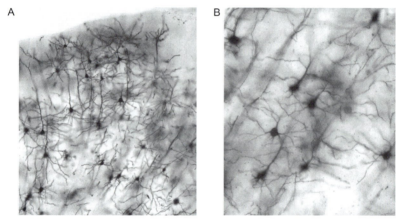

FIGURE 6.4 The Golgi stain reveals neuronal structure in fine detail. Random neurons are selectively stained in this (**A**) 100X view of a Golgi-stained mouse cortex, and (**B**) a closer view. (Courtesy of Dr. Jocelyn Krey).

Intracellular and Juxtacellular Labeling

The preceding methods label cell bodies and/or axons after a brain has already been fixed and sectioned. It is also possible to label individual neurons during an experiment for later histological analysis so that a neuron's activity or function can be coupled to the neuron's structure and location in the brain. During electrophysiology experiments, a scientist can use a glass pipette to fill a neuron of interest with a chemical that can later be visualized using histological methods. This chemical is usually a variant of a molecule called **biotin**, such as biocytin or neurobiotin, which has properties that make it useful for subsequent histology (Box 6.2). A scientist can intracellularly label cells *in vivo* or in slices. During an extracellular recording session, the electrode tip can be apposed to the neuron's membrane for juxtacellular labeling. This labeling can completely fill a neuron's dendrites or axonal arborizations.

BOX 6.2 Methods of Labeling Probes

A probe, such as an antibody in immunohistochemistry (IHC) or antisense strand in in situ hybridization (ISH), is not visible by itself. In order to be visualized, the probe must be conjugated to a label that is visible on its own or that reacts with other chemicals to create a visible product. The use of a particular label depends on a number of factors, including the degree of specificity needed, whether other methods will be used in combination, and what types of microscopes are available. Following are some common types of labels (fluorescent, chromogenic, radioactive, and colloidal gold) and how they can be detected (Table 6.2).

TABLE 6.2 Common Labels for Detection

Label Type	Comments	Visualization Method
Fluorescent dyes	Green dyes: FITC, Alexa 488, Cy2 Red dyes: Rhodamine, Texas Red, Alexa 594, Cy3 Infrared dyes: Cy5	Fluorescence microscopy
Quantum dots	Spectral selectivity based on size	Fluorescence microscopy
Alkaline phosphatase (AP)	NBT/BCIP substrate produces blue/purple product	Usually light microscopy, though fluorescent substrates are available
Horseradish peroxidase (HRP)	DAB substrate produces brown/black product	Light or electron microscopy of colored DAB product
β-galactosidase (encoded by *lacZ* gene)	X-gal or IPTG substrate produces blue/purple product	Usually light microscopy, though fluorescent substrates are available
Biotin	Biotinylated antibodies or proteins are detected using avidin conjugated to another label described (fluorescent, chromogenic, or gold)	Light, fluorescent, or electron microscopy depending on type of label conjugated to avidin
Digoxygenin	Used to label ISH probes, detected with antidigoxygenin probe conjugated to another label, frequently AP	Light or fluorescence microscopy, depending on substrate for label
Radioactive isotope (^{35}S, ^{33}P, ^{3}H, ^{125}I)	ISH probes: ^{35}S, ^{33}P, ^{3}H Proteins: ^{3}H, ^{125}I	Light microscopy
Gold	Electron-dense label used for detecting immunostaining in EM	Electron microscopy

FLUORESCENT LABELS

Fluorescent molecules are ubiquitous in biological research. Fluorescent proteins, such as green fluorescent protein (GFP), can be genetically encoded and serve as reporter proteins. Other nonprotein **fluorophores** can be attached to a variety of molecular probes. Because these fluorescent probes have different excitation and emission spectra (Chapter 5), scientists can use multiple fluorescent labels to visualize distinct genes, proteins, and structures in a neural specimen.

There are a variety of fluorophores available for labeling probes. Commonly used fluorescent labels include the organic cyanine and Alexa fluor dyes, which are commercially available and often used to label antibodies. **Quantum dots** are a recently developed type of fluorophore. They are semiconductor nanocrystals that are extremely bright and resistant to photobleaching, making them particularly useful for live-cell imaging over long time periods.

Fluorescence is a sensitive method of detection, as bound fluorophores will emit light against a dark background. Attaching multiple fluorophores to a probe can amplify the signal to enhance detection. However, this also makes it trickier to quantify exact levels of the probed substance, since multiple fluorescent molecules can be attached to single probes; it is not a one-to-one mapping of label-to-probed substance, as it can be with other methods, such as radioactivity.

Fluorescently labeled structures can be made visible to light and electron microscopes through a process called **photoconversion**. Fluorophores can photoconvert the chemical DAB (diaminobenzidine), a substrate commonly used for chromogenic enzymes. Exciting certain fluorophores with the appropriate wavelength of light in the presence of DAB forms a dark-colored precipitate.

CHROMOGENIC/COLORIMETRIC LABELS

Chromogenic or **colorimetric** labels are enzymes that react with a substrate to produce a colored product that is visible under the light microscope. The most commonly used combinations are alkaline phosphatase (AP) with the substrates NBT/BCIP to produce a blue to purple stain; horseradish peroxidase (HRP) with the substrate DAB to produce a brown to black stain; and β-galactosidase (genetically encoded by *lacZ*) with X-gal or IPTG substrate to produce a blue to purple stain. Commercially available substrates that produce fluorescent rather than colored products are also available, though colorimetric substrates are more frequently used.

Though not a chromogenic label itself, **biotin** is commonly conjugated to antibodies or incorporated into ISH oligonucleotide probes. The biotinylated probe can then be detected using the biotin-specific binding partner avidin/streptavidin conjugated to a chromogenic label like AP or HRP. Avidin and biotin can form large complexes and greatly enhance the signal compared to using a probe alone. **Digoxigenin** is another common molecule that is incorporated into ISH probes as an alternative to radioactive labels. Antidigoxigenin antibodies conjugated to chromogenic enzymes are used to detect the presence of the dig-labeled probe.

RADIOACTIVE LABELS

Radioactive isotopes, or radioisotopes, have an unstable nucleus that randomly disintegrates to produce a different atom. In the course of disintegration, these isotopes emit either energetic subatomic particles, such as electrons, or radiation, such as gamma rays. By using chemical synthesis to incorporate one or more radioactive atoms into a small molecule of interest such as a sugar, amino acid, or neurotransmitter, the fate of that molecule can be traced during any biological reaction.

Radioisotopes are visualized using autoradiography. Specimens are coated with a photographic emulsion and left in the dark while the radioisotope decays. Each decay event causes a silver grain to precipitate in the emulsion, so the position of the radioactive probe is indicated by the position of the developed silver grains. Because it is a direct chemical reaction, each silver grain represents one decay event. Therefore, radioactive labels allow quantitative studies of biological probes.

Primarily used to label ISH oligonucleotide probes, ^{35}S and ^{33}P are radioisotopes incorporated into the probe sequence, allowing sensitive detection, even for genes expressed at low levels. However, it can be tricky to calibrate the amount of time needed for exposing the emulsion to get the right signal-to-noise ratio, and the exposure may require weeks to months to see a signal.

^{3}H and ^{125}I are radioisotopes that are used to label proteins and nucleic acids to investigate a number of processes, such as receptor localization, proliferation, and protein trafficking. Radioactively labeled ligands added to a specimen can bind to active receptors, allowing autoradiographic detection for the presence of those receptors. For a description of autoradiography in proliferation and protein trafficking tracking, see Chapter 7.

GOLD LABELS

Colloidal gold labels are conjugated to antibodies for probe detection in electron microscopes (Chapter 5). Gold provides an electron-dense material that can be visualized as a dark spot in EM. Avidin/streptavidin can also be conjugated to gold for detection of biotinylated probes in EM. Gold particles of different sizes can be used to detect multiple targets in the same sample.

VISUALIZING GENE AND PROTEIN EXPRESSION

A scientist can classify a cell using its morphology, location in the brain, or its gene and protein expression profile. The genes expressed by a neuron determine whether it is excitatory or inhibitory, what kinds of neurotransmitters it releases and responds to, its physiological characteristics, and many other functional properties. Therefore, determining the genes and proteins expressed by a cell provides

FIGURE 6.5 *In situ* **hybridization shows where a gene is transcribed. (A)** Labeled oligonucleotide probes are synthesized with a complementary sequence to a gene of interest so they hybridize
to the mRNA when applied to a brain section. **(B)** This example ISH highlights the nuclei of the
specific subsets of neurons that express the mRNA for a particular enzyme. The probe was conjugated to digoxigenin and visualized with a colorimetric reaction (Courtesy of Dr. Pushkar Joshi).

vital clues to its role in the brain. Conversely, a scientist may identify a gene or
protein of interest, and then want to know where in the brain it is expressed. The
following methods provide tools an investigator can use to visualize gene and
protein expression in the nervous system.

In Situ Hybridization

In situ hybridization (ISH) is used to visualize the expression of nucleic
acids, usually mRNA, allowing a scientist to determine when and where a specific gene is expressed in the nervous system. This technique is good for identifying which neurons express a gene, although it does not reveal where the
functional protein product is expressed within the cell.

To perform an *in situ* hybridization experiment, a scientist must first identify the genetic sequence of an mRNA to be studied. Next, the scientist creates a single-stranded nucleic acid probe that has a complementary base-pair
sequence. The probe is tagged with radioactive nucleotides or other molecules
that allow detection (Box 6.2). A tissue sample is exposed to the tagged probes
such that the complementary strands can form a probe-mRNA duplex, marking
the location of neurons that produce the mRNA sequence (Figure 6.5). This
tissue sample is almost always a brain section, although it is possible to perform *in situ* hybridization on cultured cells or even un-sectioned, whole-mount
brains and animals.

In situ hybridization techniques require specific control experiments. Because
an investigator uses a complementary ("antisense") strand of mRNA as a probe,
a common negative control is a probe with the exact same ("sense") sequence
of the gene of interest. This control probe should not produce any signal, as no

TABLE 6.3 Histochemical Stains for Neuroanatomy

Stain	Use	Appearance	Comments
Cresyl violet	Cell nuclei, Nissl stain	Blue to purple	Useful for examining cytoarchitecture; stains each type of neuron slightly differently
Hematoxylin	Cell nuclei	Blue to blue-black	Often used in combination with eosin; known as H&E
Eosin Y	Cytoplasm	Pink to red	Counterstain with hematoxylin; acidophilic stain
Thionine	Cell nuclei, Nissl stain	Blue to purple	
Methylene blue	Cell nuclei	Blue	Can be perfused through the brain before fixation
Toluidine blue	Cell nuclei	Nucleus is stained blue; cytoplasm light blue	Often used to stain frozen sections
DAPI	Cell nuclei	Fluorescent blue	Fluorescent DNA intercalating agent; excited by UV illumination
Hoechst (*bis*-benzamide)	Cell nuclei	Fluorescent blue	Fluorescent DNA intercalating agent; excited by UV illumination
Propidium iodide (PI)	Cell nuclei	Fluorescent red	Fluorescent DNA intercalating agent; excited by green light illumination
Weigert	Myelin	Normal myelin is deep blue; degenerated myelin is light yellow	Combines hematoxylin with other chemicals to selectively stain myelin
Weil	Myelin	Black	Combines hematoxylin with other chemicals to selectively stain myelin
Luxol fast blue (LFB)	Myelin	Blue	
Golgi stain	Fills neuron cell bodies and processes	Black	Stains individual neurons at random

hybridization should occur. Another good control to ensure that the antisense strand of mRNA is specific to the gene of interest is to perform additional experiments with other antisense strands that hybridize to a different region of the same mRNA.

Though it is possible to use fluorescent molecules to label probes for *in situ* hybridization, **fluorescent *in situ* hybridization (FISH)** generally refers to a specific technique used to examine the location of DNA within chromosomes rather than tissue. A scientist can use this method to visualize chromosomal abnormalities, such as missing, mutated, or dislocated genetic segments.

Immunohistochemistry

Immunohistochemistry (IHC; *histo* = "tissue") is used to visualize the expression of proteins and other biomolecules. IHC applied to cells is often referred to as **immunocytochemistry** (ICC; *cyto* = "cells"), and IHC using fluorescent reagents is referred to as **immunofluorescence** (IF). The root *immuno-* refers to the fact that these techniques depend on antibodies to recognize and bind to specific proteins (antigens). By exploiting the immune system of other animals, such as mice, sheep, rabbits, and goats, scientists and commercial suppliers have generated a wide number of antibodies for many proteins. Table 6.3 lists a few of the many antibodies that are commonly used in neuoscience. See Chapter 14 for details about antibody production.

To perform an IHC experiment, a scientist incubates a sample with an antibody that recognizes a specific antigen. This **primary antibody**, the antibody that binds to the protein, can be conjugated directly with a fluorescent molecule or chromogenic enzyme. This procedure is known as **direct IHC** (Figure 6.6A). Alternatively, in **indirect IHC**, a scientist adds a **secondary antibody** that recognizes the primary antibody (Figure 6.6B). Because multiple secondary antibodies can bind to the primary antibody, the signal is amplified. The secondary antibody will only react with the primary antibody if it is directed against the correct immunoglobulin molecule of the animal species in which the primary antibody was raised. For example, if the primary antibody was produced in a rabbit, then the secondary antibody must be antirabbit and produced in a different animal. The secondary antibody is conjugated with a fluorescent molecule or chromogenic enzyme that allows the scientist to visualize the expression of the protein.

It is critical that an investigator determines whether the result of an IHC experiment is specific for the protein of interest. Using both a positive control, a specimen known to contain the antigen of interest, and a negative control, a specimen known *not* to contain the antigen of interest, is useful for determining whether the antibody works properly. Another good control experiment for primary antibody specificity is to preincubate the primary antibody with a known solution of its antigen: this process should form antibody/antigen complexes so there is no free antibody to bind to a sample. In indirect IHC, incubating the specimen with the secondary antibody but no primary antibody is important to test for nonspecific binding of the labeled secondary antibody.

While IHC is straightforward in theory, there are a number of steps that must be optimized for IHC to work in practice. A scientist may need to adjust the method of fixation, antibody concentration, length of time the antibody is

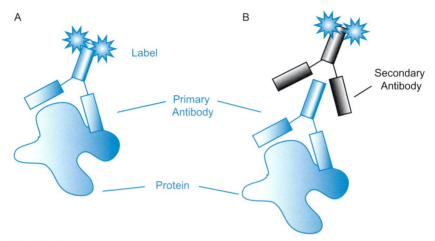

FIGURE 6.6 Immunohistochemistry. (A) Direct IHC uses primary antibodies conjugated directly to a label that can be visualized, while **(B)** indirect IHC uses labeled secondary antibodies that recognize the primary antibodies.

incubated with tissue, and accessibility to the antigen for good results. Each antibody is different, even antibodies directed against the same antigen. Therefore, an investigator should optimize experimental conditions for each new antibody before performing IHC experiments on multiple samples.

Enzymatic Histochemistry

An enzyme is a biomolecule capable of catalyzing a biological reaction. Often, these enzymes are detected using immunohistochemistry. Another method of detecting enzymes is using enzymatic histochemistry—the process of using an enzyme's endogenous activity to create a visible reaction product. The scientist incubates a brain section in a solution containing a chromogenic chemical capable of serving as a substrate for the enzyme. A colored by-product indicates the presence of the enzyme. For example, endogenous acetylcholinesterase (the enzyme that breaks down acetylcholine) is often used as a marker for acetylcholine in brain sections. Another commonly detected enzyme is cytochrome oxidase, which aids in visualizing metabolically active neurons, such as whisker barrels in rodent sensory cortex (Figure 6.7).

Reporter Genes

A reporter gene is a nonendogenous gene encoding an enzyme or fluorescent protein whose expression is controlled by a promoter for a separate gene of interest. Therefore, it is possible to examine the spatial and temporal expression of the gene by measuring the expression of the reporter. For example, imagine that a scientist is interested in studying the expression pattern of Gene X. The

TABLE 6.4 Commonly Used Antibodies that Mark Distinct Nervous System Cell Types

Cell Type	Antibodies
Progenitors/radial glia	Nestin; Pax6; RC2; vimentin; NF (neurofilament)
Young neurons	Doublecortin (DCX); NeuroD
Neurons	Tuj1 (neuron-specific β-tubulin); NeuN
Dendrites	MAP2
Axons	Tau-1, L1, Tag-1
Synapses	PSD95, synapsin
Neuronal subtypes	GAD (GABAergic neurons); vGLUT (glutamatergic); TH (dopaminergic); 5-HT (serotonergic); AChE (cholinergic)
Glia	GFAP (astrocytes); MBP (oligodendrocytes, myelin); PLP
Oligodendrocyte progenitor cells (OPC)	NG2, A2B5, O4 (late progenitor)

BOX 6.3 Array Tomography

Array tomography is a relatively new technique that uses immunohistochemistry on ultrathin serial sections, improving spatial resolution and eliminating out-of-focus fluorescence during imaging. This method allows investigators to create highly detailed three-dimensional images of protein expression in the nervous system.

An ultramicrotome is used to cut and collect ribbons of ultrathin (50–200 nm) serial sections on glass microscope slides. A scientist can perform immunohistochemistry to detect the presence of an antigen, but then the antibody can be removed and another IHC reaction performed. As many as 10 IHC reactions can be performed on the same section! Because the sections are so thin, antibodies have greater access to antigens in the section. The serial sections can then be reconstructed to provide high-resolution, three-dimensional information about the spatial relationships of these proteins. In addition, the sections can be used for electron microscopy, allowing an unprecedented investigation of protein expression combined with powerful magnification and resolution.

scientist can create a reporter construct in which the reporter gene is placed under the control of the promoter for Gene X. Alternatively, the scientist can fuse the DNA sequence of the reporter with the DNA sequence for Gene X and place both under the control of the promoter for Gene X. In either case, the reporter gene provides the scientist the ability to study the spatial and temporal expression of Gene X by measuring expression of the reporter (Figure 6.8).

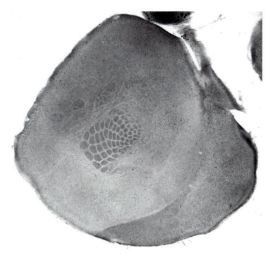

FIGURE 6.7 Enzymatic histochemistry. In this example, Rodent whisker barrels are enriched in the enzyme cytochrome oxidase, so incubating sections with the substrate (cytochrome c) and a colorimetric indicator (e.g. DAB) allows the whisker barrels to be visualized (Courtesy of Dr. Pushkar Joshi.)

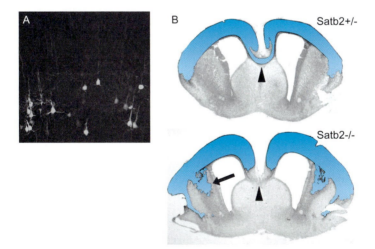

FIGURE 6.8 Using reporter genes to examine expression patterns. **(A)** In this BAC transgenic mouse line (see Chapter 11), the promoter for *neurotensin* was used to drive expression of the GFP reporter gene. This allows visualization of the specific neurons in which neurotensin is normally expressed. **(B)** Reporter genes can also be used to replace a gene when creating a knock-out animal (see Chapter 12). Here, *LacZ* gene expression (shown by β-galactosidase staining) replaces the gene *Satb2*, showing where Satb2 is normally expressed. Mutants have altered axonal projections (arrowheads). (A, Reprinted with permission from The Gene Expression Nervous System Atlas (GENSAT) Project, NINDS Contracts N01NS02331 & HHSN271200723701C to The Rockefeller University (New York, NY); B, Reprinted from Alcamo E.A. et al. (2008) Satb2 regulates callosal projection neuron identity in the developing cerebral cortex, *Neuron* 57(3): 364–377, with permission From Elsevier).

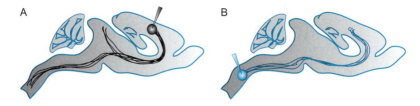

FIGURE 6.9 **Anterograde and retrograde tracers.** (A) Anterograde tracers show efferent projections, revealing regions that *receive* projections from cells in the labeled area, while (**B**) retrograde tracers show afferent projections, indicating regions that *project to* the labeled area.

Common reporters include GFP (green fluorescent protein) and its derivatives, *lacZ* (which encodes the enzyme β-galactosidase), and luciferase (the firefly protein). The expression pattern of reporter genes can reveal information about the gene on multiple levels, from subcellular localization within a neuron to patterns of organization in a whole animal.

VISUALIZING CIRCUITRY

So far, this chapter has reviewed methods of investigating cells based on their morphology, location, and gene/protein expression profiles. Of course, a neuron never functions in isolation; its role in the brain is governed by its unique combination of inputs and outputs and the neural circuits to which it belongs. Therefore, scientists often want to know the specific cells in the brain that provide synaptic input to a neuron of interest, as well as the specific cells to which a neuron projects. As just described, a Golgi stain can provide initial insight into the connections formed by neurons, but this method is limited by the random staining pattern of cells. Other methods allow scientists to more specifically investigate a neuron's direct and indirect connections with other neurons in the nervous system.

Anterograde and Retrograde Tracers

A neuronal **tracer** is a chemical probe that labels axon paths to illuminate connectivity in the nervous system. Tracers are described by the direction they travel in a neuron. **Anterograde** refers to transport from the cell body through the axon to the presynaptic terminal; **retrograde** refers to tracing in the opposite direction—transport from the synaptic terminal back to the cell body (Figure 6.9). Some tracers are strictly capable of only anterograde or retrograde transport. Others work in both directions. Common tracers include horseradish peroxidase, biotinylated dextran amine (BDA), diamidino yellow, Fluoro-Gold, and fast blue (Table 6.5). These tracers may be fluorescent or produce a colorimetric product that can be seen with light microscopy (see Box 6.2).

TABLE 6.5 Common Tracers

Tracer	Direction	Comments
Horseradish peroxidase (HRP)	Retrograde	Produces brown precipitate after reaction with hydrogen peroxide and DAB (diaminobenzidine)
Fluorescent microspheres	Retrograde	Available in many different colors; nontoxic
Fluoro-Gold	Retrograde	Widely used, rapid labeling
Diamidino yellow	Retrograde	Produces yellow fluorescence
Fast blue	Retrograde	Stable, rapid labeling; produces blue fluorescence
Cholera toxin, subunit B (CTB)	Retrograde	May also be anterograde
DiI, DiO	Anterograde and retrograde	Lipophilic dye crystals
Biotinylated dextran amine (BDA)	Anterograde and retrograde	Widely used; direction of transport depends on molecular weight and pH; can be visualized by EM
Phaseolus vulgaris-leucoagglutinin (PHA-L)	Anterograde	Plant lectin; can be visualized by EM
Tritiated amino acids (^{3}H-proline, ^{3}H-leucine)	Transsynaptic (anterograde)	Detected using autoradiography
Wheat-germ agglutinin (WGA)	Transsynaptic	Plant lectin; anterograde and retrograde transport possible; often conjugated to HRP for detection; transgene encoding WGA can be used to label genetically defined neural circuits
Tetanus toxin, fragment C (TTC)	Transsynaptic (retrograde)	Transgene encoding TTC can be used to label genetically defined neural circuits; nontoxic fragment
Pseudorabies virus (PRV)	Transsynaptic	Does not infect primates, including humans; bartha strain most commonly used for tracing studies; less virulent, only retrograde transport
Herpes simplex virus (HSV)	Transsynaptic	Broad host range, including humans

To use an anterograde or retrograde tracing technique, a scientist almost always injects the tracer into the brain of a live animal. After a waiting period (1–7 days), the brain can be removed, processed, and examined for presence of the tracer. An anterograde tracer should be present in the cell bodies and presynaptic terminals of the injected neurons, highlighting areas to which those

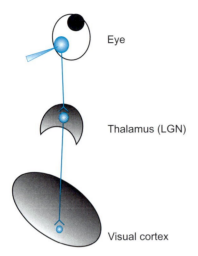

Eye

Thalamus (LGN)

Visual cortex

FIGURE 6.10 Transsynaptic labeling highlights neural circuits. Transsynaptic tracers can label functionally connected cells, such as those in the visual circuit where retinal ganglion cells project to the lateral geniculate nucleus of the thalamus which then project to primary visual cortex. Radioactive amino acids injected in the eye were used to show the presence of ocular dominance columns in the visual cortex.

neurons project. A retrograde tracer should be present in the cell bodies of neurons that project to the injection site.

Tracing experiments are often combined with other techniques, such as electrophysiology or immunohistochemistry. For example, an investigator might wish to know which cells in the brain send synaptic input to a neuron immediately following an *in vivo* electrophysiology recording. The scientist could inject a retrograde tracer, such as horseradish peroxidase, wait 3–7 days for transport to take place, and then perfuse the animal. By performing histology and surveying brain sections for labeled cells, the scientist could identify which cells projected to the cell(s) at the recording site.

Transsynaptic Labeling

Traditional anterograde or retrograde tracing experiments label only direct connections in a neural circuit. However, functional brain circuits are not simply composed of direct one-to-one connections, but also contain complicated pathways involving second- or third-order neurons. Scientists wishing to study these multisynaptic circuits can use a variety of **transsynaptic tracers**—tracers that can cross synapses and label multiple neurons in a circuit (Figure 6.10).

Radioactive Amino Acids

Radioactive amino acids, such as ^{3}H-proline or ^{3}H-leucine (^{3}H is pronounced "tritiated"), can be used for anterograde transsynaptic labeling. A scientist injects the amino acids into the local environment of the primary neurons, some of which are taken up by cells via amino acid transporters. Once inside the cell, they are incorporated into proteins, some of which are shuttled down axons by endogenous anterograde transport. Some of these proteins are secreted at the synaptic terminal and received by a second-order neuron. This process may repeat itself so that a third-order neuron is labeled. The labeled tracts can be detected using autoradiography (see Box 6.2).

Plant Lectins

Lectins are proteins that exhibit extremely high binding affinities for specific sugars. Some plant lectins have affinities for specific glycoproteins on neuronal plasma membranes. After binding, they can be internalized and transported within the neuron. Some examples of plant lectins that can also be transported across synapses include wheat germ agglutinin (WGA) and tetanus toxin fragment C (TTC). These lectins can be detected using antibodies against the lectin itself or by conjugation with a chromogenic enzyme (see Box 6.2). In addition to the standard method of injecting the tracer into an area of interest, researchers can create transgenic animals that express transneuronal lectin tracers under the control of neuron-specific promoters to examine cell-type specific circuits (Chapter 11).

Viruses

The use of viruses as transsynaptic tracers exploits the ability of viruses (especially viruses that naturally infect neural tissue) to invade neurons and then produce infectious progeny. Some strains of virus release progeny capable of crossing synaptic junctions and subsequently infecting downstream neurons in a circuit. The virus acts as a self-amplifying tracer that progressively labels neurons in a circuit as they become infected. The most common viruses used to label transneuronal circuits are pseudorabies virus (PRV) and herpes simplex virus (HSV).

The major limitation to using viruses as transsynaptic tracers is that these viruses can actually kill neurons and lead to generalized infection in the animals. The longer the animal is infected, the more likely higher-order neurons in a circuit will be labeled, but the greater the likelihood that the first- and second-order neurons will die as the infection proceeds.

CONCLUSION

The organization and structure of the nervous system provide the first clues about function and how particular structures or classes of cells can give rise to behaviors. This chapter has surveyed methods of investigating the morphology, expression patterns, and connectivity of cells. In the next chapter, we will survey methods that visualize the electrical and biochemical activity within neurons.

SUGGESTED READING AND REFERENCES

Books

Chandler, D. E., Roberson, R. W. (2009). *Bioimaging: Current Concepts in Light and Electron Microscopy*. Jones and Bartlett Publishers, Sudbury, MA.

Kiernan, J. A. (2008). *Histological and Histochemical Methods: Theory and Practice*, 4th ed.. Cold Spring Harbor Laboratory Press, Cold Spring Harbor, NY.

Polak, J. M., Van Noorden, S. (2003). *Introduction to Immunocytochemistry*, 3rd ed.. BIOS Scientific Publishers, Oxford.

Yuste, R., Konnerth, A. (2005). *Imaging in Neuroscience and Development: A Laboratory Manual.* Cold Spring Harbor Laboratory Press, Cold Spring Harbor, NY.

Yuste, R., Lanni, F., Konnerth, A. (2000). *Imaging Neurons.* Cold Spring Harbor Laboratory Press, Cold Spring Harbor, NY.

Review Articles

Enquist, L. W. & Card, J. P. (2003). Recent advances in the use of neurotropic viruses for circuit analysis. *Curr Opin Neurobiol* **13**, 603–606.

Jones, E. G. (2007). Neuroanatomy: Cajal and after Cajal. *Brain Res Rev* **55**, 248–255.

Kobbert, C., Apps, R., Bechmann, I., Lanciego, J. L., Mey, J. & Thanos, S. (2000). Current concepts in neuroanatomical tracing. *Prog Neurobiol* **62**, 327–351.

Massoud, T. F., Singh, A. & Gambhir, S. S. (2008). Noninvasive molecular neuroimaging using reporter genes: part I, principles revisited. *AJNR Am J Neuroradiol* **29**, 229–234.

Primary Research Articles—Interesting Examples from the Literature

Alcamo, E. A., Chirivella, L., Dautzenberg, M., Dobreva, G., Farinas, I., Grosschedl, R. & McConnell, S. K. (2008). Satb2 regulates callosal projection neuron identity in the developing cerebral cortex. *Neuron* **57**, 364–377.

Braz, J. M. & Basbaum, A. I. (2008). Genetically expressed transneuronal tracer reveals direct and indirect serotonergic descending control circuits. *J Comp Neurol* **507**, 1990–2003.

Chen, B., Schaevitz, L. R. & McConnell, S. K. (2005). Fezl regulates the differentiation and axon targeting of layer 5 subcortical projection neurons in cerebral cortex. *Proc Natl Acad Sci USA* **102**, 17184–17189.

Micheva, K. D. & Smith, S. J. (2007). Array tomography: a new tool for imaging the molecular architecture and ultrastructure of neural circuits. *Neuron* **55**, 25–36.

Takeuchi, A., Hamasaki, T., Litwack, E. D. & O'Leary, D. D. (2007). Novel IgCAM, MDGA1, expressed in unique cortical area- and layer-specific patterns and transiently by distinct forebrain populations of Cajal-Retzius neurons. *Cereb Cortex* **17**, 1531–1541.

Protocols

Card, J. P. & Enquist, L. W. (2001). Transneuronal circuit analysis with pseudorabies viruses. *Curr Protoc Neurosci*, Chapter 1, Unit 1.5.

Gerfen, C. R. (2003). Basic neuroanatomical methods. *Curr Protoc Neurosci*, Chapter 1, Unit 1.1.

Volpicelli-Daley, L. A. & Levey, A. (2004). Immunohistochemical localization of proteins in the nervous system. *Curr Protoc Neurosci*, Chapter 1, Unit 1.2.

Wilson, C. J. & Sachdev, R. N. (2004). Intracellular and juxtacellular staining with biocytin. *Curr Protoc Neurosci*, Chapter 1, Unit 1.12.

Websites

IHC World http://www.ihcworld.com/

The Antibody Resource Page http://www.antibodyresource.com

Chapter 7

Visualizing Nervous System Function

After reading this chapter, you should be able to:

- Describe techniques for visualizing activity and function in fixed tissue
- Describe nonelectrophysiological methods of visualizing and manipulating neuronal activity
- Describe techniques for visualizing and manipulating protein function

Techniques Covered

- **Static measures of activity and function**: immediate early genes, measuring cell proliferation with thymidine analogs, measuring protein trafficking with pulse-chase labeling
- **Visualizing neural activity**: voltage sensors (voltage-sensitive dyes, genetically encoded voltage sensors), calcium indicators (calcium indicator dyes, genetically encoded calcium sensors), synaptic transmission sensors (FM dyes, synapto-pHluorin)
- **Optical manipulation of neural activity**: molecule uncaging, light-activated channels
- **Visualizing protein function**: reporter genes, fluorescence resonance energy transfer (FRET), bimolecular fluorescence complementation (BiFC), fluorescence recovery after photobleaching (FRAP), photoactivation/ photoconversion
- **Optical manipulation of protein function**: molecule uncaging, CALI

Photographs in textbooks and other publications can sometimes create the impression that an individual cell is a small, static blob—a microscopic collection of organelles in a cytoplasmic soup that sits silently within a tissue. These individual snapshots are extremely misleading. The living cell is a dynamic wonder—an exquisitely complicated machine full of activity that a scientist cannot see with the naked eye, especially not in a single photograph. A neuron can fire electrical impulses in a steady tonic rhythm or phasic bursts, depending on the activity in other cells. Genes are constantly transcribed, with mRNA shuttling out of the nucleus and ribosomes quick to translate genetic sequences into functional proteins. Many proteins dash about the cell, changing their function and activity depending on the other proteins they encounter and the cellular signals transduced from the plasma membrane.

Methods to visualize the structure and localization of cells and proteins, as described in the last chapter, offer a great introduction to studying the activity that takes place in a neuron. However, these methods do not fully reveal the dynamic electrical and biochemical activity found within the cell. As a metaphor, pretend you know nothing about soccer (or football if you are outside the United States). In order to learn more about the sport, you find several photographs of people playing soccer, mostly photographs taken from a bird's-eye view of the field. While these photographs would no doubt teach you some aspects of the game, the only way to determine the rules, how individual players move, and how the players work as a team is to watch movies of the game. The same is true for biological systems; while photographs reveal the structure and localization of cells and proteins at a snapshot in time, a scientist can learn much more by observing the dynamic events that take place within neurons.

The purpose of this chapter is to survey methods used to visualize electrical and biochemical activity in neurons. We begin with a description of how we can visualize functional processes in fixed tissue. Then, we will survey the major methods of visualizing and manipulating electrophysiological properties of neurons using optical methods. Finally, we survey the growing collection of methods that can be used to visualize and manipulate biochemical dynamics. Many of these methods are based on the use of fluorescent dyes and proteins that make specific components of living cells visible under the microscope.

STATIC MARKERS OF ACTIVITY

Although it is ideal to measure neural activity as it occurs, it is not always possible or practical. Most structures in the brain are too deep to visualize *in vivo* with currently available methods, and it may simply be impractical to watch certain specialized activities because modern dynamic probes may not reveal any more information than static images. Certain cellular functions can be examined through the use of activity markers in fixed, histological brain sections. There are two fundamental approaches to examining neural activity after it has already occurred: (1) look for the presence of activity indirectly by measuring byproducts that accumulate during specific processes in active neurons; and (2) during certain cellular functions, incorporate a marker into cells that can subsequently indicate the presence of activity during future histological examination. In both methods, a scientist examines a snapshot of activity, a fixed representation of one moment in a dynamic process.

Assaying Neural Activity in Fixed Tissue

Neural activity leads to the transient and rapid (within minutes) transcription of a group of genes known as **immediate early genes (IEGs)**. These genes encode a diverse range of proteins, including transcription factors (c-Fos, c-myc, c-jun), cytoskeletal-interacting proteins (Arc), and growth factors (BDNF), that play roles in neural plasticity.

By using histological methods to stain for these markers, a scientist can indirectly determine which regions were active just prior to the time the brain was fixed (Figure 7.1). A scientist can use *in situ* hybridization (ISH) to show expression of the IEGs or immunohistochemistry (IHC) to show expression of their protein products (Chapter 6).

Because IEG expression patterns and levels vary due to a number of uncontrollable factors, investigators do not typically examine IEGs as stand-alone indicators of activity. The most reliable evidence compares IEG expression between a group of experimental animals and a control group—for example, demonstrating that there is significantly higher expression of IEGs in a specific neuron population in animals that received a stimulus compared to control animals that did not receive the stimulus. IEG patterns can also be used to screen neurons for the presence of activity that correlate with specific behaviors. For example, some brain regions thought to promote sleep states in the brain were discovered by examining IEGs in animals that were asleep prior to fixation. In these types of studies, IEGs only provide the first clue as to the function of a brain region. It is necessary to follow up these studies with other, more reliable measurements of neural activity to obtain meaningful results.

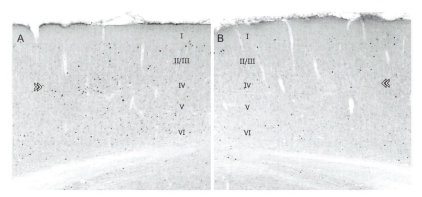

FIGURE 7.1 Measuring activity in a fixed brain sample. Overall activity levels can be detected in fixed tissue by labelling for IEGs using immunohistochemistry or *in situ* hybridization techniques. Here, c-Fos immunoreactivity shows that patterned visual stimulation evokes more activity in the visual cortex contralateral (**A**) to the stimulated eye compared to the ipsilateral (**B**) visual cortex. (Reprinted from Dotigny F., et al. (2008). Neuromodulatory role of acetylcholine in visually-induced cortical activation: Behavioral and neuroanatomical correlates. *Neuroscience*, 154 (4): 1607–1618, with permission from Elsevier.)

Assaying Cellular Function in Fixed Tissue

To measure certain functional processes in fixed tissue, a scientist can introduce a marker while the process occurs in live tissue, and then detect this marker in subsequent histological experiments. Two biological functions commonly assayed using these methods include cell proliferation and protein trafficking.

Assaying Cell Proliferation with Thymidine Analogs

As a cell divides, it progresses through different stages of the cell cycle. Cells in the DNA synthesis phase of mitosis can be marked by exposing them to tagged DNA base pair analogs. BrdU (bromo-deoxyuridine), a synthetic analog of the DNA base thymidine, or radioactive tritiated thymidine (^{3}H-thymidine), can be injected into an animal or introduced into cells or tissue in culture media. Cells that are synthesizing DNA will incorporate BrdU or ^{3}H-thymidine into their newly made DNA in place of endogenous thymidine molecules. In subsequent histology experiments, BrdU can be detected using IHC, and ^{3}H-thymidine can be detected using autoradiography (Box 6.2).

The presence of BrdU or ^{3}H-thymidine indicates that a cell was dividing around the time of injection. However, these markers do not indicate whether the cell continued to proliferate or stopped dividing to become a functional, differentiated cell. To answer these questions, a scientist performs additional IHC experiments for proteins that are present only during specific stages of the cell cycle or in certain types of cells. For example, Ki67 and PCNA are proteins that are present during the active, proliferating stages of the cell cycle but not during the resting stage. Therefore, investigators can compare the number of proliferating cells that differentiated into postmitotic cells to the number of proliferating cells that continued to divide by combining BrdU/^{3}H-thymidine pulses with IHC for other markers, such as Ki67 (Figure 7.2).

Assaying Protein Trafficking with Pulse-Chase Labeling

Pulse-chase labeling is used to observe the movement of a substance through a biochemical or cellular pathway. A labeled probe is injected into animals or added to cultured cells for a brief period (the pulse) and then washed away and

BrdU   Ki67

FIGURE 7.2 Measuring proliferation in a fixed brain sample. Cells that took up BrdU when it was injected into a live animal can be detected using immunohistochemistry against BrdU. BrdU+ cells can be compared to cells that are positive for the protein Ki67, which indicates a cell is still proliferating. (Courtesy of Dr. Sandra Wilson).

replaced by unlabeled molecules (the chase). By following changes in the localization of a marker in different subcellular compartments over time, a scientist can observe protein trafficking pathways. For example, multiple proteins can be pulsed at the same time with a radioactively labeled amino acid. By fixing specimens at different time points, a scientist can use autoradiography to determine the location of labeled probes at regular intervals to observe changes in the localization of the incorporated label. Pulse-chase experiments have been important in elucidating many biochemical pathways, including synthesis and release of neurotransmitters.

VISUALIZING NEURAL ACTIVITY

In Chapter 1, we discussed methods of visualizing activity in the whole brain with a spatial resolution of millimeters (about 100,000 neurons) and a temporal resolution of several seconds. In Chapter 4, we discussed electrophysiological methods, which can record neural activity with the spatial resolution of a single neuron (depending on the type of electrode and location in neural tissue) and a temporal resolution of milliseconds. The methods of visualizing neural activity described here can be considered a compromise between whole-brain imaging and electrophysiology: it is possible to visualize activity with single-neuron resolution or in thousands of neurons at once while simultaneously examining the spatial relationships of those neurons. The temporal resolution is typically on the order of milliseconds to seconds, depending on the specific technique. These methods can be used to visualize activity in dissociated cells in culture, tissue slices in bath media, and even intact brains in living organisms.

Visualizing neural activity depends on specialized fluorescent probes that report changes in membrane potential, calcium concentration, or synaptic vesicle fusion. These probes can be organic dyes that a scientist adds to a neural system prior to performing an experiment or a genetically encoded fluorescent protein stably expressed in a transgenic animal. Dyes tend to exhibit better temporal properties and signal-to-noise characteristics than proteins. However, genetically encoded proteins can be targeted to specific cell types in the brain, allowing a scientist to observe activity in molecularly defined circuits.

Imaging Voltage

Techniques that visualize changes in membrane potential are the closest analogs to electrophysiological recordings, as they report voltage changes in neuronal membranes with a fine temporal resolution. **Voltage-sensitive dye imaging (VSDI)** is currently the primary method by which scientists visualize changes in voltage. However, the active development of better genetically encoded voltage sensors may make their use more widespread in the future.

Voltage-Sensitive Dyes

Voltage-sensitive dyes shift their absorption or emission fluorescence based on the membrane potential, allowing a scientist to gauge the global electrical state of a neuron. Unlike extracellular electrophysiology techniques, it is possible to detect subthreshold synaptic potentials in addition to spiking activity. These dyes allow activity measurements in large populations of neurons at once, a task that would otherwise require impractically large electrode arrays.

There are a variety of voltage-sensitive dyes available, each differing in signal duration, intensity, signal-to-noise ratio, and toxicity. Most dyes exhibit small signal changes; the **fractional intensity change** ($\Delta I/I$ or $\Delta F/F$, I = intensity and F = fluorescence) is in the range of 10^{-4} to 10^{-3} (0.1%), though some dyes may reach 6% for a large 100 mV change in potential. Thus, there must be very little noise to detect a reliable signal. Furthermore, activity-dependent changes in the intrinsic optical absorption and reflection properties of the brain itself can interfere with the voltage-sensitive dye measurements.

Results from voltage-sensitive dye imaging tend to look like electrophysiology data, showing changes in $\Delta I/I$ over a certain period of time to report membrane potential changes during that same time. This information can then be encoded to show the magnitude of changes in pseudo-color maps overlaying the specimen, providing spatial information about the signal (Figure 7.3).

Genetically Encoded Voltage Sensors

To examine molecularly defined subsets of neurons, a scientist may use a genetically encoded voltage sensor. However, these probes are currently inadequate

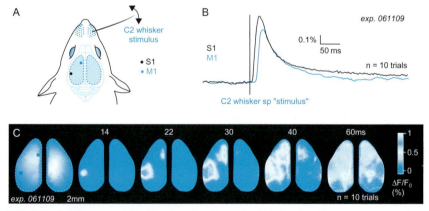

FIGURE 7.3 Voltage-sensitive Dye Imaging Example. (A) Craniotomies over somatosensory (S1) and motor (M1) cortex in anesthetized mice were stained with the voltage sensitive dye RH1691. **(B)** Moving the C2 whisker evokes transient increases in fluorescence, first in S1 (black), then M1 (blue). **(C)** The spatiotemporal dynamics of voltage changes across the brain can be observed after deflecting the C2 whisker. (Reprinted from Ferezou I. *et al.* (2007). Spatiotemporal dynamics of cortical sensorimotor integration in behaving mice. *Neuron*, 56 (5): 907–923, with permission from Elsevier).

for most studies, especially compared with VSDI. Over the past decade, many laboratories have attempted to develop genetically encoded sensors by fusing a fluorescent protein (e.g., GFP) to a voltage-gated ion channel so that the protein could act as a sensor for membrane potential changes. The first genetically encoded voltage sensor, FlaSh ("fluorescent Shaker," not to be confused with FlAsH, which is discussed later, as a method to cause photoinactivation), was created by fusing GFP to the *Drosophila* potassium channel Shaker. At depolarizing membrane potentials, FlaSh fluorescence intensity decreases. Unfortunately, this probe does not work well in mammalian systems due to poor membrane targeting and high background fluorescence. However, a number of research groups are actively working on designing better fluorescent protein voltage sensors, with encouraging results.

Imaging Calcium

Intracellular calcium is central to many physiological processes, including neurotransmitter release, ion channel gating, and second messenger pathways. In neurons, calcium dynamics link electrical activity and biochemical events. Thus, changes in calcium concentration can indirectly indicate changes in electrical activity. As with voltage-sensitive probes, fluorescent calcium indicators exist as dyes, such as Fluo-4 and Fura-2, and fluorescent proteins, such as aequorin and variants of GFP.

Data for calcium imaging experiments typically show changes in fluorescence intensity or the ratio of fluorescence intensity over time, normalized to initial levels of fluorescence (F_0). These traces are then used to create pseudocolored maps to show spatial information about the changes, with hot colors (yellow-red) representing large changes and cool colors (blue-purple) representing small changes. Compared to VSDI, calcium imaging produces larger, more robust visible signal changes ($\Delta F/F = 1–20\%$) but with lower temporal resolution (milliseconds rather than microseconds).

Calcium Indicator Dyes

Calcium indicator dyes, organic molecules that change their spectral properties when bound to Ca^{2+}, can be categorized as ratiometric or nonratiometric dyes. **Ratiometric dyes** are excited by or emit at slightly different wavelengths when they are free of Ca^{2+} compared to when they are bound to Ca^{2+}. Thus, they can report changes in Ca^{2+} through changes in the ratio of their fluorescence intensity at distinct wavelengths. A common ratiometric indicator is Fura-2, which emits at 510 nm but can change the wavelength at which it is excited from 340 nm to 380 nm in response to calcium binding. Thus, monitoring the ratio of the fluorescence intensity emitted by Fura-2 at 340 nm to the fluorescence emitted at 380 nm reveals changes in calcium concentration (Figure 7.4). Ratiometric dyes allow investigators to correct for background changes in fluorescence unrelated to calcium dynamics, such as artifacts related to photobleaching, variations

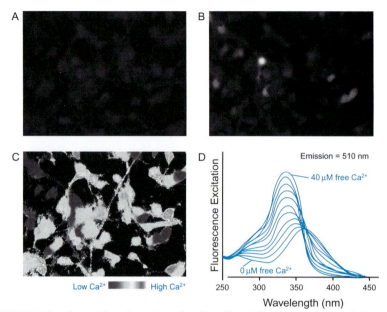

FIGURE 7.4 Imaging calcium dynamics using the ratiometric dye Fura-2. Fura-2 changes its emission spectra when bound to calcium. To examine calcium dynamics, images are collected by exciting cells loaded with Fura-2 at (**A**) 340 nm, dim at resting levels of calcium and (**B**) 380 nm, brighter at resting levels of calcium, and collecting emission at 510 nm. The ratio of 340/380 indicates cells with high (bright) or low (dim) concentrations of calcium (**C**). The ratio can be converted into known concentrations of free calcium levels (**D**). (Courtesy of Dr. Jocelyn Krey.)

in illumination intensity, or differences in dye concentration. The primary disadvantage of using ratiometric dyes compared to nonratiometric dyes is that data acquisition and measurements are more complicated.

Nonratiometric dyes report changes in Ca^{2+} directly with changes in either excitation or emission fluorescence intensity. The common nonratiometric indicators Fluo-3 and Fluo-4 exhibit predictable increases in emission fluorescence intensity with increases in calcium concentration. While the direct relationship between fluorescence intensity and calcium concentration is sensitive for detecting changes due to calcium binding, the measurement is also prone to detecting changes based on dye concentration and experiment-specific conditions. However, nonratiometric dyes tend to be easier to use and quantify.

Because dye-indicators bind to Ca^{2+}, they contribute to buffering of cellular Ca^{2+} and can alter Ca^{2+} homeostasis. Thus, quantitative measurements of Ca^{2+}-signaling require careful understanding of the indicator properties, experimental conditions and potentially interfering binding partners (e.g., Mg^{2+}).

Genetically Encoded Calcium Sensors

The bioluminescence of the jellyfish *Aequorea victoria* is due to the presence of both GFP and a chemiluminescent protein called **aequorin**. GFP

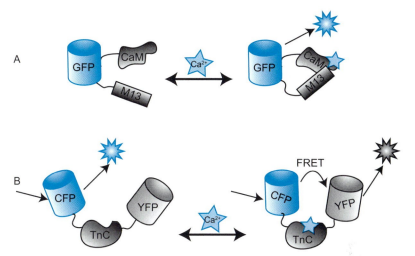

FIGURE 7.5 Genetically-encoded calcium sensors use proteins with calcium-dependent conformational changes to affect fluorescent protein emission. (A) Some sensors (e.g. Pericam) use a form of GFP whose fluorescence intensity increases when calcium binds to a calmodulin (CaM) recognition element. The Ca^{2+}-dependent conformational change is magnified by binding to the M13 peptide. **(B)** Others (e.g. TN-L15) use a conformational change in troponin-C (TnC) to induce FRET.

has received much attention and is ubiquitous in biological research. However, aequorin has been used for calcium imaging since the late 1960s. This protein emits light when bound to calcium, without requiring illumination for excitation. Unfortunately, the measurable signal is extremely low and requires amplification. Thus, it has mostly been superseded by the use of fluorescent dyes and more robust genetically encoded calcium sensors.

Genetically encoded calcium sensors take advantage of the conformational changes that occur in certain endogenous calcium-binding proteins when they bind to calcium. Cameleons, the first genetically encoded Ca^{2+} indicators, use FRET (see the following section) to cause changes in the emission spectra after binding to calcium. Cameleons can be used as a ratiometric indicator by comparing fluorescence intensity of the unbound CFP (cyan fluorescent protein) emission wavelength to fluorescence intensity of the bound YFP (yellow fluorescent protein) emission wavelength. Other genetically encoded calcium sensors, such as Camgaroo, G-CaMP, or Pericam, are used nonratiometrically through direct changes in fluorescence intensity caused by calcium-sensitive changes in the structure of the fluorophore. Cameleons, Camgaroo, G-CaMP, and Pericam all use the calcium-binding protein calmodulin, while TN-L15, another FRET-based calcium sensor, uses troponin C (Figure 7.5). Like calcium dyes, genetically encoded sensors also contribute to buffering of cellular Ca^{2+} levels.

Imaging Synaptic Transmission

A critical component of neural activity is the release of neurotransmitters from vesicles into the synaptic junction. Electrophysiology first revealed the fusion and reuptake of synaptic vesicles, but imaging fluorescent dyes and proteins that report synaptic vesicle activity provide greater details about the dynamic process of synaptic transmission.

FM Dyes

FM dyes are lipophilic styryl dyes that fluoresce when bound to membranes. They are useful for reporting the exocytosis and endocytosis events that occur during synaptic vesicle recycling and can stain nerve terminals in an activity-dependent manner. Typically done in cultured cell preparations, a scientist adds dye to the medium to label the membrane surface. For activity-dependent labeling, neurons are stimulated so that all synaptic vesicles undergo exocytosis and endocytosis with the stained membrane. The stained membranes are internalized during endocytosis. The rest of the dye is washed away, leaving behind clusters of stained recycled synaptic vesicles. Then, the investigator observes the destaining process that occurs as the stained synaptic vesicles release dye during further exocytosis. FM dyes have been used to characterize neurotransmitter release mechanisms, measure the kinetics of vesicle recycling, and observe vesicle movements.

pH-Sensitive Fluorescent Proteins

Neurotransmitter release can also be examined using the genetically encoded synaptic transmission reporter **synapto-pHluorin**, which uses pH-sensitive mutants of GFP. The interior of synaptic vesicles are acidified to a pH of approximately 5.7, an environment that keeps the pHluorins in an off state. When the vesicle fuses with the plasma membrane during exocytosis, the pH rises to extracellular levels, switching the pHluorin on and causing it to fluoresce (Figure 7.6). One of the advantages of using synapto-pHluorins is that the signal regenerates through multiple rounds of vesicle release and recycling, which permits vesicle recycling to be imaged in addition to synaptic transmission.

Table 7.1 compares various methods of visualizing neural acivity.

OPTICALLY MANIPULATING NEURAL ACTIVITY

In addition to observing activity using optical methods, a scientist can also stimulate or inhibit neural activity using optical methods. There are many advantages to using optical methods of manipulation over electrophysiological methods: optical probes can be genetically targeted to specific cell types within the brain, large populations of neurons can be simultaneously controlled, and off-target effects from neighboring brain regions are easier to avoid.

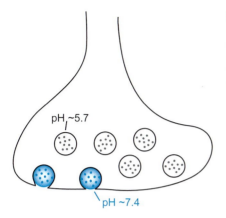

FIGURE 7.6 Synapto-pHluorin reports synaptic transmission dynamics. Synapto-pHluorin does not fluoresce when in acidified (pH ~5.7) synaptic vesicles, but does fluoresce when the vesicles exocytose and expose the reporter to physiological pH (pH ~7.4).

Stimulation through Uncaging of Molecules

A variety of different compounds can be "caged" by adding a photoremovable protecting group that chemically masks their active function. Illuminating the caged molecule removes the protecting group, releasing the active molecules into the illuminated area (Figure 7.7). Uncaging occurs rapidly (microseconds to 1 millisecond). Using this method, investigators can control the timing and location of release by controlling when and where molecules are illuminated and the size of the illuminated area. Calcium and neurotransmitters, such as glutamate, are commonly used for uncaging experiments to induce neuronal activity.

Light-Activated Channels

The ability to control neuronal activity on a natural time scale is important in determining exactly how activity relates to functional output. Both electrical microstimulation and optical stimulation of light-sensitive ion channels allow investigators to control neural activity with rapid time scales. However, optical stimulation has some advantages over electrical microstimulation. Stimulation of neurons using light eliminates artifacts that can be caused by electrical stimulation. Also, because light-activated channels are genetically encoded, investigators have the ability to selectively target specific cell types, enhancing the ability to distinguish the contribution of activity from particular populations of neurons.

SPARK (synthetic photoisomerizable azobenzene-regulated K^+) channels are potassium channels fused to a molecular tether containing an ion that can block the channel pore (Figure 7.8C). The molecular tether changes conformation depending on the wavelength of light shined on it: long wavelengths of light (500 nm, green) allow the ion to block the channel pore, turning the channel off, while shorter wavelengths of light (380 nm, blue) shorten the tether, preventing the ion blocker from reaching the channel and allowing efflux of potassium ions out of the cell. Thus, by shining the short (blue) wavelength of light, the neuron becomes hyperpolarized, with its activity silenced.

TABLE 7.1 Commonly Used Activity Sensors

Category	Examples	Advantages	Disadvantages
Voltage-sensitive dyes	Di-8-ANEPPS, RH 414	High temporal resolution (microsecond response time).	Low signal intensity changes (order of 0.1%). Promiscuous labeling.
Genetically encoded voltage sensors	FlaSh (Fluorescent Shaker K^+ channel), SPARC (Na^+ channel)	Targeting specificity.	Lower temporal resolution than dyes (milliseconds). High background. Not as useful in mammalian systems.
Nonratiometric calcium indicator dyes	Fluo-3, Fluo-4, Calcium Green-2	Direct fluorescence intensity changes correlated with $[Ca^{2+}]$.	Prone to artifacts from loading concentration, photobleaching, and other experiment-specific conditions. Cannot be calibrated. Buffering of intracellular Ca^{2+} can alter signaling pathways if present at high concentration.
Ratiometric calcium indicator dyes	Fura-2, Indo-1	Large signal changes (10%). Resistant to experiment-specific artifacts, like loading concentration and photobleaching. Enables calibration.	More complicated data acquisition and measurement than nonratiometric dyes. Buffering of intracellular Ca^{2+} can alter signaling pathways if present at high concentrations.
Genetically encoded calcium sensors	FRET-based: Cameleon, TN-L15 Other: Camgaroo, G-CaMP/Pericam	Targeting specificity.	Can have long signal decay time, giving low temporal resolution. Buffering of intracellular Ca^{2+} can alter signaling pathways if present at high concentration.
Dye-based synaptic vesicle markers	FM dyes (FM1-43, AM4-64)	Can vary stimulation and exposure to dye to control number of vesicles labeled.	Labels all membrane surfaces so background can be high and difficult to wash out of brain slices.
pH-sensitive fluorescent proteins	Synapto-pHluorin	Allows study of presynaptic release.	Many active release sites can make imaging of specific synapses difficult. Vesicle is only visible while fused for exocytosis, but becomes invisible after endocytosis.

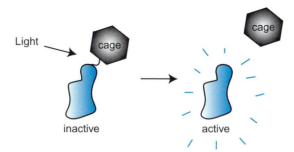

FIGURE 7.7 Molecular uncaging. A protective chemical group ("cage") can be added to a molecule to block its activity until light of a specific wavelength is applied that cleaves off the "cage" and releases active molecules into the illuminated region.

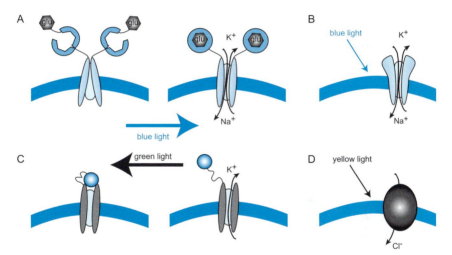

FIGURE 7.8 Optical control of neural activty. Blue light (380 nm) can be used to depolarize a cell expressing **(A)** LiGluR or **(B)** ChR2 channels, or hyperpolarize a cell expressing **(C)** SPARK channels. Green light (500 nm) reverses the ligand-gated activation of LiGluR and SPARK channels. **(D)** Yellow light (580 nm) can be used to hyper polarize cells expressing NpHR.

A light-gated ionotropic glutamate receptor (LiGluR) complements the SPARK channel, allowing short wavelengths of light to activate the glutamate receptor and thus depolarize the cell (Figure 7.8A). LiGluR works in a similar manner to the SPARK channel, but rather than a function-blocking ion, LiGluR has an agonist ligand attached to the tether that will lead to activation when bound to the receptor.

Without relying on conformational changes or external ligands, optically controlled channelrhodopsin-2 (ChR2) and halorhodopsin (NpHR) molecules allow fast neuronal activation and inhibition, respectively, permitting extremely precise control over activity (Figure 7.8B, D). ChR2 is a cation channel that allows sodium ions to enter the cell upon illumination by ~470 nm blue light, while NpHR is a chloride pump that activates upon illumination by ~580 nm yellow light. Both constructs rapidly activate upon illumination with the appropriate wavelength of light, and rapidly turn off without proper illumination, making the stimulation fully reversible and highly temporally precise.

VISUALIZING PROTEIN FUNCTION

Scientists can use fluorescent probes to visualize the activity of specific proteins within a cell. Using these methods, it is possible to track the subcellular localization of proteins over time, as well as to detect the interaction between two or more proteins. New tools are continually being developed to reveal the movements and dynamic interactions of proteins.

Time-Lapse Imaging with Reporter Genes

Reporter genes were described in the previous chapter as useful markers to localize protein expression. Fusing a fluorescent reporter gene, such as GFP or its variants, to a protein of interest allows a scientist to observe the location and trafficking of the protein in live cells and tissues (Figure 7.9). In most cases, the GFP fusion protein behaves in the same way as the original protein. The GFP fusion protein strategy has become a standard way to determine the distribution and dynamics of many proteins of interest in living cells. For example, a transcription factor fused to GFP can reveal changes in localization in response to growth factors, moving from cytoplasm to nucleus.

Fluorescence Resonance Energy Transfer (FRET)

Fluorescence (or Förster) resonance energy transfer (FRET) is used to monitor interactions between proteins (Figure 7.10A). Two molecules of interest are each labeled with a different fluorophore, chosen so the emission spectrum of one fluorophore (donor) overlaps with the absorption spectrum of the other (acceptor). If the two proteins bind, bringing their fluorophores into close proximity (closer than about 2 nm), the energy of the absorbed light can be transferred directly from the donor to the acceptor. Thus, when the complex is illuminated at the excitation wavelength of the first fluorophore, light is emitted at the emission wavelength of the second. FRET is measured as the ratio of fluorescence intensity of the donor to fluorescence intensity of the acceptor. This method can be used with two different spectral variants of GFP as fluorophores to monitor processes such as the interaction of signaling molecules with their receptors. FRET has traditionally been exploited to assess distances and orientations between separate molecules or different sites within a single macromolecule. This technique can also be used to report any substrate modification that leads to a conformational change resulting in the interaction of donor and acceptor fluorophores.

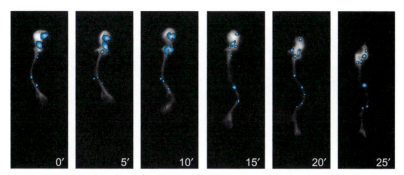

FIGURE 7.9 Example of time-lapse imaging with reporter proteins. This migrating neuron is expressing two different reporter proteins with different excitation and emission spectra: a cytoplasmic fluorescent protein that allows visualization of the cell's morphology (white) and a GFP fusion protein that labels early endosomes (blue). Observing changes in the position of the endosomes over time can reveal intracellular trafficking dynamics.

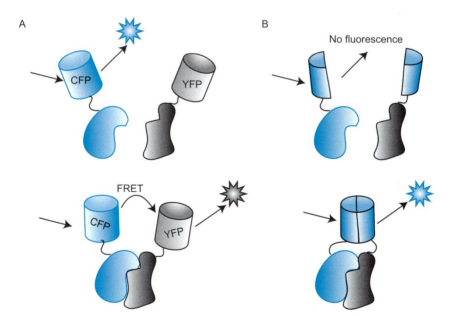

FIGURE 7.10 FRET and BiFC allow protein-protein interactions to be monitored. (A) Two proteins expressing a FRET pair—blue protein conjugated to a FRET donor and grey protein conjugated to the FRET acceptor—are excited at a wavelength that excites the donor (CFP). If the proteins do not interact, donor (CFP) fluorescence will be emitted, but if the two proteins interact closely, FRET can occur and acceptor (YFP) fluorescence will be emitted. **(B)** BiFC is similar but uses proteins attached to two halves of a single fluorophore. When the proteins do not interact, no fluorescence is detected. If the proteins do interact, the fluorophore should be able to fold and emit fluorescence.

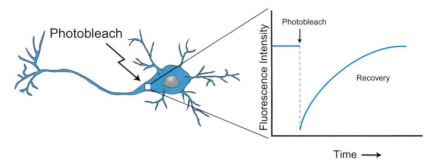

FIGURE 7.11 **FRAP reveals information about reporter protein kinetics.** Intense illumination is used to bleach a small region of a cell expressing a fluorescent reporter protein and the photobleached area is then observed. The kinetics of protein trafficking or turnover can then be measured by observing the dynamics of fluorescence returning to the bleached area.

Bimolecular Fluorescence Complementation (BiFC)

Bimolecular fluorescence complementation (BiFC) is similar to FRET in that it relies on the interactions and proximity of two proteins or parts of a protein to produce a visible change. Rather than two separate fluorophores, however, two halves of a single fluorophore (e.g., GFP) are split. Each half of the fluorophore is fused to one protein and does not fluoresce. When the two proteins are close enough in proximity, the two halves of the fluorophore can fold and form the functional fluorophore (Figure 7.10B).

Fluorescence Recovery after Photobleaching (FRAP)

One limitation to using fluorophores is the phenomenon of photobleaching and decrease in fluorescence intensity over time. However, investigators can exploit this phenomenon to examine protein turnover rates or trafficking. **Fluorescence recovery after photobleaching (FRAP)** uses strong laser illumination to bleach the fluorescent molecules in a particular region. The investigator then measures the timecourse of fluorescence returning to the bleached region. This can reveal the kinetics of a fluorescently tagged protein's diffusion, binding, or dissociation, or active transport into the bleached region (Figure 7.11).

Photoactivation and Photoconversion

Some fluorescent proteins have been engineered so they are not fluorescent until hit with a specific wavelength of light (often UV) that **photoactivates** the fluorophore (Figure 7.12A). Photoactivatable fluorescent proteins can be fused to a protein and later activated, allowing selected subpopulations of the protein to be labeled and followed as they move around a cell. This is particularly useful for observing the behavior and dynamics of a highly abundant protein that may be difficult to track in a large population of fully fluorescent molecules. Some sophisticated photoactivatable proteins allow precise control over their fluorescence. For example, the photoactivatable protein Dronpa can be reversibly

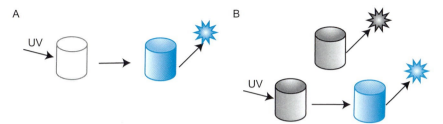

FIGURE 7.12 Photoactivation/Photoconversion. (A) Some fluorophores do not emit light until activated by a flash of UV light. **(B)** In photoconversion, a fluorophore emitting at one wavelength (grey) can be converted to emit at a different wavelength (blue) after UV stimulation.

photoactivated; UV light increases fluorescence intensity, while blue light will quench the fluorescence.

Photoconversion serves a similar purpose as photoactivation, but rather than activating a previously nonfluorescent molecule, a pulse of light causes a fluorophore to change its emission spectra from one color to another (Figure 7.12B). This can be used to examine the timing of protein behaviors. Kaede is an example of a photoconvertible fluorescent protein that initially emits green light but emits red light after UV illumination.

OPTICALLY MANIPULATING PROTEIN FUNCTION

In addition to optical tools that allow a scientist to observe proteins and biochemical processes with light, other tools allow the manipulation of protein activity. The development of tools to both visualize and manipulate function at the same time is a rapidly expanding field.

Photoactivation/Photo-uncaging

Earlier in this chapter, we described the process of uncaging molecules, such as calcium or neurotransmitters, to induce neuronal activity. This technique can be used more generally for optical control over many molecules to control biochemical activity as well as neural activity (Figure 7.7). For example, the function of ligand-gated receptors can be probed by photo-uncaging ligand molecules, the function of molecules in a signaling cascade can be probed through photo-uncaging of second messengers, and the function of many proteins can be probed through photorelease of agonists or antagonists.

Photoinactivation

Chromophore-assisted laser inactivation (CALI) is a method used to inhibit the activity of specific proteins in precise subcellular regions. Some dye molecules generate reactive oxygen species that can damage a protein. By using a laser to activate targeted dye molecules, specific proteins can be rendered nonfunctional (Figure 7.13). In order to target specific proteins, a scientist

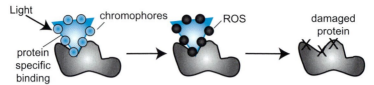

FIGURE 7.13 CALI is used to disrupt protein function. Chromophores that generate reactive oxygen species (ROS) can be conjugated to molecules that bind specific proteins (blue triangle, e.g. an antibody). When illuminated, the chromophores generate ROS that damage and disable the bound protein.

TABLE 7.2 Comparison of Techniques to Optically Probe Protein Function

	Description
Reporter Genes	Proteins or organelles can be fluorescently tagged and observed over time. Useful for observing behavior, localization, or movements of tagged protein.
FRET (Fluorescence/ Förster Resonance Energy Transfer)	A donor fluorophore with an emission spectra that overlaps with the excitation spectra of an acceptor fluorophore transfers energy to the acceptor when the donor and acceptor get close enough to each other. Measure shift in the ratio of donor emission intensity to acceptor emission intensity. Useful for observing interactions between two tagged proteins or two parts of a single protein.
BiFC (Bimolecular Fluorescence Complementation)	Two different proteins can be tagged with half a fluorophore. The fluorophore will fluoresce only if the two halves come near enough, reporting a close interaction between the tagged proteins. Measure change in fluorescence intensity to indicate whether the two proteins are near each other.
FRAP (Fluorescence Recovery After Photobleaching)	High-intensity light is used to photobleach a region of fluorescence. Measure kinetics of fluorescence intensity returning to photobleached region. Useful for observing dynamics of trafficking, diffusion, or binding/dissociation of fluorescently tagged proteins.
Photoactivation/ Photoconversion	Photoactivation causes a dramatic increase in fluorescence intensity after stimulation by light in specialized fluorophores. PA-GFP (photoactivatable GFP) becomes fluorescent after stimulation with UV light. Photoconversion uses light to change the emission spectra of a fluorophore. Kaede normally emits green fluorescence but when illuminated by UV light, irreversibly converts to red emission. Useful for tracking localization changes of small populations or individual cells, organelles, or proteins.

TABLE 7.2 (Continued)

	Description
Photoactivation/ Photo-uncaging	Light removes an inhibiting chemical group from a "caged" compound, releasing a now active molecule. Rapid acting with experimenter control over spatial and temporal release of compound. Many different compounds can be caged. Glutamate and calcium uncaging allow optical control for inducing neuronal activity.
CALI (Chromophore-Assisted Light Inactivation)	Light stimulates a chromophore to produce reactive oxygen species that destroy targeted proteins. Proteins can be targeted through the use of antibodies or genetic tags. FlAsH/ReAsH are dyes that bind to tetracysteine tags incorporated into targeted proteins and generate reactive oxygen species that inactivate the protein upon illumination.

conjugates a protein-specific antibody to a phototoxic dye or uses dyes that recognize genetically encoded tags on proteins.

The membrane-permeable biarsenical dyes, green FlAsH (not to be confused with FlaSh, the voltage sensor described previously) and red ReAsH, bind to proteins with a tetracysteine tag that can be genetically encoded. High-intensity illumination of bound FlAsH and ReAsH forms reactive oxygen species that inactivate the tagged protein. Molecular specificity is due to the genetic addition of the tetracysteine tag to a protein of interest. Spatial and temporal specificity come from high-precision control over laser illumination of the sample, and the timing and location of the FlAsH or ReAsH addition. Table 7.2 summarizes the techniques used to probe protein function.

CONCLUSION

The development of fluorescent probes to observe and manipulate neural activity and protein function allows scientists to view dynamic processes in living cells, organs, and whole animals. Taken together, these techniques, along with complementary methods to visualize structure (Chapter 6), record from neurons (Chapter 4), or investigate intracellular signaling cascades (Chapter 14), constitute a fantastic arsenal of methods to investigate how proteins determine the properties of neurons, as well as how neurons determine the properties of circuits and behavior.

SUGGESTED READING AND REFERENCES

Books
Yuste, R. & Konnerth, A. (2005). *Imaging in Neuroscience and Development: A Laboratory Manual.* Cold Spring Harbor Laboratory Press, Cold Spring Harbor, New York.

Yuste, R., Lanni, F. & Konnerth, A. (2000). *Imaging Neurons.* Cold Spring Harbor Laboratory Press, Cold Spring Harbor, New York.

Review Articles
Airan, R. D., Hu, E. S., Vijaykumar, R., Roy, M., Meltzer, L. A. & Deisseroth, K. (2007). Integration of light-controlled neuronal firing and fast circuit imaging. *Curr Opin Neurobiol* **17**, 587–592.

Barth, A. L. (2007). Visualizing circuits and systems using transgenic reporters of neural activity. *Curr Opin Neurobiol* **17**, 567–571.

Ellis-Davies, G. C. (2007). Caged compounds: photorelease technology for control of cellular chemistry and physiology. *Nat Methods* **4**, 619–628.

Giepmans, B. N., Adams, S. R., Ellisman, M. H. & Tsien, R. Y. (2006). The fluorescent toolbox for assessing protein location and function. *Science* **312**, 217–224.

Guzowski, J. F., Timlin, J. A., Roysam, B., McNaughton, B. L., Worley, P. F. & Barnes, C. A. (2005). Mapping behaviorally relevant neural circuits with immediate-early gene expression. *Curr Opin Neurobiol* **15**, 599–606.

Lippincott-Schwartz, J., Altan-Bonnet, N. & Patterson, G. H. (2003). Photobleaching and photoactivation: following protein dynamics in living cells. *Nat Cell Biol* Suppl, S7–S14.

Miesenbock, G. & Kevrekidis, I. G. (2005). Optical imaging and control of genetically designated neurons in functioning circuits. *Annu Rev Neurosci* **28**, 533–563.

Zhang, F., Aravanis, A. M., Adamantidis, A., de Lecea, L. & Deisseroth, K. (2007). Circuit-breakers: optical technologies for probing neural signals and systems. *Nat Rev Neurosci* **8**, 577–581.

Zhang, J., Campbell, R. E., Ting, A. Y. & Tsien, R. Y. (2002). Creating new fluorescent probes for cell biology. *Nat Rev Mol Cell Biol* **3**, 906–918.

Primary Research Articles—Interesting Examples from the Literature
Airan, R. D., Meltzer, L. A., Roy, M., Gong, Y., Chen, H. & Deisseroth, K. (2007). High-speed imaging reveals neurophysiological links to behavior in an animal model of depression. *Science* **317**, 819–823.

Airan, R. D., Thompson, K. R., Fenno, L. E., Bernstein, H. & Deisseroth, K. (2009). Temporally precise *in vivo* control of intracellular signalling. *Nature* **458**, 1025–1029.

Angevine, J. B., Jr. & Sidman, R. L. (1961). Autoradiographic study of cell migration during histogenesis of cerebral cortex in the mouse. *Nature* **192**, 766–768.

Chenn, A. & Walsh, C. A. (2002). Regulation of cerebral cortical size by control of cell cycle exit in neural precursors. *Science* **297**, 365–369.

Dulla, C., Tani, H., Okumoto, S., Frommer, W. B., Reimer, R. J. & Huguenard, J. R. (2008). Imaging of glutamate in brain slices using FRET sensors. *J Neurosci Methods* **168**, 306–319.

Marek, K. W. & Davis, G. W. (2002). Transgenically encoded protein photoinactivation (FlAsH-FALI): acute inactivation of synaptotagmin I. *Neuron* **36**, 805–813.

Matsuzaki, M., Honkura, N., Ellis-Davies, G. C. & Kasai, H. (2004). Structural basis of long-term potentiation in single dendritic spines. *Nature* **429**, 761–766.

Murphy, L. O., Smith, S., Chen, R. H., Fingar, D. C. & Blenis, J. (2002). Molecular interpretation of ERK signal duration by immediate early gene products. *Nat Cell Biol* **4**, 556–564.

Ohki, K., Chung, S., Ch'ng, Y. H., Kara, P. & Reid, R. C. (2005). Functional imaging with cellular resolution reveals precise micro-architecture in visual cortex. *Nature* **433**, 597–603.

Outeiro, T. F., Putcha, P., Tetzlaff, J. E., Spoelgen, R., Koker, M., Carvalho, F., Hyman, B. T. & McLean, P. J. (2008). Formation of toxic oligomeric alpha-synuclein species in living cells. *PLoS ONE* **3**, e1867.

Patterson, G. H. & Lippincott-Schwartz, J. (2002). A photoactivatable GFP for selective photolabeling of proteins and cells. *Science* **297**, 1873–1877.

Plath, N., Ohana, O., Dammermann, B., Errington, M. L., Schmitz, D., Gross, C., Mao, X., Engelsberg, A., Mahlke, C., Welzl, H., Kobalz, U., Stawrakakis, A., Fernandez, E., Waltereit, R., Bick-Sander, A., Therstappen, E., Cooke, S. F., Blanquet, V., Wurst, W., Salmen, B., Bosl, M. R., Lipp, H. P., Grant, S. G., Bliss, T. V., Wolfer, D. P. & Kuhl, D. (2006). Arc/Arg3.1 is essential for the consolidation of synaptic plasticity and memories. *Neuron* **52**, 437–444.

Reiff, D. F., Ihring, A., Guerrero, G., Isacoff, E. Y., Joesch, M., Nakai, J. & Borst, A. (2005). *In vivo* performance of genetically encoded indicators of neural activity in flies. *J Neurosci* **25**, 4766–4778.

Reijmers, L. G., Perkins, B. L., Matsuo, N. & Mayford, M. (2007). Localization of a stable neural correlate of associative memory. *Science* **317**, 1230–1233.

Schroder-Lang, S., Schwarzel, M., Seifert, R., Strunker, T., Kateriya, S., Looser, J., Watanabe, M., Kaupp, U. B., Hegemann, P. & Nagel, G. (2007). Fast manipulation of cellular cAMP level by light *in vivo*. *Nat Methods* **4**, 39–42.

Teather, L. A., Packard, M. G., Smith, D. E., Ellis-Behnke, R. G. & Bazan, N. G. (2005). Differential induction of c-Jun and Fos-like proteins in rat hippocampus and dorsal striatum after training in two water maze tasks. *Neurobiol Learn Mem* **84**, 75–84.

Wang, J. W., Wong, A. M., Flores, J., Vosshall, L. B. & Axel, R. (2003). Two-photon calcium imaging reveals an odor-evoked map of activity in the fly brain. *Cell* **112**, 271–282.

Protocols

Barreto-Chang, O. L. & Dolmetsch, R. E. (2009). Calcium imaging of cortical neurons using Fura-2 AM. JoVE 23. http://www.jove.com/index/details.stp?id=1067, doi: 10.3791/1067

Carlson, G. C. & Coulter, D. A. (2008). *In vitro* functional imaging in brain slices using fast voltage-sensitive dye imaging combined with whole-cell patch recording. *Nat Protoc* **3**, 249–255.

Fuger, P., Behrends, L. B., Mertel, S., Sigrist, S. J. & Rasse, T. M. (2007). Live imaging of synapse development and measuring protein dynamics using two-color fluorescence recovery after photo-bleaching at *Drosophila* synapses. *Nat Protoc* **2**, 3285–3298.

Gaffield, M. A. & Betz, W. J. (2007). Imaging synaptic vesicle exocytosis and endocytosis with FM dyes. *Nat. Protocols,* **1**, 2916–2921.

Hatta, K., Tsujii, H. & Omura, T. (2006). Cell tracking using a photoconvertible fluorescent protein. *Nat Protoc* **1**, 960–967.

Royle, S. J., Granseth, B., Odermatt, B., Derevier, A. & Lagnado, L. (2008). Imaging phluorin-based probes at hippocampal synapses. *Methods Mol Biol* **457**, 293–303.

Chapter 8

Identifying Genes and Proteins of Interest

After reading this chapter, you should be able to:

- Understand basic concepts in genetics, such as transcription and translation
- Compare and contrast the strengths and limitations of commonly used genetic model organisms
- Describe methods for identifying genes and proteins important to a biological phenotype, including genetic, *in silico*, and molecular screening techniques

Techniques Covered

- **Genetic screens:** forward and reverse genetic screens
- ***In Silico* screens:** BLAST, Ensembl
- **Molecular screens:** cDNA microarray, RNAi screen

Virtually every cell in your body contains a copy of your entire **genome**, the complete set of genetic information embedded in DNA that is necessary for your development, plasticity, metabolism, and normal physiology. The functional unit of the genome is the **gene**, a segment of DNA that indirectly codes for a functional protein product. **Proteins** are the molecular machines responsible for virtually all of the cell's structural and functional properties. In order to understand how the brain works, a neuroscientist must understand the anatomy and physiology of neurons—and in order to understand the anatomy and physiology of neurons, a neuroscientist must understand the genes and proteins that give rise to their properties.

One of the overarching questions in neuroscience is how an animal's **genotype**, the genetic constitution of an animal, determines that animal's **phenotype**, an observable trait or set of traits. In other words, what genes influence a trait, such as proper development of an axon toward its target, or appropriate behavioral response following a stimulus? Each month, scientists publish a variety of studies either identifying a novel gene and demonstrating its important role in an animal's phenotype or showing that a previously discovered gene also contributes to a separate phenotype. How do scientists identify these genes? How do they identify the genetic and molecular substrates that underscore development, physiology, and behavior?

The purpose of this chapter is to describe methods useful for identifying which genes or proteins are important for a specific biological phenotype. These methods usually involve using a type of screen to find a handful of genes among thousands that contribute to a neurobiological phenomenon. In previous chapters, we have examined techniques that generally study whole brains, behaviors, or the activity of cells. Starting with this chapter, we begin to study genetic and molecular neuroscience, the contribution of genes and molecules to development, physiology, and behavior. Therefore, we start the chapter with a brief introduction to studying genes and proteins, as well as a survey of commonly studied model organisms. The second half of the chapter describes various methods for identifying genes or proteins of interest: using genetic screens, *in silico* screens, and molecular screens. Once these genes are identified, scientists can perform other experiments to further understand their contribution to a phenotype, as described in later chapters.

A BRIEF REVIEW OF GENES AND PROTEINS

The number of genetic and molecular techniques used in the literature can often be intimidating to neuroscientists without formal backgrounds in genetics or molecular biology. Fortunately, the background information necessary to understand these techniques and why they are used is not complex, and can be summarized in terms of the flow of information within each cell. Highly detailed descriptions of this information flow can be found in other textbooks. Here, we provide the essential information that a neuroscientist needs to know in order to understand genetic and molecular neuroscience, the subject of the remaining chapters in this book.

The Central Dogma of Molecular Biology

The central dogma of molecular biology is a model of the flow of information within a cell. Described by Francis Crick in 1958, this model posits that information in a nucleic acid, encoded in specific DNA or RNA bases, can be transferred to another nucleic acid or to a protein, but information in a protein, encoded in specific amino acids, can never be transferred back to another protein or to a nucleic acid. Many of us learned the simplest version of the central dogma suggesting a unidirectional flow of information—specific sequences of DNA code for molecules of RNA—and these RNA molecules, in turn, code for proteins (Figure 8.1). This model lays the foundation for all of molecular biology research. While scientists have elaborated on this model over the past 40 years and found more complex situations that contradict the simple version of the central dogma, such as the discovery of microRNAs and other methods of gene regulation, it continues to serve as the framework from which all molecular neuroscientists design and execute experiments.

DNA

DNA is the ultimate source of all genetic information within a cell. Residing within the cell nucleus, a DNA molecule is a long polymer made up of repeating units called **nucleotides** (Figure 8.2). These nucleotides are composed of a phosphate group, a sugar molecule (2-deoxyribose—what gives DNA its name), and a nitrogenous base. The phosphate and sugar molecules form the backbone of a DNA polymer. There are four types of nitrogenous bases found in DNA that define the properties of the nucleotide: adenine (abbreviated A), thymine (T), guanine (G), and cytosine (C).

In all eukaryotic organisms, DNA exists as a tightly associated pair of two long strands that intertwine, like two handrails of a spiral staircase, forming the shape of a **double helix**. The two strands of DNA are stabilized by hydrogen bonds between the nitrogenous bases attached to the two strands. A base on one strand forms a bond with only one type of base on the opposite strand: A only bonds with T, and C only bonds with G. The arrangement of two nucleotides binding together across the double helix is referred to as a **base pair**. Strands of DNA that form matches among base pairs are called **complementary strands**. Hydrogen bonds, unlike covalent bonds, are relatively weak and can

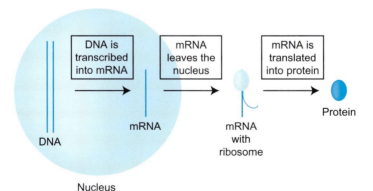

FIGURE 8.1 The flow of information within a cell. DNA resides in the nucleus and codes for relatively short sequences of mRNA. mRNA leaves the nucleus, where it is translated by ribosomes into proteins. Thus, the molecular flow of information starts with nucleic acids and ends with amino acids.

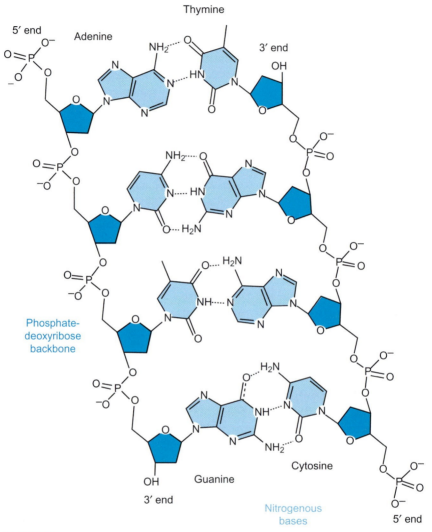

FIGURE 8.2 The molecular structure of DNA. DNA is composed of a sugar-phosphate back-bone on the outside and nitrogenous bases on the inside. The four nitrogenous bases form hydro-gen bonds in the interior of the DNA molecule and hold the two complementary strands together. The two strands are oriented in opposite directions, with one strand running in the 5' to 3' direc-tion and the other running 3' to 5'.

be broken and rejoined relatively easily. This allows other enzymes to unwind DNA during DNA replication or gene transcription.

In addition to being complementary, the two strands of DNA are also anti-parallel, oriented in opposite directions. The directionality of a DNA strand is defined by the carbon atom in the 2-deoxyribose molecule that attaches to another phosphate molecule in a nucleotide monomer. One strand is said to be oriented in the 5′ to 3′ direction, while the other strand is oriented in the 3′ to 5′ direction (Figure 8.2). This orientation has important consequences in DNA synthesis, as DNA can only be synthesized in the 5′ to 3′ direction. This orientation also has consequences for RNA synthesis, as RNA can also only be synthesized in the 5′ to 3′ direction.

Transcription

Transcription is the synthesis of a messenger RNA (mRNA) molecule from a DNA template. There are three major differences between an RNA molecule and a DNA molecule: (1) RNA molecules use the sugar ribose instead of 2-deoxyribose; (2) RNA molecules are single-stranded; and (3) RNA molecules use the nitrogenous base uracil (U) instead of thymine.

In transcription, only one of the two strands of DNA, called the template strand, is transcribed. The other strand, the coding strand, has a sequence that is the same as the newly created RNA transcript (with U substituted for T). An enzyme called RNA polymerase reads the template strand in the 3′ to 5′ direction and synthesizes the new mRNA in the 5′ to 3′ direction (Figure 8.3).

Transcription begins when various proteins called transcription factors interact with regions of the genome called promoters. A promoter is a specific sequence of DNA that facilitates the transcription of a specific gene. These promoters are often located adjacent and upstream of the genes they regulate, towards the 5′ region of the coding strand. When a transcription factor binds to a promoter, multiple proteins are recruited to begin the transcription process.

After transcription terminates, a newly formed RNA molecule receives further processing called RNA splicing before leaving the nucleus. At this point, the RNA is composed of exons and introns flanked by untranslated regions (UTRs) at either end (Figure 8.4). An exon is a sequence of an RNA strand that remains after splicing occurs, while an intron is a sequence that is lost. The 5′ and 3′ UTRs are sequences that cap the RNA molecule and are not translated into amino acids. RNA splicing is catalyzed by a macromolecular complex known as a spliceosome, which cuts exon and intron sequences apart and then joins the exons back together to form a single strand. After RNA splicing, a mature mRNA molecule leaves the nucleus and migrates to the cytoplasm where it can be translated into a protein.

Translation

Translation is the process of using an mRNA molecule as a template to produce a protein. This process is performed by specialized macromolecules in the cytoplasm called ribosomes. A ribosome reads an mRNA molecule in the 5′ to 3′ direction and uses this template to guide the synthesis of a chain of amino acids to form a protein. Amino acids are the building blocks of all proteins

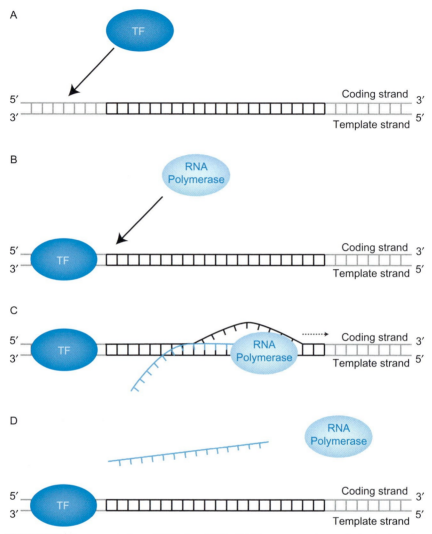

FIGURE 8.3 The transcription of DNA into RNA. (A) The process of transcription begins when specific transcription factors interact with promoter sequences on the genome. These factors recruit other proteins that are necessary to attract RNA polymerase. **(B)** RNA polymerase binds to the complex between transcription factors and DNA. Once bound, the enzyme begins prying the two strands of DNA apart. **(C)** RNA polymerase proceeds down the length of the DNA, simultaneously unwinding DNA and adding nucleotides to the 3′ end of the growing RNA molecule. **(D)** When transcription is finished, RNA polymerase dissociates from the DNA and the two DNA strands reform a double helix conformation. A new mRNA is produced that is complementary to the DNA template strand.

and are assembled as a long strand, much like nucleotides are assembled as a long strand in a nucleic acid molecule. Each sequence of three nucleotides of an mRNA molecule code for one amino acid. The three nucleotides collectively form a **codon**, a translation of genetic information into an amino acid monomer.

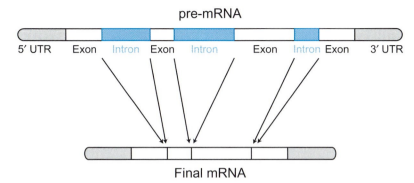

FIGURE 8.4 RNA splicing. Before mRNA leaves the nucleus, macromolecular complexes called spliceosomes remove noncoding intron sequences and rejoin exon sequences. Noncoding sequences remain at the 5′ and 3′ ends.

There is a precise genetic code for the conversion of codons into amino acids (Figure 8.5), with each codon precisely coding for one specific amino acid. One codon, "AUG" codes for the amino acid methionine, which is the start of any protein sequence. Other codons ("UAA," "UAG," and "UGA") serve as "stop" codons and terminate translation. After translation ends, the newly formed polypeptide chain folds to assume its native protein conformation. These proteins can be further processed by other organelles, such as the endoplasmic reticulum, or receive **post-translational modifications** (PTMs) from other proteins.

This incredibly brief survey of DNA, transcription, and translation is no match for much more detailed descriptions found in genetics or basic biology textbooks. But this information is the foundation on which most molecular neuroscience techniques are based.

GENETIC SCREENS

Genetic screens fall into two categories: forward or reverse screens. A **forward genetic screen** is a screen used to identify genes important for a biological phenotype. In these types of screens, an investigator selects a scientifically interesting phenotype and seeks to determine which genes are necessary for the generation of this phenotype. A **reverse genetic screen** is the opposite: an investigator selects an interesting gene and seeks to determine which phenotypes result from its absence. Forward genetic screens often test thousands of genes at once, while a reverse genetic screen tests only a single gene. Forward genetic screens have been compared to "fishing" because a scientist is never quite sure which genes he or she will catch; reverse genetic screens have been compared to "gambling" because a scientist places all of his or her hopes on that gene producing an interesting phenotype. Forward genetic screens have been said to be "hypothesis generating," as they typically produce multiple candidate genes that

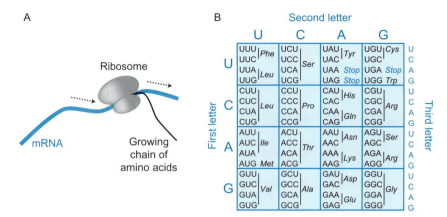

FIGURE 8.5 **The translation of mRNA into proteins. (A)** A ribosome translates the genetic sequence of an mRNA molecule into an amino acid sequence that forms a polypeptide. mRNA sequences are processed in the 5′ to 3′ direction. Each series of three nucleotides, called codons, translates into a single amino acid. **(B)** The genetic code for the conversion of codons into amino acids. Each codon translates into a specific amino acid (abbreviated here in three letters, such as *Phe*, which is the abbreviation for phenylalanine). Because there are more possible codons than amino acids, some codons translate into the same amino acids, such as UUU and UUC. The AUG codon translates into methionine and signals the start of translation. The UAA, UAG, and UGA codons do not translate into amino acids, instead serving as a stop signal and terminating translation.

may be important for a phenotype; reverse genetic screens have been said to be "hypothesis testing," as a scientist can test specific hypotheses about the role of a gene in a phenotype. Here we briefly describe the process of performing a forward or reverse genetic screen.

Forward Genetic Screen

In a forward genetic screen, a scientist mutagenizes animals to produce thousands of lines that may have a mutation in a single gene, and then screens each individual to identify animals with an abnormal phenotype. These screens are typically performed in flies and worms, although they have also rarely been performed in mice (Box 8.1). There are six main steps in performing a forward genetic screen:

- **Design an assay to measure a phenotype.** The first step in determining which genes may be important for a phenotype is to design a specific, quantitative assay in order to discriminate between wild-type individuals and animals with an aberrant phenotype. This requires fully characterizing the wild-type phenotype and choosing measurable parameters to identify individuals with an abnormal phenotype. For example, a scientist performing a forward genetic screen to determine genes important for axon guidance

BOX 8.1 Genetic Model Organisms

Each week, hundreds of studies are published in peer-reviewed journals that use genetic tools to investigate the nervous system. Interestingly, nearly all of these studies use the same five species as model organisms: the worm (*Caenorhabditis elegans*), the fly (*Drosophila melanogaster*), the zebrafish (*Danio rerio*), the mouse (*Mus musculus*), and the human (*Homo sapiens*). There are occasional exceptions, such as the frog (*Xenopus laevis*) or the sea slug (*Aplysia californica*), but these species collectively represent less than 1% of published studies in the literature. It is obvious why neuroscientists would want to learn more about the genes of humans, as the ultimate goal of neuroscience is to better understand our own species. But why did the other four species become such tractable organisms for genetic studies? Why use roundworms and fruit flies? Why not flatworms, bees, goldfish, or gerbils? Many of these other organisms have the same short breeding time, high number of offspring, and other factors that might make them attractive model species.

The answer can essentially be summarized in a single word: history. There really isn't a good reason why scientists started using mice instead of gerbils, but when multiple laboratories started performing genetic experiments on these animals, it made sense for other labs to follow. It is useful for multiple laboratories to study the same species: when dozens or hundreds of labs each study the same organism, a community develops that allows researchers to better share results, reagents, and genetic tools.

There is no better example of a community forming around a genetic model organism than the *Drosophila* community of scientists. Ever since Thomas Hunt Morgan began using the fly as a model organism to study genetics in 1910, a growing number of scientists learned more and more about ideal conditions and practices for using flies in experiments. As the knowledge about fly maintenance increased, so did the number of genetic tools amenable to fly genetic studies. In the late 1960s, Seymour Benzer used *Drosophila* to study the link between genes and behavior. In the 1980s and 1990s, scientists developed dozens of molecular genetic tools for carefully dissecting genetic circuits in flies, and in 2000, the fly became the second eukaryotic organism to have its genome sequenced.

A similar history developed for the worm: the prominent scientist Sydney Brenner began using *C. elegans* as a model organism in the 1960s. After years of study, scientists have given all 302 neurons specific names and mapped out the connections of all 7000 synapses. There is now an energetic *C. elegans* community just as there is a vibrant *Drosophila* community, with many established methods for maintaining and manipulating genetic strains of worms.

Table 8.1 summarizes some of the traits and benefits of these commonly used genetic organisms. We mention these species not only because they are highly represented in the literature, but also because virtually all novel genes and proteins are discovered in these species. The methods described below to identify functionally relevant genes depend on the decades-long development of these organisms for biological research.

TABLE 8.1 Relative Advantages and Disadvantages of Different Organisms Used for Genetic Studies

Species	Advantages	Disadvantages
Worm (*Caenorhabditis elegans*)	Simple, multicellular organism Known genome Short lifespan Relatively easy to manipulate genes and screen for genes of interest through mutagenesis Can be frozen All neurons and their connections are known	Invertebrate Primitive nervous system
Fruit fly (*Drosophila melanogaster*)	Complex, multicellular organism Known genome Short lifespan, rapid reproduction rate Relatively easy to manipulate genes Many mutant lines available	Invertebrate Primitive nervous system
Zebrafish (*Danio rerio*)	Vertebrate Can use powerful invertebrate techniques like Gal4/UAS system Good for imaging studies during development because eggs are clear Large number of offspring	Not a mammal
Mouse (*Mus musculus*)	Complex, higher-order organisms Known genome Possible to manipulate genes in whole organism or specific tissues Nervous system is homologous to humans	Relatively long lifespan and reproductive time Genetic manipulations are lengthy and costly Greater genetic redundancy than invertebrates
Human (*Homo sapiens*)	Complex, higher-order organism Known genome Can observe naturally occurring genetic mutations and polymorphisms	Long lifespan Not tractable for experimental genetic manipulation

might choose a specific axon and characterize its normal growth and development over time. When screening individuals, the scientist could look for axons that mistarget to abnormal locations in the brain. Note that this screen requires the scientist to identify the same axon in thousands of individuals. Alternatively, a scientist could choose a behavioral phenotype, characterizing the normal behavioral response to a stimulus and then examining individuals with an abnormal behavior. The abnormal behavior would have to be statistically different from the normal variation in behavior between individuals.

- **Mutagenize eggs/larvae.** There are three methods of developing mutant lines of organisms. In **chemical mutagenesis**, a scientist applies a mutagenizing chemical, such as ethyl methane sulfonate (EMS) or N-ethyl-N-nitrosourea (ENU), to thousands of eggs/larvae, which statistically creates lines of animals with mutations in a single gene in the genome. In **irradiation mutagenesis**, a scientist achieves the same goal with high-intensity UV light, which damages DNA at a rate such that animals statistically have a mutation in a single gene in the genome. Finally, in **insertional mutagenesis**, a scientist uses various methods to insert mobile genetic elements called **transposons** into the genomes of offspring. These transposons insert at random locations in the genome, occasionally disrupting an endogenous gene. Therefore, they can be used to test thousands of animals, each with a potential loss-of-function of a single gene.

- **Screen for abnormal phenotypes**. After hundreds or thousands of mutant lines are produced, a scientist screens each individual in a phenotypic assay, one mutant at a time. This is easily the most time-consuming aspect of a forward genetic screen. Any mutant lines that show an abnormal phenotype are maintained and bred for future experiments. The vast majority of mutant lines that do not show aberrant phenotypes are either used for other genetic screens or discarded.

- **Perform complementation analysis.** Once a genetic mutant is isolated, investigators often perform a **complementation test** to determine if the mutation is unique or has been previously described. This approach is used when a mutation results in the same phenotype as a previously described mutation. The two strains could have mutations in the same gene or in two different genes. In order to differentiate between these two possibilities, the two strains are mated and the phenotypes of the offspring are identified. If the offspring exhibits the wild-type phenotype, a scientist concludes that each mutation is in a separate gene. However, if the offspring does not express the wild-type phenotype but rather a mutant phenotype, the scientist concludes that both mutations might occur in the same gene.

- **Map the gene.** Just because a scientist has discovered a mutation in a gene that alters a phenotype doesn't mean that the scientist knows the molecular identity of that gene. The only way to determine the molecular sequence of the gene and its location in the genome is to map the gene.

If mutagenesis was caused by chemical or irradiation methods, investigators typically perform a **linkage analysis** to map the gene. This technique is used to identify the relative position of a novel gene in relation to the location of known genes. To map the gene, a scientist mates the mutant animals with animals of a separate mutant phenotype. The two traits are then assessed in the offspring to determine if they are inherited together at relatively high rates. When two traits are closely "linked" in future offspring, they appear together more frequently and there is a greater likelihood that they are located close to each other on a chromosome. Because many genomes have been sequenced, a scientist can identify potential regions on a chromosome where the mutant gene may be found. If mutagenesis was caused by insertional mutagenesis methods, it is much easier to identify the molecular identity of the mutated gene. A scientist simply uses the transposon's genetic sequence to design primers for DNA sequencing, and then sequences DNA in either direction of the transposon to identify the region of the genome where it landed (Chapter 9).

• **Clone the gene.** The final step in the process of identifying novel genes for a phenotype is to clone the DNA encoding the gene. This allows a scientist to perform future genetic experiments that manipulate the gene *in vitro* or *in vivo*. We will discuss methods of cloning genes in Chapter 9.

Reverse Genetic Screen

In a reverse genetic screen, a scientist identifies a gene of interest and then perturbs the gene to determine its role in various phenotypes. For example, if a scientist hypothesizes that a certain gene is necessary for proper formation of neuromuscular junctions in the mouse, the scientist could make a knockout mouse, removing the gene from the genome and examining whether proper synapses form between peripheral nerves and motor units. We describe the process of performing these kinds of experiments in Chapters 11 and 12.

IN SILICO SCREENS

It is possible for a scientist to identify a gene or protein of interest by using bioinformatics or genomic databases. An *in silico* **screen** is a computer-based method that identifies and compares similar DNA and/or protein sequences from multiple species. Let's say that a scientist identifies a gene encoding a novel transmembrane receptor. The scientist may be able to identify and discover other receptors by looking for similar sequences in the genome. Alternatively, if a gene is discovered in one species, a scientist can use bioinformatics to identify orthologous sequences in other species. These genes may have the same function or very different functions. Two examples of publicly available genetic databases are BLAST and Ensembl.

BLAST

The **Basic Local Alignment Search Tool (BLAST)** is a tool used to compare the nucleotide sequences of DNA or the amino acid sequences of proteins. The power of BLAST is the ability to compare a DNA or protein sequence with the entire library of known sequences from all species. The user can set a threshold for the degree to which two sequences form a match. Therefore, if a scientist wants to identify genes with high sequence similarity to a gene of interest, the BLAST program can identify similar genes with high sequence similarity, and potentially, similar functions. Many genes in the mouse genome have been identified due to their similarity with other structures, especially neuropeptides and ion channels. The best way to learn about BLAST is to try it: http://blast.ncbi.nlm.nih.gov/Blast.cgi.

Ensembl

Ensembl is a bioinformatics browser that functions much like BLAST but with a greater emphasis on genomic analysis. This tool provides the actual genomic sequences for a huge diversity of animal models, allowing comparative genomic searches to be performed. Additionally, this program can be used for analyzing sequence alignment, to investigate gene homology, and to obtain specific gene sequences to be used for subsequent experiments. Ensembl has the complete genomes and rough drafts of genomes for dozens of animals, including standard model organisms, as well as relatively esoteric organisms, such as the bush baby, the bat, and the platypus. Again, the best way to learn about this genetic tool is to try it: http://www.ensembl.org/index.html.

MOLECULAR SCREENS

A molecular screen uses molecular biology techniques to identify genes important for a specific biological process. Unlike standard genetic screens, molecular screens use high-throughput strategies and thousands of nucleic acid probes to identify the molecular components of a phenotype.

cDNA Microarray Screen

A **microarray** is a device used to measure and compare gene expression profiles from biological samples. It can be used to analyze gene expression from a single sample or to compare gene expression between two different samples. Because gene expression is tightly regulated, genes are only expressed at times when they serve a useful function. Therefore, a scientist can potentially identify a gene of interest based on how its expression changes over time or if its expression changes in response to a stimulus or environmental condition. The power of a microarray is that it is possible to assay all genes in the genome at the same time in a relatively small device that can fit in the palm of your hand.

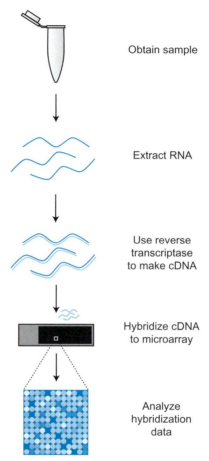

Obtain sample

Extract RNA

Use reverse
transcriptase
to make cDNA

Hybridize cDNA
to microarray

Analyze
hybridization
data

FIGURE 8.6 A cDNA microarray. A microarray is used to analyze gene expression in a biological sample and compare expression with other samples. mRNA is extracted from a sample and
reverse transcribed to produce cDNA. The cDNA is labeled and then hybridized to a microarray
chip containing thousands of potentially complementary strands. A computer detects the amount
of signal from each specific sequence. Thus, a microarray can be used to detect which specific
mRNA sequences were expressed in the original sample.

The device consists of an array of thousands of tiny microscopic spots of
DNA sequences, often sequences that are parts of genes (Figure 8.6). A scientist takes a biological sample, extracts the RNA, and then uses a special
enzyme called **reverse transcriptase** to make complementary DNA (cDNA)
from the RNA strands. The cDNA is labeled and incubated with the surface of the
DNA array. The cDNA is allowed a time period to hybridize with the sequences
on the array. A computer subsequently scans the entire array and detects hybridization between cDNA in the sample and spots on the array. Because the computer contains a database of every spot on the array, a scientist can learn which
genes were present in a sample and which genes were undetected.

A microarray can be used to identify novel genes for a neurobiological phenotype. For example, suppose a scientist wishes to identify genes that respond to light/dark cycles. The scientist can collect samples from a part of the brain known to play a role in regulating light/dark cycles, such as the suprachiasmatic nucleus. Some samples can be collected from animals during the day, while other samples are collected at night. A comparison in the gene expression profile between day/night genes will produce a list of genes that show differential expression during those two periods. No doubt, a great number of genes will appear on this list. However, this list may provide a scientist with a great number of candidates to study in subsequent experiments.

RNAi Screen

RNA interference (RNAi) is a technique used to knock down the expression of mRNA molecules and will be discussed in Chapter 12. RNAi screens use thousands of RNAi molecules to disrupt individual gene function in a variety of organisms, most notably worms, or cultured cells. For example, in a multi-well-based RNAi screen, hundreds of unique RNAi molecules are each added to the bottom of 384 well plates, with a separate RNAi molecule in each well. Cells are added to the wells and the scientist observes changes in various parameters, such as morphology, proliferation, or death rate. Thus, an investigator can determine which RNAi molecules, and therefore which genes, are necessary for a cellular process to occur.

CONCLUSION

This chapter serves as the first step in identifying new genes and their protein products. These techniques are often the first step in discovering the role of a gene in a biological process or phenotype. Other methods are necessary to validate this discovery, such as methods that perturb the expression of the gene to identify its necessity and/or sufficiency for a phenotype. The remaining chapters of this book use molecular and genetic methods to gain insight into newly discovered genes as to their role *in vitro* and *in vivo*.

SUGGESTED READING AND REFERENCES

Books
Griffiths, A. J. F. (2008). *Introduction to Genetic Analysis*, 9th ed. Freeman, NY.
Lewin, B. (2008). *Genes IX*, 9th ed. Jones and Bartlett, Sudbury, MA.

Review Articles
Gahtan, E. & Baier, H. (2004). Of lasers, mutants, and see-through brains: functional neuroanatomy in zebrafish. *J Neurobiol* **59**, 147–161.
Goldowitz, D., Frankel, W. N., Takahashi, J. S. *et al.* (2004). Large-scale mutagenesis of the mouse to understand the genetic bases of nervous system structure and function. *Brain Res Mol Brain Res* **132**, 105–115.

Paddison, P. J., Silva, J. M., Conklin, D. S. *et al.* (2004). A resource for large-scale RNA-interference-based screens in mammals. *Nature* **428**, 427–431.

Ryder, E. & Russell, S. (2003). Transposable elements as tools for genomics and genetics in *Drosophila*. *Brief Funct Genomic Proteomic* **2**, 57–71.

Wu, S., Ying, G., Wu, Q. & Capecchi, M. R. (2007). Toward simpler and faster genome-wide mutagenesis in mice. *Nat Genet* **39**, 922–930.

Primary Research Articles—Interesting Examples from the Literature

Crick, F. H. (1958). On protein synthesis. *Symp Soc Exp Biol* **12**, 138–163.

Crick, F. H. (1970). Central dogma of molecular biology. *Nature* **227**, 561–563.

Schuldiner, O., Berdnik, D., Levy, J. M. *et al.* (2008). piggyBac-based mosaic screen identifies a post-mitotic function for cohesin in regulating developmental axon pruning. *Dev Cell* **14**, 227–238.

Tracey, W. D., Jr., Wilson, R. I., Laurent, G. & Benzer, S. (2003). Painless, a *Drosophila* gene essential for nociception. *Cell* **113**, 261–273.

Ule, J., Ule, A., Spencer, J. *et al.* (2005). Nova regulates brain-specific splicing to shape the synapse. *Nat Genet* **37**, 844–852.

Walsh, T., McClellan, J. M., McCarthy, S. E. *et al.* (2008). Rare structural variants disrupt multiple genes in neurodevelopmental pathways in schizophrenia. *Science* **320**, 539–543.

Zarbalis, K., May, S. R., Shen, Y., Ekker, M., Rubenstein, J. L. & Peterson, A. S. (2004). A focused and efficient genetic screening strategy in the mouse: identification of mutations that disrupt cortical development. *PLoS Biol* **2**, E219.

Protocols

Bökel, C. (2008). EMS screens : from mutagenesis to screening and mapping. *Methods Mol Biol* **420**, 119–138.

Lehner, B., Tischler, J. & Fraser, A. G. (2006). RNAi screens in Caenorhabditis elegans in a 96-well liquid format and their application to the systematic identification of genetic interactions. *Nat. Protocols* **1**, 1617–1620.

Nolan, P. M., Kapfhamer, D. & Bucan, M. (1997). Random mutagenesis screen for dominant behavioral mutations in mice. *Methods* **13**, 379–395.

Ramadan, N., Flockhart, I., Booker, M., Perrimon, N. & Mathey-Prevot, B. (2007). Design and implementation of high-throughput RNAi screens in cultured *Drosophila* cells. *Nat. Protocols* **2**, 2245–2264.

Chapter 9

Molecular Cloning and Recombinant DNA Technology

After reading this chapter, you should be able to:
- Describe tools used to manipulate DNA
- Describe the steps involved in creating novel DNA constructs

Techniques Covered
- **Isolating DNA fragments**: restriction enzymes, PCR, DNA synthesis
- **Cloning DNA**: vectors, ligation, DNA libraries, recombineering, transformation
- **Purifying DNA**: gel electrophoresis, mini-/midi-/maxipreps
- **Identifying DNA**: DNA sequencing, nucleic acid hybridization techniques (*in situ* hybridization, Southern and northern blots)

Watson and Crick's discovery of DNA structure led to the understanding of the physical basis of genetics and heredity. Since then, rapid advances in genetic engineering technology have enabled important discoveries and the ability to manipulate the functional properties of DNA molecules. Genetic engineering permits scientists to probe the connection between the molecules of life and the structure and function of the nervous system, providing insight into how small differences in DNA, even single nucleotide changes, can lead to profound changes in behavior.

Modern molecular biologists manipulate DNA much like classic filmmakers edit movies. If a filmmaker wishes to take a scene from one film reel and add it to another film reel, he can locate the scene, physically cut it out, splice it together into the new film reel, and run the film to ensure that the sequence has integrated correctly. Likewise, a molecular biologist may want to cut out one DNA sequence from a specialized DNA storage system called a vector,

207

and splice it into a new vector. A scientist can do this in the same way as the filmmaker, with specific molecular tools that take the place of scissors, tape, and a sequence identification system.

Identifying and cutting a piece of DNA out of the genome is referred to as **molecular cloning**. Cloning allows the isolation and amplification of a specific DNA sequence. This sequence can then be manipulated using the tools of recombinant DNA technology to form novel sequences, mutate endogenous genes, tag genes with reporter constructs, and much more. In other words, modern molecular techniques allow a scientist to manipulate DNA in just about any way that he or she would like.

In particular, two molecular cloning tools that have revolutionized a scientist's ability to manipulate DNA form the foundation for recombinant DNA technology: (1) the capacity to cut, alter, and join DNA molecules *in vitro*; and (2) host/vector systems that allow recombinant DNA molecules to be copied and mass produced.

The purpose of this chapter is to describe these tools and methods and how they are used to clone and manipulate DNA constructs. First, we will discuss the discoveries and advances in technology that allow the isolation of DNA fragments from the genome. Then, we will describe molecular cloning: how those isolated fragments are used to create new recombinant DNA constructs, as well as how to purify either DNA fragments or recombinant constructs. Finally, we describe ways to examine the sequence identity of DNA. After creating a novel DNA construct, a scientist can deliver this construct into various cell types (Chapter 10) or use the construct to produce a transgenic or knock-out animal (Chapters 11 and 12).

ISOLATING DNA FRAGMENTS

In Chapter 8, we discussed methods to identify a particular stretch of DNA that is important to a neural phenotype. Once that sequence has been identified, isolating fragments of the DNA sequence is critical for subsequent experiments. The first step is to purify the DNA fragment away from the rest of the genomic DNA. There are two primary methods used to create DNA fragments: (1) restriction digests that cut DNA into small pieces, and (2) the polymerase chain reaction (PCR), which amplifies only the region of interest.

Restriction Enzymes

The discovery of **restriction enzymes** (also called **restriction endonucleases**) that recognize and cut specific DNA sequences is one of the most important advances in biology that makes recombinant DNA technology possible. Restriction enzymes are like targeted molecular scissors, cutting double-stranded DNA at unique base pair sequences called **recognition** (or **restriction**) **sites**.

These enzymes are naturally produced by bacteria, which make these enzymes to cleave foreign DNA molecules in their environment. Molecular biologists have identified and purified hundreds of different restriction enzymes from many species of bacteria, and now these enzymes can be easily ordered from a variety of manufacturers.

Each enzyme recognizes a unique sequence of about 4–8 nucleotides and cuts at specific locations within these recognition sites, leaving behind different types of ends—either smooth or jagged. Some restriction enzymes leave **blunt ends** after digesting DNA (Figure 9.1A). Others produce staggered cuts that leave **sticky** or **cohesive ends** with short single-stranded tails hanging over the ends of each cut fragment (Figure 9.1B). Each tail can form complementary base pairs with the tail at another end produced by the same enzyme. This allows any two DNA fragments to be easily joined together, as long as the fragments were generated with the same restriction enzyme (or another enzyme that leaves the same sticky ends). These different ends can be thought of as puzzle pieces whose shape is determined by the particular enzyme used. If the same enzyme is used to cut two different pieces of DNA, then there will be two possible matching complements: the original parts that were separated, but also a mix-and-match between the different parts with the complementary shape.

In addition to its use in creating short fragments of DNA, **restriction digests**, the process of cutting pieces of DNA with restriction enzymes, can also be used to deduce structural information about a piece of DNA without knowing its precise sequence. **Restriction mapping** uses restriction digests and separation of digested fragments by gel electrophoresis (described later in this chapter) to create a restriction digest map of the locations of various recognition sites. These maps facilitate cloning strategy planning and can allow the comparison of different DNA fragments without knowing exact sequences.

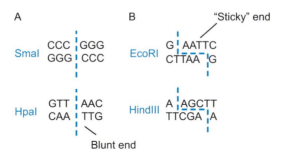

FIGURE 9.1 Restriction enzymes. Acting like molecular scissors, restriction enzymes recognize specific sequences and cut at specific locations within those sequences. They can either **(A)** leave blunt ends, such as the enzymes SmaI and HpaI, or **(B)** make uneven cuts that result in sticky ends, such as the enzymes EcoRI and HindIII.

This is a particularly useful, quick way to examine recombinant DNA products to test whether they are correct.

Polymerase Chain Reaction (PCR)

The **polymerase chain reaction (PCR)** is a biochemical reaction that uses controlled heating and cooling in the presence of DNA synthesizing enzymes to exponentially amplify a small DNA fragment. It requires only tiny amounts of DNA to be used as a template from which a selected region will be amplified a billion-fold, effectively purifying this DNA fragment away from the rest of the template DNA.

Short single-stranded oligonucleotides (about 20–30 nucleotides in length) called **primers** are essential ingredients in PCR. Primers specify what fragment will be amplified by binding to the template in the region flanking the sequence of interest (Figure 9.2). Primers are used in PCR to amplify and modify DNA sequences. Where do these primers come from? They are synthesized *in vitro* using what is called the "solid-phase" method. In this method, short DNA strands are synthesized by adding activated monomers to a growing DNA chain linked to an insoluble support. DNA chains of as many as 100 nucleotides can be synthesized using this automated method. This type of chemical DNA synthesis can also be used to create labeled oligonucleotides for nucleic acid hybridization techniques. DNA sequences can also be modified using PCR either by adding restriction sites to the ends of a DNA fragment or by introducing mutations to characterize the function of the DNA fragment. Labs do not often chemically synthesize DNA fragments themselves but rather send the sequence they want synthesized off to a company that specializes in oligonucleotide synthesis. These companies can also add special tags or make mutant versions of the synthesized DNA.

Standard PCR

Standard PCR starts with a DNA template, often from a genomic or cDNA library (Box 9.1), and uses primers designed to target a specific DNA fragment of interest to selectively amplify that fragment. A solution used for PCR amplification contains: (1) a nucleic acid template with the target sequence to amplify; (2) a pair of oligonucleotide primers that hybridize with the sequences flanking the target DNA; (3) all four deoxyribonucleoside triphosphates (dNTPs—A, T, G, C); and (4) a heat-stable DNA polymerase, an enzyme that adds nucleotides onto the end of a forming DNA strand.

A PCR cycle consists of three steps (Figure 9.2): (1) *Strand separation*—the two strands of the template/parent DNA molecule are separated by heating the solution to about 95° C for 15 seconds so each strand can serve as a template. (2) *Hybridization of primers*—the solution is abruptly cooled to about 50–55° C to allow each primer to hybridize to a DNA strand. One primer hybridizes to the 3'-end of the target on one strand, and the other primer hybridizes to the

BOX 9.1 DNA Libraries

A **DNA library** is a comprehensive collection of cloned DNA fragments from a cell, tissue, or organism. DNA libraries can be used to isolate a specific gene of interest, as they generally include at least one fragment that contains that gene. Libraries can be constructed using any of a variety of different vectors (plasmid, phage, cosmid, or artificial chromosomes) and are generally housed in a population of bacterial cells.

Plasmid DNA libraries are created using the same process used to clone genes into expression vectors (Figure 9.4). Plasmid and cellular DNA are both cut with restriction enzymes and annealed via their sticky ends to form recombinant DNA constructs. These recombinant molecules containing foreign DNA inserts are then covalently sealed with DNA ligase, and bacteria are transformed using these recombinant DNA plasmids. Each bacterial cell that was initially transfected contains, in general, a different foreign DNA insert, replicated in all the clones of the bacterial colony.

For many years, plasmids were used to clone DNA fragments of 1–30 kb. Larger DNA fragments are more difficult to handle and thus were harder to clone. Then researchers began to use larger DNA vectors, **yeast artificial chromosomes (YACs)**, which could handle very large pieces of DNA (100–2000 kb). Today, new plasmid vectors based on the naturally occurring F plasmid of the bacteria *E. coli* are used to clone DNA fragments of 150–350 kb. Unlike smaller bacterial plasmids, the F plasmid—and its derivative, the **bacterial artificial chromosome (BAC)**—is present in only one or two copies per *E. coli* cell. The fact that BACs are kept in such low numbers in bacterial cells may contribute to their ability to maintain large stably cloned DNA sequences: with only a few BACs present, it is less likely that the cloned DNA fragments will become scrambled due to recombination with sequences carried on other copies of the plasmid. Because of their stability, ability to accept large DNA inserts, and ease of handling, BACs are now the preferred vector for building DNA libraries of complex organisms—including those representing the human and mouse genomes.

Genomic DNA Libraries

Cleaving the entire genome of a cell with a specific restriction enzyme and cloning each fragment is sometimes called the "shotgun" approach to gene cloning. This technique can produce a very large number of DNA fragments that will generate millions of different colonies of transformed bacterial cells. Each of these colonies is composed of cell clones derived from a single ancestor cell and therefore harbors many copies of a particular stretch of the fragmented genome. Such a plasmid is said to contain a genomic DNA clone, and the entire collection of plasmids is called a **genomic DNA library**. But because the genomic DNA is cut into fragments at random, only some fragments contain genes. Many of the genomic DNA clones obtained from the DNA of a higher eukaryotic cell contain only noncoding DNA, which makes up most of the DNA in such genomes. Thus, genomic DNA clones contain many DNA sequences that may not be particularly relevant for a cell of interest. By using mRNA, investigators can bypass DNA that is not transcribed to make cDNA libraries that contain the DNA sequences relevant and active for the cell or area of interest.

cDNA Libraries

By starting with mRNA rather than genomic DNA, only DNA sequences that correspond to protein-encoding genes are cloned. This is done by extracting the mRNA from cells and then using RT-PCR to make a cDNA (complementary DNA) copy of each mRNA molecule present, and inserting this cDNA into a vector. Each clone obtained in this way is called a cDNA clone, and the entire collection of clones derived from one mRNA preparation constitutes a **cDNA library**. There are important differences between genomic DNA clones and cDNA clones. Genomic clones represent a random sample of all the DNA sequences in an organism and, with very rare exceptions, are the same regardless of the cell type used to prepare them. By contrast, cDNA clones contain only those regions of the genome that have been transcribed into mRNA. Because cells of different tissues produce distinct sets of mRNA molecules, a distinct cDNA library is obtained for each type of cell used to prepare the library.

The use of a cDNA library for gene cloning has several advantages. DNA encoding proteins that are important for a specific cell type will be highly enriched in a library made from that cell type. This makes it possible to compare the genes and proteins expressed in different cell types. By far the most important advantage of cDNA clones is that they contain the uninterrupted coding sequence of a gene—only the exons that code for proteins, free from noncoding intron sequences. Thus, when the aim of cloning is either to deduce the amino acid sequence of a protein from the DNA sequence or to produce the protein in bulk by expressing the cloned gene in a bacterial or yeast cell, it is preferable to start with cDNA. However, genomic DNA libraries allow noncoding regulatory sequences to be examined.

3'-end on the complementary target strand so that DNA extension from the primers will be directed toward each other. This amplifies the region between the primers. Parent DNA duplexes do not form, because the primers are present in large excess. (3) *DNA synthesis*—the solution is heated to about 70–75° C, the optimal temperature for *Taq* DNA polymerase. This heat-stable polymerase comes from *Themus aquaticus*, a thermophilic bacterium that lives in hot springs. The polymerase elongates both primers in the direction of the target sequence because DNA synthesis is in the 5' to 3' direction; new dNTPs are added to the growing 3' end of the primer. DNA synthesis takes place on both strands and extends beyond the target sequence.

These three steps (strand separation, hybridization of primers, and DNA synthesis) constitute one cycle of PCR amplification and can be carried out repetitively just by changing the temperature of the reaction mixture. Nothing much is produced in the first cycle of DNA synthesis; the power of the PCR method is revealed only after repeated cycles. Every cycle doubles the amount of DNA synthesized in the previous cycle. With each round of DNA synthesis, the newly generated fragments serve as templates in their turn, and within a few cycles the predominant product is a single species of DNA fragment whose length

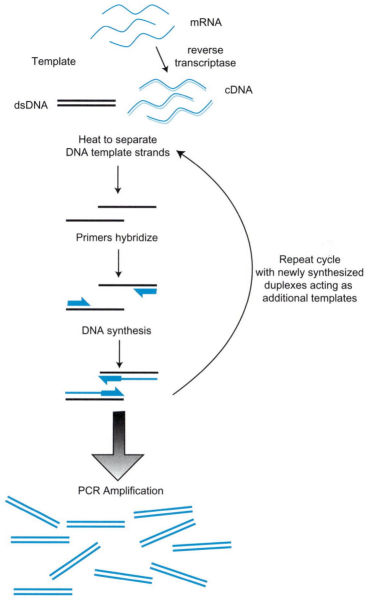

FIGURE 9.2 PCR and RT-PCR. PCR can either begin with a double-stranded DNA template or use the enzyme reverse transcriptase to create a cDNA template from mRNA (RT-PCR).

corresponds to the distance between the two original primers. In practice, 20–30 cycles of a reaction are required for effective DNA amplification. A single cycle requires only about 5 minutes, and the entire procedure can be easily automated using modern PCR machines, also known as thermocyclers, that control the cycling temperature changes required for each step in a PCR cycle.

Reverse Transcription PCR (RT-PCR)

Reverse transcription PCR (RT-PCR) can use mRNA rather than DNA as the starting template, amplifying **complementary DNA (cDNA)**. The enzyme **reverse transcriptase** synthesizes a DNA chain on an RNA template, and DNA polymerase converts the single-stranded DNA molecules into double-stranded DNA molecules that can further be used as templates (Figure 9.2, top). Then a PCR reaction is run to amplify a fragment of DNA to indirectly detect an mRNA species present in the original sample. Thus, the value of RT-PCR is to amplify a cDNA sequence based on an mRNA template, either to identify the presence of mRNA or to clone a cDNA molecule for future manipulation.

Quantitative Real-Time PCR (qRT-PCR or qPCR)

In addition to amplifying pieces of DNA for later manipulation or use, a slight modification of standard PCR methods allows quantitative comparisons of the amount of DNA or RNA present in a sample. A very sensitive method that requires only a tiny amount of either DNA or RNA template, **quantitative real time PCR** (**qRT-PCR**, not to be confused with RT-PCR described above) is based on detecting a fluorescent signal produced in proportion to the PCR product during amplification (Figure 9.3). A fluorogenic probe within the region to be amplified contains a fluorescent reporter tag on one end and a fluorescence

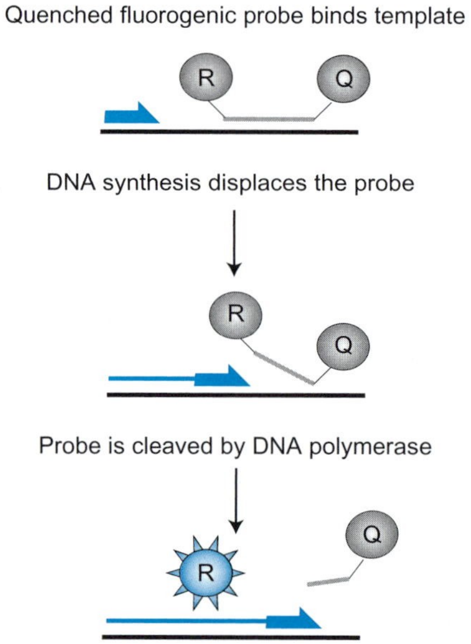

Quenched fluorogenic probe binds template

DNA synthesis displaces the probe

Probe is cleaved by DNA polymerase

FIGURE 9.3 Quantitative Real-Time PCR (qRT-PCR). A probe containing a fluorescent reporter (R) that is initially quenched (Q) binds to a specific sequence. During PCR, the probe is cleaved, removing the quencher so that each round of amplification produces fluorescence that can be quantified to reveal the amount of nucleic acid template present in a sample.

quencher on the other end. During each cycle of amplification, DNA polymerase cleaves the fluorescent probe molecule, removing the quencher and allowing the reporter to emit fluorescence. The increase in fluorescence is proportional to the amplification during the exponential phase of amplification. These changes in fluorescence are read out by a sensitive camera and comparisons can be made among different samples. Nucleic acid hybridization techniques such as Southern (DNA) or northern (RNA) blots are alternative methods to quantitatively compare amounts of DNA or RNA present in samples.

Modifying DNA Sequences Using PCR

Though the primary purpose of PCR is to make exact copies of a DNA sequence, it can also be used to modify a sequence and make copies of the modified sequence. This is primarily done by creating primers that contain the desired sequence modifications, either by adding the base pairs of specific restriction enzyme recognition sites to the ends or by altering a single base pair to introduce a point mutation. Though the primers may not hybridize as tightly for the first few PCR cycles, with each successive cycle, the added or mutated sequences in the primers will be copied and incorporated into the amplified PCR product.

CLONING DNA

The core of molecular cloning is creating many copies of a particular DNA fragment (Box 9.2). In order to clone a DNA fragment that has been isolated using restriction enzyme digests or PCR, it must be put into a form that can be copied. Bacterial host cells are often used as factories to mass-produce clones of the DNA of interest. It is these colonies of identical bacterial clones that lend their name to the technique of "molecular cloning." For bacteria to mass produce DNA of interest, it must be in a form that can be replicated and passed

---

BOX 9.2 Defining the Word *Clone*

The word *clone* has many definitions, which can make terminology in molecular biology rather confusing. The general definition involves making copies, but the term also describes distinct though related topics. Animal or cell clones are so called because of the shared copies of genomic DNA. Mapping the location of a gene discovered through a forward genetic screen (Chapter 8) is also known as **positional cloning,** which draws on molecular cloning techniques to map the chromosomal location of an unidentified, disrupted gene. And when researchers say they are "cloning a gene," they are most likely isolating and identifying a particular stretch of DNA to define an unknown gene, again using molecular cloning techniques.

   Molecular cloning literally refers to the creation of many identical copies of a DNA molecule. The term *cloning* arises from the colonies of identical host cells (the so-called clones) that are produced to make these copies of a particular DNA sequence.

on by the dividing cells. Thus, isolated DNA fragments must be put into a storage device: the vector. If the DNA fragment comes from another vector, the process of transferring the desired sequence into a new vector is called **subcloning**. Box 9.5 describes an example of a subcloning experiment, going through all the steps that will be described in the following section.

Vectors

A **DNA vector** is a carrying vehicle that can hold an isolated DNA sequence of interest. The essential features of a vector are that it can replicate autonomously in an appropriate host, and it can be combined with other pieces of DNA. Often, the inserted DNA fragment comes from a different organism than the vector DNA, and the fusion is called a recombinant DNA construct. Two commonly used vectors are **plasmids**, naturally occurring circles of DNA that act as independently replicating accessory chromosomes in bacteria, and **bacteriophage** or **phage**, a virus that can deliver its genetic cargo to bacteria. Many plasmids and bacteriophages have been ingeniously modified to enhance the delivery of recombinant DNA molecules into bacteria and to facilitate the selection of bacteria harboring these vectors.

Because there is a size limit to the number of base pairs that can be stably inserted into a plasmid or phage vector (1–25 kb, kilobase pairs), other vectors that can hold larger amounts of DNA have been utilized: cosmids and artificial chromosomes. A **cosmid** is a combination plasmid that contains phage sequences that allow the cosmid to be packaged and transmitted to bacteria like a phage vector. Cosmids are more stable than regular plasmids, and thus they can hold larger inserts (30–50 kb). **Artificial chromosomes** containing telomeric and centromeric regions can hold even larger inserts. Because cosmids and artificial chromosomes can stably hold large DNA inserts, they are often used to create DNA libraries (Box 9.1).

Common Vector Features

Useful vectors contain three main features: (1) an origin of replication, (2) restriction sites, and (3) a selectable marker. An origin of replication is a specific DNA sequence at which DNA replication is initiated, and thus is necessary for the reproduction of the vector. Restriction sites are needed to place a DNA sequence of interest into a vector. Because vectors are designed to be a storage system that carries inserted pieces of DNA, a standard feature is the presence of a **multiple cloning site (MCS)**, a region with multiple restriction sites to make a compatible digest of the vector and the DNA insert more easily. Both the vector and the isolated DNA fragment are cut using the same restriction enzymes so that sticky ends from the DNA fragment match cut sticky ends on the vector. Then, the two molecules can be stitched together using an enzyme called DNA ligase (described below). The recombinant DNA molecule comprised of the vector and the DNA fragment, also known as the insert, joined together is referred to as a **construct**.

A selectable marker is an important feature to be able to distinguish which host cells contain the construct. Naturally occurring plasmids often confer antibiotic resistance, so this is the most common feature used to select host cells carrying the construct. Only cells containing the vector with antibiotic resistance will be able to survive when grown in an environment containing the antibiotic.

Types of Vectors

Cloning vectors provide a basic backbone for the DNA insert to be reproduced and generally have the common features just described, but these vectors are useful only for cloning and not for expressing a protein product. The DNA insert in a cloning vector can be copied but not translated into a functional protein product.

For a gene to give rise to its protein product, an expression vector must be used that contains the appropriate pieces for a host cell to produce the protein. In the case of a mammalian cell, a standard mammalian expression vector will still contain the origin of replication, multiple cloning site, and selectable marker, but it will also need a promoter found in mammalian cells that can drive the expression of the gene, as well as other standard features found in genes. Expression vectors need to be transcribed and translated, so they need the polyadenylation tail that normally appears at the end of transcribed pre-mRNA to protect and stabilize it, a minimal untranslated region (UTR) so there is not an excess of DNA that will not be translated, and a sequence that attracts the ribosome for translation (Box 9.3).

Ligation

Inserting a piece of DNA into a vector requires using restriction enzymes as molecular scissors and pasting the ends with DNA ligase, an enzyme normally used by cells for DNA repair (Figure 9.4). Cutting the insert and the vector with

BOX 9.3 Internal Ribosome Entry Site (IRES)

One special feature present in certain expression vectors is an internal ribosome entry site (IRES). An IRES is a sequence recognized by a ribosome, so it can be used to drive the translation of two different proteins off a single mRNA transcript. It is most commonly used for the expression of a reporter protein like GFP or β-galactosidase. This allows both the gene of interest and the reporter to be regulated by the same promoter and transcribed as a single mRNA, so the reporter faithfully recapitulates the expression of the gene of interest. This is an alternative to creating a fusion protein; if fusing an additional protein to the gene of interest interferes with normal function, an IRES can be added before the reporter gene. This will create a single mRNA transcript, but because the IRES recruits another ribosome, two proteins will be created in the same location and time, driving the expression of two different genes under the same promoter.

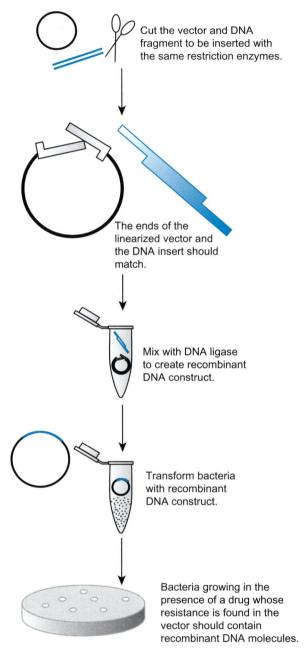

FIGURE 9.4 How DNA gets into a vector: making recombinant DNA. This diagram depicts the most frequently used method of creating a recombinant DNA molecule.

BOX 9.4 Cloning through Recombineering

Recombineering ("recombination-mediated genetic engineering" or "recombinogenic engineering") is an emerging method to create constructs more quickly without the use of restriction enzymes and DNA ligase. It uses a bacteriophage's homologous recombination proteins to directly modify DNA within a bacterial cell. This method is efficient and decreases the amount of time it takes to create many types of complicated constructs because it eliminates the need to find or create appropriate and unique restriction-enzyme cleavage sites in both cloning vectors and genomic DNA. This method of *in vivo* cloning uses homologous recombination between short (40–50 base pairs) regions of homology to insert a DNA sequence of interest into a vector.

the same restriction enzyme produces compatible ends that will stick to each other. By using two different restriction enzymes to produce different compatible ends on the 5′ and 3′ ends of the insert and vector, scientists influence the insert to stick in the proper orientation. This is why it is generally better to use enzymes that produce sticky ends. Blunt ends can cause problems such as multiple inserts, inserts in the wrong orientation, or no inserts if the vector recloses on itself. Some vectors, such as BACs, are so large that it can be very difficult to produce recombinant DNA constructs using restriction enzymes and DNA ligase. An alternative to this molecular cutting and pasting is recombineering (Box 9.4).

Transformation

In a step called transformation (for bacteria) or transfection (more generally), the construct is introduced into **competent cells**, host cells that have been made transiently permeable to DNA. We will discuss methods of getting constructs into neurons and other mammalian cells in the next chapter. Competent cells are most frequently bacteria. Bacteria can be used to grow large quantities of a single recombinant DNA species. As bacterial host cells grow and divide, doubling in number every 30 minutes, the recombinant plasmids also replicate to produce an enormous number of copies of DNA circles containing the foreign DNA (Figure 9.4). This insert is inherited by all the progeny cells of a single bacterium, which together form a small colony in a culture dish (the clones). Yeast can also be used, particularly in cloning situations in which very large pieces of DNA need to be cloned.

PURIFYING DNA

Once DNA fragments have been generated through restriction digests or PCR amplification, they must be purified and separated away from other fragments that are not of interest. This is done through one of the most ubiquitous tools

in molecular biology: gel electrophoresis. Other procedures, called DNA
purification preparations, are used to purify vector DNA from the proteins and
genomic DNA of host cells.

Isolation and Characterization of DNA Fragments Using Gel Electrophoresis

Gel electrophoresis is used to isolate, identify, and characterize properties of
DNA fragments in many different situations and at many different points during the cloning process. A small amount of DNA can be loaded into a well at
one end of a gel in an apparatus that allows a current to be run through the gel.
Because DNA is negatively charged, it will migrate to the cathode (the positive
charge) and away from the anode (the negative charge). In most types of gels,
the mobility of a DNA fragment in an electric field is inversely proportional to
the logarithm of the number of base pairs (up to a certain limit). Thus, smaller
fragments of DNA can move faster than larger fragments, and gel electrophoresis can isolate fragments of DNA based on size. A DNA ladder—DNA that
has been digested, producing fragments with known base pair sizes—should
be run next to unknown DNA samples so that the sizes of the unknown samples can be determined.

After creating smaller fragments of DNA from the source DNA, the length
and purity of those DNA molecules can be determined using gel electrophoresis (Figure 9.5). The appearance of the DNA bands in the gel can indicate the

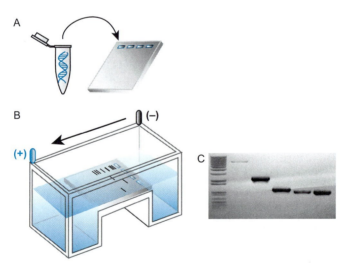

FIGURE 9.5 **Gel electrophoresis. (A)** DNA samples are loaded into individual wells of an agarose gel. **(B)** When an electric current is passed through the gel, negatively charged DNA travels
away from the anode (–) and toward the cathode (+) in a size-dependent manner. **(C)** A ladder
(far left) of DNA fragments of known sizes is usually run next to samples to identify the size of
the sample fragments.

purity of DNA fragments in a sample. For example, if an investigator believes he or she only has two fragments of DNA in a sample but finds several bands of DNA after electrophoresis, the sample may have been contaminated by other DNA.

The length of the DNA fragments that can be analyzed using gel electrophoresis depends on the type of gel used. Polyacrylamide gels can separate fragments containing about 1000 base pairs, whereas more porous agarose gels are used to resolve mixtures of larger fragments (up to 20 kb). In a technique called pulsed-field gel electrophoresis, it is possible to separate entire chromosomes containing millions of nucleotides.

The DNA bands on agarose or polyacrylamide gels are invisible unless the DNA is labeled or stained in some way. One sensitive method of staining DNA is to expose it to the dye ethidium bromide, an intercalating agent that fluoresces under ultraviolet light when it is bound to DNA. Alternatively, bands or spots of radioactive DNA in gels can be visualized by autoradiography (Box 6.2). The intensity of a DNA band is based on the amount and size of the fragment, and thus the quantity can be estimated by comparing the intensity of the band to a known quantity in the ladder.

Purifying DNA from Host Cells

After transforming bacterial host cells to grow many copies of DNA, the recombinant DNA plasmid must be purified from the host cell DNA and proteins. This is done by growing up many more copies of the bacterial clone, lysing them, and chemically purifying the DNA. Depending on the volume of bacteria that was grown, different amounts can be purified—from micrograms to grams—and there are many kits now available to aid in separating the plasmid DNA from bacterial DNA and proteins. This procedure is known as the miniprep (for small amounts), midiprep, or maxiprep (for large amounts). After separating the DNA of interest from other DNA and proteins in this manner, the next step is to sequence the DNA to make sure that its identity is intact and correct.

BOX 9.5 Walkthrough of a Subcloning Experiment

Let's say you have identified an interesting gene called *geneX* from microarray analysis (Chapter 8) to screen for genes that decrease over the course of neural development. You have determined the sequence of this gene and know that it is highly expressed in neurons. You would like to be able to examine its subcellular behavior in neurons using time-lapse imaging, so you will create a reporter construct (Chapter 7) by making a GFP-fusion expression construct for *geneX*—GFP-*geneX*. Once you've created this expression construct, you can introduce GFP-*geneX* into cells or animals using the transfection methods described in Chapter 10.

First, you must figure out an appropriate vector and restriction enzymes to use. Because you want to create a GFP-fusion construct, you could use a commercially available GFP-containing plasmid vector called pEGFP-N1. This vector contains a promoter that constitutively drives gene expression (called CMV), the EGFP (enhanced green fluorescent protein) coding sequence, antibiotic resistance to kanamycin, and a multiple cloning site (MCS), all contained within a circular plasmid. The MCS is located in front of the EGFP coding sequence, meaning the *geneX* coding sequence will be attached to the N-terminus of EGFP (this is what the N1 in the vector's name refers to).

An MCS contains many unique restriction sites to choose from, so compatible restriction enzymes can be used on both the vector and the insert. The primary factor to consider when choosing the appropriate restriction enzyme to use is whether the site is unique. You do not want to be cutting up your DNA of interest in unexpected or undesired areas. Because you will be using the same restriction enzyme to open up the vector that you will be using to generate the ends of your insert, you must consider the sequences of both items. After using software that tells you where various restriction sites are located in your DNA of interest, you end up choosing the common enzymes *Eco*RI and *Bam*HI, because *geneX* does not contain those recognition sites within its coding sequence.

The next major step is to generate a *geneX* fragment with the appropriate restriction sites to prepare the insert to go into the vector. There are multiple ways to generate the small fragment of DNA that represents *geneX*. If you already had a vector containing the segment of *geneX* you want, you might be able to directly cut it out of this vector to paste into a new vector. Alternatively, you could use PCR to amplify the appropriate fragment from a cDNA template, adding the recognition sequences for *Eco*RI and *Bam*HI to the ends of the *geneX* fragment.

After choosing a vector and preparing the insert, you will need to make them compatible by cutting them with the same restriction enzymes. You need to insert the *geneX* fragment in the proper direction and the same reading frame as the EGFP so that they form a single protein together. You have done this by choosing two different restriction enzymes (*Eco*RI and *Bam*HI) that leave "sticky" ends after digestion and by making sure that the frame of the *geneX* PCR product after digestion is the same as the EGFP coding sequence within the vector.

To isolate your digested products, you can run the digested vector and insert on an agarose gel. You should be able to identify the appropriate bands based on the size. Cut out those bands of DNA contained within the gel, and then purify the DNA from the gel. The gel can be dissolved away to extract the cut vector and cut insert.

Mix the purified digested products with DNA ligase, which will recognize those restriction sites and "glue" them back together. The *Eco*RI digest on the 5' end of the *geneX* insert matches the *Eco*RI digest on the 3' end of the plasmid, while the *Bam*HI digest on the 3' end of the *geneX* insert matches the *Bam*HI digest on the 5' end of the plasmid. Thus, the *geneX* sequence should be inserted in the proper direction into the plasmid.

Mix the ligation product with competent cells and then spread the cells on plates containing nutrients—to help cells grow—and kanamycin, the antibiotic that cells containing your construct should be resistant against.

The next morning, you should see clonal colonies of bacteria growing on the plate. Select individual colonies whose member cells should contain identical copies of the construct, and grow more clones in a larger volume of growth media.

Purify the plasmid DNA from each of the clones that was grown up using a "miniprep" kit. Basically, you lyse the bacteria, extract the DNA from the lysed bacteria, and purify the plasmid DNA from the bacterial DNA. You can use multiple methods to verify that the purified DNA from the clones you chose contains the right construct.

The purified plasmid DNA can be examined using restriction enzyme analysis. Items that can be checked using restriction digests include the presence of the insert, whether the insert was ligated in the correct direction, and whether there is more than one insert present. These can be examined by choosing appropriate restriction enzymes, which will not necessarily be the ones used to join the insert and vector. Finally, you will want to verify that the sequence in your construct matches the DNA sequence of *geneX* perfectly and no mistakes were introduced. You can do this by performing sequencing reactions (Figure 9.6).

Now that you've got your GFP-geneX construct in an expression vector, you can introduce this into cells using techniques that will be covered in the next chapter.

IDENTIFYING DNA

A critical step in characterizing the information encoded by a stretch of DNA is specifying the nucleotides that spell out that information. Other tools for manipulating or amplifying DNA of interest rely on knowing exact sequences, as do hybridization techniques that can reveal the presence, location, and amounts of DNA or RNA.

DNA Sequencing

DNA sequencing is used to identify the string of nucleotides making up the DNA region of interest. This sequence can be and is used in a number of steps to identify DNA, such as comparing known genomic sequences to identify a gene or verifying that a recombinant DNA product is correct after manipulations.

The **Sanger dideoxy chain-termination method** can determine the sequence of nucleotides in any purified DNA sample. This method reports the sequence with high fidelity for a stretch of about 200–500 base pairs. Based on *in vitro* DNA synthesis, the Sanger method synthesizes short pieces of DNA in the presence of nucleotide bases to which more bases cannot be added: chain-terminating dideoxyribonucleoside triphosphates. These chain-terminating nucleotides are labeled and mixed in with regular bases so that fragments of DNA will be created at many different lengths, being randomly stopped by the addition of a chain-terminating nucleotide. Thousands of fragments of different lengths will be run on a gel. When this method was originally performed, and for many years thereafter, each chain-terminating nucleotide was labeled with a radioactive probe, so scientists had to perform four separate reactions (one for each nucleotide) and run a gel with each reaction in a different lane. Modern techniques label each chain-terminating nucleotide with a different colored fluorescent dye. This way, only a single reaction containing each of the distinctly labeled chain-terminating nucleotides needs to be performed, and an automated fluorescence detector can quickly scan the gel to "read" the identity of the last, terminating nucleotide at that position. In this way, the order of nucleotide bases can be determined (Figure 9.6). Modern molecular biology labs tend not to run sequencing reactions themselves, as it is typically faster and less expensive to send a fragment of DNA off to a dedicated facility for sequencing.

Nucleic Acid Hybridization Techniques

The term *hybridization* refers to the process whereby two complementary nucleic acid strands bind to each other. Any fragment of DNA with a known base pair sequence can be detected by labeling a probe strand with the complementary base pairs of DNA with radioactivity or a biochemical tag that can react to give a colorimetric or fluorescent signal (Box 6.2). These complementary probes can be created using DNA synthesis or PCR amplification.

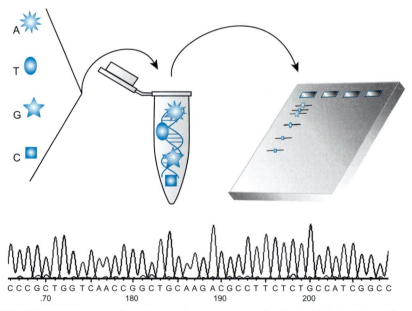

FIGURE 9.6 DNA sequencing. Each chain-terminating nucleotide is labeled with a different probe, and many fragments of DNA are synthesized. Each position may incorporate a labeled chain-terminating nucleotide. When all of the synthesized DNA fragments are separated using gel electrophoresis, they will be positioned by size and a sensitive camera can automatically detect which label is at each position producing output based on fluorescence intensity (bottom).

In situ *Hybridization (ISH)*

Discussed in Chapter 6, *in situ* hybridization uses a complementary strand to hybridize with a sequence of mRNA present in tissue to reveal spatial information about the expression of a particular gene.

Southern Blot

Developed by Edwin Southern, a **Southern blot** can reveal if a known DNA sequence is present by using a specific probe to identify the sequence in a large mixture of other DNA fragments. In a Southern blot, DNA of interest, generally extracted genomic DNA, is digested with restriction enzymes, and the mixture of fragments produced is separated by gel electrophoresis, denatured to form single-stranded DNA, and transferred to a nitrocellulose sheet (Figure 9.7).

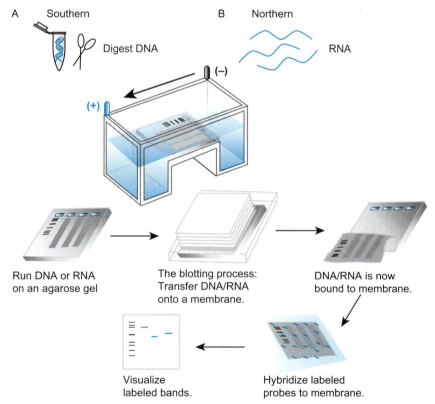

FIGURE 9.7 The Southern and northern blot processes detect specific sequences of nucleic acids. (A) The Southern blot uses restriction enzymes to digest DNA, and **(B)** a northern blot uses RNA molecules directly. Both processes separate out nucleic acids using gel electrophoresis and then transfer the nucleic acid onto a membrane. This membrane is incubated with labeled probes that will detect the presence and quantity of a specific sequence of nucleic acid.

The positions of the DNA fragments in the gel are preserved on the nitrocellulose sheet, where they are exposed to a radioactive ^{32}P-labeled single-stranded DNA probe. The probe hybridizes with a restriction fragment having a complementary sequence, and autoradiography then reveals the position of the restriction fragment-probe duplex. In this way, a particular fragment in the midst of a million others can be identified like using a magnet to find a needle in a haystack. Southern blots are often used to demonstrate a successful gene knock out by using a probe against the region of the gene that was knocked out. PCR is a more sensitive and faster alternative to identify a DNA fragment, but these techniques are often used together to confirm a result.

Northern Blot

An analogous technique for the analysis of RNA has been humorously named the **northern blot**, in reference to the Southern blot. Northern blots directly compare the relative amounts of mRNA levels between different samples on the same blot, making it the preferred method for looking at transcript size (indicated by position on the gel) and for detecting alternatively spliced transcripts. Intact mRNA molecules are separated on the basis of their size using gel electrophoresis, and a procedure identical to that in the Southern blot is performed (Figure 9.7). Even sequences with only partial homology can be used as probes, so homologous mRNA from a different species or genomic DNA fragments that may contain an intron can still be used. Doing experiments with RNA typically requires more care in protecting the samples, as RNases (enzymes that degrade RNA) are ubiquitously present, and RNA samples are much less stable than DNA molecules. Perhaps one of the trickiest and most critical steps is isolating the RNA. Alternative methods for examining mRNA levels include RT-PCR, RNase protection assays, and microarrays. Though it takes longer to perform a northern blot and is less sensitive than RT-PCR, the ability to use multiple types of probes and higher specificity still make the northern blot a commonly used technique. An analogous technique used to detect proteins with an antibody is called the **western blot** (Chapter 14).

CONCLUSION

The tools of recombinant DNA technology have provided scientists with the ability to manipulate genes and, thus, the proteins they encode. These tools allow us to alter DNA sequences, cutting and pasting them in novel ways, amplifying them to produce many copies, and sequencing them to confirm their identity. The power of recombinant DNA technology is complemented by the ability to deliver novel DNA constructs to cells and create transgenic and mutant organisms, as will be discussed in the next several chapters.

SUGGESTED READING AND REFERENCES

Books

Drlica, K. (2004). *Understanding DNA and Gene Cloning: A Guide for the Curious*, 4th ed. Wiley, Hoboken, NJ.

Griffiths, A. J. F. (2008). *Introduction to Genetic Analysis*, 9th ed. Freeman, NY.

Nicholl, D. S. T. (2002). *An Introduction to Genetic Engineering*, 2nd ed. Cambridge University Press, Cambridge.

Review Articles

Campbell, T. N. & Choy, F. Y. (2002). Approaches to library screening. *J Mol Microbiol Biotechnol* **4**, 551–554.

Copeland, N. G., Jenkins, N. A. & Court, D. L. (2001). Recombineering: a powerful new tool for mouse functional genomics. *Nat Rev Genet* **2**, 769–779.

Marziali, A. & Akeson, M. (2001). New DNA sequencing methods. *Annu Rev Biomed Eng* **3**, 195–223.

Primary Research Articles—Interesting Examples from the Literature

Agate, R. J., Choe, M. & Arnold, A. P. (2004). Sex differences in structure and expression of the sex chromosome genes CHD1Z and CHD1W in zebra finches. *Mol Biol Evol* **21**, 384–396.

Bacon, A., Kerr, N. C., Holmes, F. E., Gaston, K. & Wynick, D. (2007). Characterization of an enhancer region of the galanin gene that directs expression to the dorsal root ganglion and confers responsiveness to axotomy. *J Neurosci* **27**, 6573–6580.

Brenner, S., Williams, S. R., Vermaas, E. H., Storck, T. et al. (2000). *In vitro* cloning of complex mixtures of DNA on microbeads: physical separation of differentially expressed cDNAs. *Proc Natl Acad Sci U S A* **97**, 1665–1670.

Hsu, S. Y., Kaipia, A., McGee, E., Lomeli, M. & Hsueh, A. J. (1997). Bok is a pro-apoptotic Bcl-2 protein with restricted expression in reproductive tissues and heterodimerizes with selective anti-apoptotic Bcl-2 family members. *Proc Natl Acad Sci U S A* **94**, 12401–12406.

Paisan-Ruiz, C., Jain, S., Evans, E. W., Gilks, W. P. *et al.* (2004). Cloning of the gene containing mutations that cause PARK8-linked Parkinson's disease. *Neuron* **44**, 595–600.

Valenzuela, D. M., Murphy, A. J., Frendewey, D. & Gale, *et al.* (2003). High-throughput engineering of the mouse genome coupled with high-resolution expression analysis. *Nat Biotechnol* **21**, 652–659.

Protocols

Choi, S. & Kim, U. J. (2001). Construction of a bacterial artificial chromosome library. *Methods Mol Biol* **175**, 57–68.

Lee, S. C., Wang, W. & Liu, P. (2009). Construction of gene-targeting vectors by recombineering. *Methods Mol Biol* **530**, 15–27.

Sambrook, J., Russell, D. W. (2001). *Molecular Cloning: A Laboratory Manual*, 3rd ed. Cold Spring Harbor Laboratory Press, Cold Spring Harbor, NY.

Websites

New England BioLabs catalog http://www.neb.com/nebecomm/tech_reference

Primer-BLAST: primer designing tool http://www.ncbi.nlm.nih.gov/tools/primer-blast/

Chapter 10

Gene Delivery Strategies

After reading this chapter, you should be able to:

- Describe common methods of delivering recombinant DNA into cells
- Compare the most efficient methods of delivering DNA into cells *in vitro* and *in vivo*

Techniques Covered

- **Physical gene delivery:** microinjection, electroporation, biolistics
- **Chemical gene delivery:** calcium phosphate transfection, lipid-based transfection
- **Viral gene delivery:** adenovirus, adeno-associated virus (AAV), lentivirus, herpes simplex virus (HSV)

The previous chapter discussed methods for making and manipulating DNA constructs using recombinant DNA technology. Of course, the ultimate goal of creating a DNA construct is to deliver the manipulated sequence into living cells so that the endogenous cellular machinery can transcribe and translate the sequence into functional proteins. A neuroscientist might want to deliver a DNA sequence into a population of cells in culture, in a brain slice, or in the brain of a living animal. The purpose of this chapter is to survey the common methods of delivering recombinant DNA into cells in each of these environments.

There are three general categories of DNA delivery: physical, chemical, and viral. **Transfection** refers to nonviral methods of delivering DNA to cells, including physical and chemical methods. **Infection** refers to viral DNA delivery, in which viruses attach themselves to cells and inject their DNA cargo. Each method has its own relative advantages and disadvantages that make it useful for delivering DNA in different experimental contexts (Table 10.1). Major factors that dictate the choice of DNA delivery method include: cellular environment, cell type, and experimental goals. Furthermore, these DNA delivery methods vary in terms of the number of cells they affect, levels of gene expression they induce, and the length of gene expression over time.

PHYSICAL GENE DELIVERY

Physical gene delivery methods deliver DNA into a cell by physically penetrating the cell membrane with force. These methods are often highly efficient (transfected cells express high levels of the delivered gene) and can be used with any cell type. However, they can sometimes disrupt the integrity of the cell membrane, traumatizing the cell and leading to its death. Therefore, care must be taken in moderating the amount of force applied to deliver DNA to cells. Three common physical methods include microinjection, electroporation, and biolistics.

Microinjection

A scientist can microinject DNA into a cell by piercing the cell membrane with a small glass needle and then applying pressure to inject DNA inside the cell (Figure 10.1). If the scientist uses an extremely thin needle and withdraws the needle carefully, the cell membrane remains intact, and the cell has an opportunity to incorporate the injected DNA into its genome. This process requires

TABLE 10.1 Categories of Gene Delivery Strategies

Method	Advantages	Disadvantages
Physical	High-efficiency gene transfer No limitations on construct size No cell type dependency	Low throughput Requires specialized equipment Can physically harm cells
Chemical	High efficiency *in vitro* No limitations on construct size Relatively easy to perform Rapid High throughput Low immunogenicity	Limited *in vivo* applications Challenge to prepare consistent solutions Efficiency depends on cell type Can be toxic to cells Transient expression
Viral	High-efficiency gene transfer Cell-specific targeting is possible Long-term expression Can be used *in vitro* and *in vivo*	Complex cloning required More expensive than transfection methods Safety concerns regarding production of infectious viruses in humans May provoke immune response Laborious preparation Limited construct size

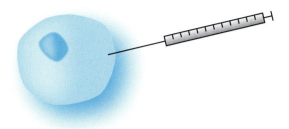

FIGURE 10.1 DNA microinjection. An extremely thin glass needle punctures the plasma membrane, allowing DNA to be injected into the cell.

a microscope to view the living cell, as well as fine micromanipulators to precisely place the needle adjacent to the cell membrane. Each cell must be injected one at a time (in contrast to other methods that deliver DNA to millions of cells at once), making microinjection laborious and low throughput. This method requires technical skill and is therefore usually performed by trained technicians or lab personnel who use these techniques regularly.

DNA microinjection is the method used to make transgenic animals. To make a transgenic mouse, a scientist microinjects the transgenic DNA construct into newly fertilized mouse eggs. In some of the new eggs, the transgene randomly incorporates into the mouse genome. These cells are injected into a female mouse to carry the egg to gestation. The full details of creating transgenic mice, as well as other transgenic animals, are discussed in Chapter 11.

In addition to DNA constructs, microinjection can also be used to deliver proteins, antibodies, and oligonucleotides into cells. Unfortunately for neuroscientists, microinjection is rarely used to transfect neurons as their small shape and sensitivity makes them more difficult to inject than other cell types. This technique is also too low throughput to transfect many neurons in culture but can be useful for situations in which only a single neuron needs to be transfected.

Electroporation

Electroporation is the process of using an electric pulse to transfect cells with DNA (Figure 10.2). Applying an electric field to cells is thought to induce temporary pores in the cell membrane, allowing the cell to take up DNA sequences. The electric field also drives negatively charged DNA strands away from the cathode (negative end) toward the anode (positive end) of an electric field. Therefore, an electric pulse causes some of the DNA to enter the cell. After the electric field is switched off, the cell membrane reseals, trapping some of the electroporated DNA within the cell.

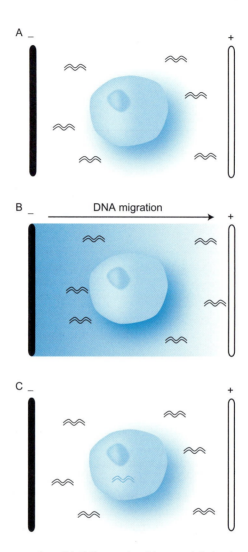

FIGURE 10.2 Electroporation. (**A**) Cells are placed in a special chamber with a solution containing the DNA to be transfected. (**B**) Applying an electric field drives the negatively charged DNA strands away from the cathode and toward the anode. The electric field also induces tiny pores in the cell membrane. (**C**) After the electric field is switched off, the cell membrane reseals, and some DNA remains in the cell.

Electroporation can be used to efficiently deliver DNA to cells in both *in vivo* and *in vitro* conditions. A scientist can electroporate dissociated cells or brain slices in special containers with electrical contacts. In a process known as *ex vivo* electroporation, a scientist removes a brain from an animal, applies electroporation techniques, and then sections the brain for slice cultures.

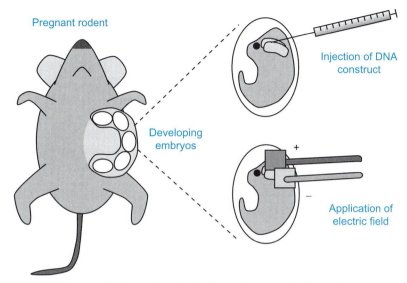

FIGURE 10.3 *In utero* **electroporation.** A scientist performs a surgical procedure on a pregnant female, exposing the embryonic pups. DNA is injected into the pups' ventricular systems, and paddles generate electric field pulses to introduce DNA into the cells lining the ventricles.

In neuroscience research, electroporation has been particularly useful for *in utero* preparations (Figure 10.3). *In utero* electroporation surgeries begin by exposing the embryonic pups of a pregnant animal, usually a rodent. DNA is injected into the pups' ventricles, and electrode paddles direct the uptake of the injected DNA construct(s) toward the anode and away from the cathode. The pups are placed back inside the mother, who is sutured and allowed to recover. Days later, pups are removed or the mother gives birth, and the neurons of the offspring express the introduced DNA construct. Similar methods can be used in the chicken model system in a process called *in ovo* electroporation. These methods are useful for delivering genes to brain regions adjacent to the nervous system's fluid-filled ventricles, as the DNA-containing solution must be injected into the ventricle. Thus, they have primarily been used in studies of chick spinal cord and rodent cerebral cortex. A scientist can control both spatial and temporal specificity by controlling the position of the electrode paddles and the timing of electroporation. For example, a scientist can target gene transfection to distinct cortical layers in the mammalian cerebral cortex by electroporating at specific embryonic time periods.

Electroporation has numerous advantages, such as the ability to transfect cells in many different *in vivo* and *in vitro* environments. Furthermore, investigators can vary expression levels of a transfected gene by varying the strength and pattern of the electric field pulses.

Biolistics

Biolistics, short for "biological ballistics" and also known as **particle-mediated gene transfer**, is the method of directly shooting DNA fragments into cells using a device called a **gene gun**.

To use a gene gun, a scientist first mixes a DNA construct with particles of a heavy metal, usually tungsten or gold. These fine particles stick to the negatively charged DNA. The DNA/metal particles are loaded onto one side of a plastic bullet (Figure 10.4). A pressurized gas, usually helium, provides the force for the gun. Gas pressure builds up until a rupture disk breaks, driving the plastic bullet down a shaft. The plastic bullet is abruptly stopped at the end of the shaft, but the DNA/metal particles emerge from the gun with great speed and force. If the gun is aimed at biological tissue, some of the metal particles will penetrate the cell membranes and deliver DNA constructs to cells.

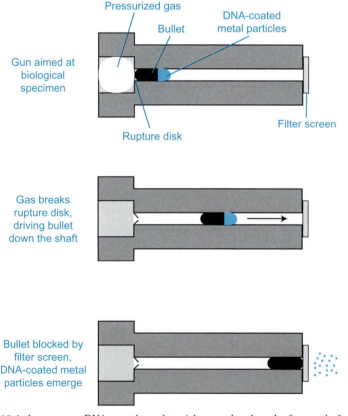

FIGURE 10.4 A gene gun. DNA-coated metal particles are placed on the front end of a bullet. High-pressured gas drives the bullet down a shaft. At the end of the shaft, the bullet is blocked, but the DNA-coated particles emerge with great speed and force.

A neuroscientist can use biolistic technology to cause efficient gene expression in neurons. This technology can produce a dispersed transfection pattern, similar to a Golgi stain, in which only individual cells receive the foreign DNA in a background of untransfected cells. Another advantage to using biolistic technology is that a gene gun can deliver DNA through relatively thick tissue, such as a tissue slice in culture. Therefore, it is possible to transfect cells within a slice that would be difficult to target using other gene delivery methods. This technique has not yet been successful in transfecting mammalian neurons *in vivo*, although it has been used *in vivo* to transfect liver and skin cells. The main disadvantage of using this technology is that it may cause physical damage to cells. Optimization is required to limit the amount of tissue damage caused by the force of impact of the projectiles.

CHEMICAL GENE DELIVERY

Chemical gene delivery is the process by which a scientist uses a chemical reaction to deliver DNA into cells. Negatively charged DNA can form macromolecular complexes with positively charged chemicals. These complexes can then interact with a cell's membrane, promoting uptake through endocytosis or fusion. These strategies are useful in that they are incredibly high-throughput and often simple to perform. However, they are generally inefficient for *in vivo* delivery and therefore mainly used for cell culture experiments (Chapter 13). Expression of the target gene is usually transient, lasting for days to weeks depending on the cell type. Common chemical gene delivery strategies include calcium phosphate transfection and lipid transfection.

Calcium Phosphate Transfection

The **calcium phosphate method** of DNA transfection is one of the simplest and least expensive methods of chemical gene delivery (Figure 10.5). This method requires two chemical solutions: calcium chloride, which serves as a source of calcium ions, and HEPES buffered saline (HBS), which serves as a source of phosphate ions. First, the calcium chloride solution is mixed with the DNA to be transfected, and then the HBS is added. When the two solutions are combined, the positively charged calcium ions and negatively charged phosphate ions form a fine precipitate. The calcium ions also cause the DNA to precipitate out of solution. After a few minutes, the solution with precipitate is directly added to cells in culture. By a process that is not entirely understood, the cells take up some of the precipitate, and with it, the DNA. It is thought that the DNA precipitate enters cells by endocytosis, but the exact mechanism remains a mystery. This method works best for a cell monolayer so that the DNA precipitate covers the cells evenly.

The advantages to using the calcium phosphate method are that it is relatively easy, reliable, and cheap. It is useful for transient expression or creating

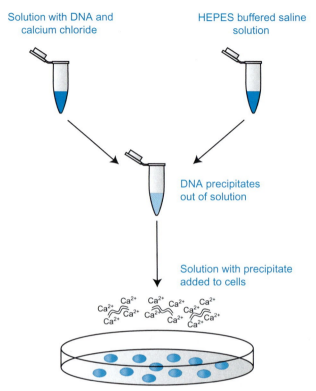

Solution with DNA and
calcium chloride

HEPES buffered saline
solution

DNA precipitates
out of solution

Solution with precipitate
added to cells

FIGURE 10.5 Calcium phosphate transfection. A scientist mixes DNA, calcium chloride, and HEPES buffered saline. The chemical reaction forms a calcium phosphate precipitate, allowing DNA to rain down on cells in the culture.

stable cell lines from immortalized cells (Chapter 13). The main disadvantage is that transfection efficiencies are low in neurons (1–3%), and this method does not work at all for transfecting neurons in intact tissue. Some immortalized cell lines, such as HEK 293T and HeLa cells (Chapter 13), exhibit high transfection efficiencies (20–100%), which is one reason for their ubiquitous use in the life sciences.

Lipid Transfection

Lipofection, also known as "lipid transfection" or "liposome-based transfection," uses a lipid complex to deliver DNA to cells. Lipids are a broad class of fat-soluble biomolecules, such as fats, oils, and waxes. The cell membranes of animal cells are composed of a bilayer of phospholipids with hydrophilic surfaces facing the cytoplasm and extracellular environment (Figure 10.6A). Lipofection technology uses tiny vesicular structures called **liposomes** that have the same composition as the cell membrane (Figure 10.6B). A scientist

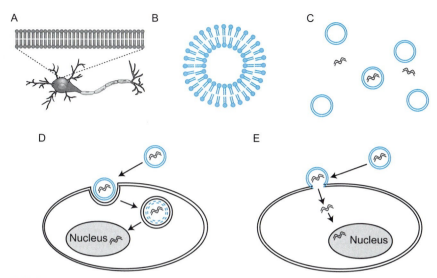

FIGURE 10.6 Lipofection. (A) The cell membrane is composed of a lipid bilayer, with a hydrophobic interior and hydrophilic exterior. **(B)** Liposomes are also composed of a lipid bilayer, arranged as a spherical shell. **(C)** A scientist performs a brief reaction that allows liposomes to form around DNA. **(D)** Cells in culture can endocytose the liposome, digesting it within vesicles to release DNA. **(E)** Alternatively, liposomes can directly fuse with the plasma membrane, directly releasing DNA into cells.

performs a simple reaction that forms a liposome around the DNA sequence to be transfected (Figure 10.6C). Depending on the liposome and cell type, the liposome can be endocytosed (Figure 10.6D) or directly fuse with the cell membrane to release the DNA construct into cells (Figure 10.6E).

The advantage to lipofection is that it works in many cell types, including cultured neurons. Commercially available kits allow transfection reactions to be performed within 30 minutes and gene expression to be assayed within hours. However, like the calcium phosphate method, lipofection is almost exclusively used in cell culture experiments.

VIRAL GENE DELIVERY

Viral gene delivery uses one of many available viral vectors to deliver DNA to cells *in vitro* or *in vivo*. A virus can be thought of as a tiny molecular machine whose entire purpose is to attach to cells and inject genetic material. In nature, this genetic material encodes the proteins necessary to make more virus, so the infected cell essentially becomes a virus-making factory. In the lab, many viruses have been manipulated so that they can no longer replicate inside a host cell on their own. Furthermore, the coding region of viral DNA is exchanged with a gene of interest to make the virus deliver an expression construct to a cell without forcing that cell to produce more virus particles.

This process of using a nonreplicating viral vector to deliver foreign DNA into a cell is called **transduction**.

Scientists can either produce virus in their own labs or outsource viral production commercially (Figure 10.7). To begin, a scientist uses recombinant DNA technology to place a DNA sequence of interest into a plasmid containing the necessary sequences to incorporate into a virus particle. Note that this plasmid is not the same as a virus itself but is simply a piece of DNA with the necessary sequences to incorporate into a virus particle, provided the other necessary proteins are present. In order to produce a batch of viruses, cultured cells are used to produce all the necessary viral components. A scientist transfects these cells with the recombinant plasmid, as well as additional DNA sequences coding for the necessary viral proteins. Because viral DNA has been engineered so that it cannot replicate more virus on its own, these viral proteins are necessary to produce functional, infectious viral particles. Immortalized cell lines, such as HEK 293T cells are the best cells to serve as virus packaging cells, as they grow quickly, are transfected relatively easily, and can produce large amounts of virus. One to two days after transfection, the **packaging cells** produce many virus particles and release them into the extracellular medium. The final step of virus production is to collect the medium, filter out cell debris, and centrifuge the medium to collect the viral pellet. The pellet is dissolved in a buffer and, depending on the needs of the investigator, either used immediately to infect the target cells of interest or frozen until needed.

Viral infections are robust, highly efficient, and can lead to long-term expression. Viruses have been used to deliver genes to cultured cells, brain slices, tissue explants, and brain regions *in vivo*. They are the tools of choice to deliver genes into the brains of animals, as most other methods of gene delivery are

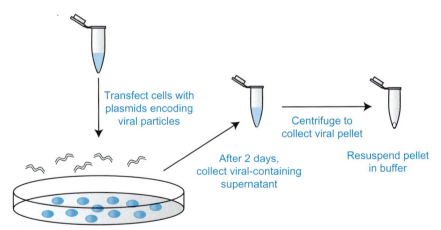

FIGURE 10.7 Virus production. A scientist transfects cultured cells with DNA encoding the necessary proteins to make virus. Over 2–3 days, the cells produce a virus and release it into the culture medium. The medium is collected and centrifuged to collect a viral pellet. The pellet is resuspended in buffer and frozen until use.

incapable of working *in vivo*. However, there are also disadvantages and limitations to using viral gene delivery. The infected cells, especially cells *in vivo*, may not begin expressing the transduced gene until 7–14 days after exposure to the virus. Also, the size of the viral vector restricts the size of the DNA construct that a scientist can transduce. Finally, scientists must take great care and safety when using viruses, as it is possible for scientists to accidentally infect themselves.

There are a variety of viral vectors that are useful to neuroscientists. They vary in terms of infection efficiency, expression levels, duration of expression, time to start of expression, host cell toxicity, and host cell preference. Table 10.2 compares the relative features of each viral vector.

Adenovirus

Adenovirus can infect both dividing and postmitotic cells in a broad range of species and cell types. It can carry DNA constructs that are 7.5–30 kb

TABLE 10.2 Viral Vectors Used in Neuroscience

Vector	Insert Size	Features
Adenovirus	7.5–30 kb	Infects dividing and nondividing cells Transient expression—doesn't integrate in the genome High transduction efficiency Onset of expression in days Triggers immune response
Adeno-Associated Virus (AAV)	<5 kb	Naturally replication-deficient—safe Infects dividing and nondividing cells Stable expression—integrates in the genome High transduction efficiency Onset of expression in weeks Minimal toxicity Labor-intensive production
Lentivirus	8 kb	Form of retrovirus that can infect nondividing cells in addition to dividing cells Stable expression—integrates in the genome Safety concerns
Herpes Simplex Virus (HSV)	>30 kb	Neuron specific Stable, long-term expression—doesn't integrate in the genome Low transduction efficiency Safety concerns Complex production Onset of expression in hours

(kilobases) long. The DNA does not integrate into the genome. Therefore, expression is transient, lasting weeks or months. It is useful for high-level transient expression in neurons, but it can often cause an inflammatory response and lead to cell death.

Adeno-Associated Virus (AAV)

Adeno-associated virus is naturally replication deficient, requiring a helper virus to replicate, so it is safer to use than other viral strains. AAV can infect both dividing and postmitotic cells, but, unlike adenovirus, does integrate into the host genome, permitting long-term gene expression. AAV also seems to be less toxic to neurons than adenovirus. However, AAV production is labor-intensive and the carrying capacity is much less than adenovirus, about 5 kb.

Lentivirus

Lentivirus belongs to a class of virus called retrovirus that has an RNA genome rather than DNA. In order to produce functional gene products, the virus also contains the enzyme reverse transcriptase which produces cDNA from the RNA template (Chapter 9). When a cell endocytoses a lentivirus particle, the RNA is released and reverse transcriptase produces cDNA. The DNA migrates to the nucleus where it integrates into the host genome.

Most retroviruses only infect dividing cells, making them useful for studying neuronal development and cell fate. However, lentivirus is capable of infecting both dividing and postmitotic cells (such as neurons) and is therefore widely used in neuroscience experiments. Lentivirus is based on the HIV virus and has an 8 kb carrying capacity. Because the DNA integrates into the genome, lentivirus delivery leads to long-term expression.

Herpes Simplex Virus (HSV)

Herpes simplex virus is a neurotropic virus that naturally targets neurons as host cells. It has a very large genome compared to other viruses (100–200 kb) that allows delivery of long DNA sequences. While the DNA does not integrate into the host genome, it exists stably in the nucleus to produce long-term (months to years) expression. The disadvantages are that it can be difficult to engineer HSV because of its large size, and it can be toxic to cells. There are also safety concerns because it is highly infectious.

CONCLUSION

This chapter has surveyed the common methods used to deliver DNA sequences into cells. A scientist chooses one method over another based on a number of factors, such as the cell type and the cell's environment, as well as the specific goals of the experiment (Table 10.3). For cell culture experiments,

TABLE 10.3 Choosing a Gene Delivery Strategy

Environment	Considerations	Commonly Used Gene Delivery Methods
Cells in the Intact Nervous System of Living Animals	Cells within the brain are notoriously resistant to most transfection methods.	Electroporation, viruses
Cells in Brain Slices	Gene delivery must be fast and efficient and may need to go through thick tissue.	Electroporation, biolistics
Cells in Dissociated Cultures	Gene delivery must be highly efficient and high throughput to transfect/ transduce thousands or millions of cells.	Electroporation, biolistics, chemical transfection, viruses
Individual Cells	Cells are often valuable, such as extracted embryos or newly fertilized eggs, so care is taken to efficiently deliver DNA to each cell.	Microinjection, electroporation

chemical, electroporation, or viral gene delivery strategies all work well. For *in vivo* DNA delivery to adult animals, viral gene delivery is the most efficient option. Now that we have surveyed these methods, the next two chapters will focus on methods to manipulate the genomes of living organisms.

SUGGESTED READING AND REFERENCES

Books

Twyman, R. M. (2005). *Gene Transfer to Animal Cells*. Garland Science/BIOS Scientific Publishers, Abingdon, Oxon, UK.

Review Articles

Bonetta, L. (2005). The inside scoop—evaluating gene delivery methods. *Nat Meth* **2**, 875–883.

Luo, L., Callaway, E. M. & Svoboda, K. (2008). Genetic dissection of neural circuits. *Neuron* **57**, 634–660.

Washbourne, P. & McAllister, A. K. (2002). Techniques for gene transfer into neurons. *Curr Opin Neurobiol* **12**, 566–573.

Primary Research Articles—Interesting Examples from the Literature

Kitamura, T., Feng, Y., Kitamura, Y. I., Chua, S. C., Jr., Xu, A. W., Barsh, G. S., Rossetti, L. & Accili, D. (2006). Forkhead protein FoxO1 mediates Agrp-dependent effects of leptin on food intake. *Nat Med* **12**, 534–540.

Kohara, K., Kitamura, A., Morishima, M. & Tsumoto, T. (2001). Activity-dependent transfer of brain-derived neurotrophic factor to postsynaptic neurons. *Science* **291**, 2419–2423.

Pekarik, V., Bourikas, D., Miglino, N., Joset, P., Preiswerk, S. & Stoeckli, E. T. (2003). Screening for gene function in chicken embryo using RNAi and electroporation. *Nat Biotechnol* **21**, 93–96.

Tabata, H. & Nakajima, K. (2001). Efficient in utero gene transfer system to the developing mouse brain using electroporation: visualization of neuronal migration in the developing cortex. *Neuroscience* **103**, 865–872.

Wilson, S. P., Yeomans, D. C., Bender, M. A., Lu, Y., Goins, W. F. & Glorioso, J. C. (1999). Antihyperalgesic effects of infection with a preproenkephalin-encoding herpes virus. *Proc Natl Acad Sci USA* **96**, 3211–3216.

Protocols

Lappe-Siefke, C., Maas, C. & Kneussel, M. (2008). Microinjection into cultured hippocampal neurons: a straightforward approach for controlled cellular delivery of nucleic acids, peptides and antibodies. *J Neurosci Methods* **175**, 88–95.

O'Brien, J. & Lummis, S. C. (2004). Biolistic and diolistic transfection: using the gene gun to deliver DNA and lipophilic dyes into mammalian cells. *Methods* **33**, 121–125.

Walantus, W., Castaneda, D., Elias, L. & Kriegstein, A. (2007). In utero intraventricular injection and electroporation of E15 mouse embryos. JoVE. 6. http://www.jove.com/index/details.stp?id=239, doi: 10.3791/239

Chapter 11

Making and Using Transgenic Organisms

After reading this chapter, you should be able to:
- Describe and categorize commonly used transgenes
- Describe the necessary components of a transgenic DNA construct
- Understand the benefits and utility of binary transgenic expression systems
- Explain the process of creating viable transgenic organisms

Techniques Covered
- **Common transgenes in neuroscience**: reporter genes, genes used to ablate neurons, genes used to measure neural activity, genes used to manipulate neural activity, genes used to disrupt endogenous genes
- **Binary transgenic systems:** Gal4/UAS, Cre/lox, Flp/Frt, Tet-off/Tet-on
- **Making transgenic organisms:** mice, flies, worms

One of the modern marvels of genetic engineering technology is the ability to take a gene from one species and place it in the genome of another species. The rich diversity of life on Earth provides an extraordinary variety of genes that not only help organisms survive unique niches on the planet, but also help neuroscientists probe the structure and function of neural systems. For example, the jellyfish *Aequorea victoria* evolved a gene capable of emitting photons of light for bioluminescence. This gene, known as green fluorescent protein (GFP), has been used in thousands of studies to examine the structure and function of the nervous system. A variety of species have evolved other genes useful for visualizing neurons and manipulating their function, such as *luciferase* from fireflies, *channelrhodopsin-2* from single-celled algae, *lacZ* from *E. coli*, or *allatostatin receptor* from insects. Genetic engineering technology provides the ability to express these genes in specific cell types at specific times in an organism's lifespan. Thus, modern genetic engineering technology allows scientists to stably express genetically encoded tools for the precise study of specific neurons and neural circuits.

Any gene expressed in an animal that does not normally carry this gene is called a **transgene**, and any animal that carries this foreign DNA is called a **transgenic organism**. The purpose of this chapter is to survey common transgenes used in modern neuroscience research and to describe how these genes can be restricted to specific cell types at specific times. Furthermore, we will examine the process of creating a transgenic animal in commonly used model organisms including mice, flies, and worms. Chapter 12 will provide additional information on gene targeting technologies used to remove genes from a genome.

TRANSGENES

There are now dozens of commonly used transgenes that can be used to answer a variety of experimental questions. Scientists can inject these transgenes into embryos to create transgenic lines of animals that pass on the transgene to their offspring. Alternatively, these transgenes can be introduced directly into cells or brain regions as described in Chapter 10. We survey some of the major categories of transgenes here and in Table 11.1, though this is not an exhaustive list.

Reporter Genes

A **reporter gene** is any gene whose protein product can report the location of a particular protein, cell type, or circuit. It can be used by itself to report the expression pattern of a specific gene, or it can be used in conjunction with another transgene to report the successful expression of the other transgene and mark cells expressing both constructs.

The most common reporter gene is GFP and its variants. When placed under the control of a cell-specific promoter, GFP can mark the location of these specific cells so they can be used in subsequent experiments (for example, electrophysiology, colocalization experiments, or identification of cells in culture). In addition to GFP, there are dozens of fluorescent proteins with different spectral properties that appear as different colors, such as red fluorescent protein (RFP), cyan fluorescent protein (CFP), and yellow fluorescent protein (YFP).

Fluorescent proteins can be manipulated so they tag proteins or organelles within a cell. For example, a scientist can attach GFP to a gene that encodes a protein of interest so that the location, expression, and trafficking of the protein can be studied *in vitro*. It is also possible to add specific sequences to fluorescent proteins so that they localize to discrete regions of the cell, such as the cell membrane. Membrane-bound fluorescent reporters can be useful for tracing axons and dendrites in addition to cell bodies.

TABLE 11.1 Examples of commonly used transgenes

	Type of Transgene	Utility	Examples
Reporter Genes	Fluorescent proteins	Marks the location of cells, subcellular structures, and organelles	Green fluorescent protein (GFP) and multicolored variants
	Fluorescent proteins to trace neural connections	Addition of membrane tags to fluorescent proteins so they are bound to the membrane	Farnesylated-GFP, Myristolated-GFP
	Bioluminescent proteins	Proteins that produce light as a by-product of a biochemical reaction	Luciferase
	Nonluminescent proteins	Proteins that mark expressing neurons after histological processing	LacZ, Alkaline Phosphatase (AP)
Genes that Cause Cell Death	Toxins	Kills all expressing neurons	Ataxin
	Toxin receptors	When the toxin is applied to the brain, only cells expressing the receptor will die	Diphtheria toxin receptor
Genes for Measuring Neural Activity	Calcium indicators	Sensors that change fluorescence in response to different concentrations of calcium ions	Pericams, Cameleons, Camgaroos, TN-L15
	Voltage sensors	Sensors that change fluorescence in response to changes in membrane potential	FlaSh, Flare, VSFP1, VSFP2, SPARC
	Synaptic activity sensors	Sensors that change fluorescence in response to changes in synaptic vesicle fusion	Synapto-pHluorin
Genes for Inducible Activation of Neural Activity	Optogenetics	Membrane-bound channels that depolarize cells when stimulated by light	Channelrhodopsin-2 (ChR2)

(Continued)

TABLE 11.1 (Continued)

	Type of Transgene	Utility	Examples
	Ligand receptors	Ligand-gated ion channels that depolarize cells when a ligand is added	Capsaicin receptor (TRPV1)
Genes for Inducible Inhibition of Neural Activity	Optogenetics	Membrane-bound pumps that hyperpolarize cells when stimulated by light	Halorhodopsin (NpHR)
	Ligand receptors	Ligand-gated ion channels that hyperpolarize cells when a ligand is added	Allatostatin receptor (Alstr)
	Shibire mutants	Temperature sensitive mutants of the *Drosophila* shibire gene (dynamin) that prevent synaptic transmission at certain temperatures	Shi^{ts}

Many investigators use *lacZ* as a reporter gene. *LacZ* is a bacterial gene that encodes the enzyme β-galactosidase. This enzyme reacts with a substrate, X-gal, to create a dark blue product. Therefore, *lacZ* can serve as a faithful reporter of gene expression after histological processing with X-gal.

Genes Used to Ablate Neurons

Historically, lesion studies have been a staple of neuroscience research to investigate the necessity of a brain region or specific group of neurons for a specific behavior or brain function, or to model a human disease that involves the death of specific neural subtypes (for example, Parkinson's disease). However, using physical or pharmacological methods to ablate neurons raises concerns about the specificity and totality of the lesions. Transgenic technology allows lesions to be targeted to specific, genetically defined subsets of neurons. Genetic ablation studies can be performed using the *ataxin* transgene, which encodes a toxin that kills all expressing neurons. It is also possible to use a transgene that encodes a toxin's receptor, such as the diptheria toxin receptor. This receptor is toxic to neurons, but only when diptheria toxin is injected into the brain. Neurons that do not express the receptor will not be sensitive to the toxin. Using a genetically encoded toxin receptor instead of a genetically encoded toxin itself allows a scientist to ablate neurons at any time-point during an animal's lifespan. By controlling when the toxin is injected, the scientist controls the timing of ablation.

Genes Used to Measure Neural Activity

Chapter 7 described methods to image calcium dynamics and the membrane potential for use as an indirect measure of neural activity. Transgenes-encoding calcium indicators, such as *Pericam* and *Cameleon*, or voltage-sensitive fluorescent indicators, such as FlaSh, allow targeted observation of neural activity in specific cell types. This specificity is an advantage over the use of dyes that label all neurons in a region.

Genes Used to Manipulate Neural Activity

Neural activity can be manipulated through nongenetic methods, such as electrical stimulation or pharmacological inhibition. However, these methods rarely stimulate or inhibit all cells of a desired subtype, and they often affect undesired subtypes. Transgenes that allow inducible activation or inhibition of neural activity solve these problems by providing control over genetically targeted cells, with no effect on cells that do not express the transgene. These transgenes use either optically controlled or ligand-gated channels to selectively stimulate or dampen neural activity.

Chapter 7 described two optically controlled transgenes used to manipulate neural activity. Channelrhodopsin-2 (ChR2), originally derived from single-celled algae, is a cation channel that can depolarize a neuron upon stimulation with blue light. Because ChR2 can be placed under cell-specific promoters, it is possible to stimulate specific cell types in a heterogeneous population of neurons by delivering blue light to those neurons. The genetically encoded membrane-bound chloride pump Halorhodopsin (NpHR) hyperpolarizes a neuron upon stimulation with yellow light. Using this transgene, it is possible to silence single action potentials or sustained trains of neuronal spiking.

Another method of manipulating neural activity using genetically encoded transgenes is to express a ligand-gated ion channel in the brain of an animal that normally does not express this channel. For example, the capsaicin receptor (TrpV1) is a ligand-gated channel that depolarizes a neuron upon exposure to capsaicin. Flies and other invertebrates do not normally express this channel, so capsaicin exposure will depolarize only those neurons that express the receptor. A similar strategy can be used to inhibit activity and hyperpolarize neurons in the presence of a ligand. For example, the insect allatostatin receptor (Alstr) is not expressed in mammals. In the presence of the insect protein allatostatin, this receptor inhibits neural activity. Thus, if Alstr is expressed in the mammalian brain, addition of allatostatin will reversibly inactivate genetically targeted populations of neurons until the ligand is washed out from the system.

Genes Used to Disrupt Endogenous Gene Function

In mice, it is possible to knockout genes from the genome using gene targeting technology, as will be discussed in Chapter 12. This technology is specific to mice and currently not possible in other model organisms. However, two methods that are commonly used to disrupt endogenous gene or protein function can be achieved using transgenic technology. The overexpression of **dominant negative** transgenic constructs directly interferes with protein function, while **RNA interference** (RNAi) technology can be used to knock down the expression of a gene by targeting mRNA before it is translated into a functional protein. We will discuss dominant negatives, RNAi, and how they work in Chapter 12, but it is worth mentioning in the context of transgenic organisms, as these techniques allow scientists to express a transgene that can disrupt the function of any gene. Thus, although it is currently impossible to create "knockout flies" in the same way as creating knockout mice, it is possible to create a transgenic fly in which a protein's function is disrupted by expressing a dominant negative construct or a gene of interest is effectively knocked down by expressing an RNAi construct.

Overexpression or Misexpression of Endogenous Genes

Some laboratories produce transgenic mice that overexpress an endogenous gene or express the gene in a cell-type that normally does not express it. For example, a scientist wishing to perform a gain-of-function experiment in which a gene of interest is expressed at a much higher level than normal could create a transgenic animal that highly overexpresses the gene. Scientists can also show that overexpression of a gene in a knockout background can rescue the gene's function. Alternatively, a scientist can show that expressing a gene is sufficient to cause a particular function. For example, he or she may hypothesize that a gene encoding a cell-surface receptor is important during development to attract axons from a different brain region. A transgenic mouse that expresses the receptor in a different cell type could be used to test whether the presence of this receptor in other cells is sufficient to cause the axons to target the wrong region.

Now that we have surveyed many commonly used transgenes, we examine how a scientist creates a transgenic construct and incorporates foreign DNA into an animal's genome.

THE TRANSGENIC CONSTRUCT

The first step in producing a transgenic organism is to use recombinant DNA technology (Chapter 9) to make a transgenic construct. This construct must contain (1) the DNA sequence encoding the transgene and (2) the necessary **promoter** sequences for the expression of the gene (Figure 11.1). Simply inserting a gene into the genome will not cause it to be expressed: endogenous cellular machinery must recognize a promoter sequence in order to transcribe the DNA into mRNA.

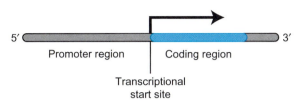

5′ Promoter region | Coding region 3′

Transcriptional
start site

FIGURE 11.1 A generic transgenic construct. A construct is composed of the transgene (coding region) and promoter region.

Promoters control the expression of genes so they are expressed in specific cells and tissues at specific times. For example, every cell in the human body contains the gene encoding the enzyme tyrosine hydroxylase (TH). However, this gene is only expressed in a subset of neurons in the brain because the promoter elements of the TH gene only recruit the **transcription factors** and other regulatory proteins needed to transcribe TH in that subset of cells. Therefore, a transgenic construct containing all the necessary promoter elements for the TH gene will allow a transgene to be expressed specifically in the subset of cells that normally transcribe TH. The promoters of endogenous genes expressed in cells throughout the entire brain can be used to drive expression of transgenes in cells throughout the entire brain. Promoters of genes expressed in specific cell types can be used to drive expression in just those cell types.

Promoter sequences are often depicted as existing just upstream of the genes they regulate (Figure 11.1). However, this is not necessarily the case, especially in mammals. Promoter elements necessary to drive a gene of interest may be located thousands or even hundreds of thousands of base pairs upstream of a gene. They may be located in the **introns** within the coding sequences of a gene or even downstream of the coding sequence of the gene. This phenomenon restricts the promoters that can be used to drive transgene expression. In general, it is easier to use promoters to target specific cell-types in invertebrate model organisms as promoters in these species are often located just upstream of the genes they regulate.

Because mammalian promoters can be difficult to isolate and place in a transgenic construct, some laboratories make transgenic mice using **BAC constructs**. BAC stands for **bacterial artificial chromosome**, a large 150–200 kb vector that can be used to create genetic libraries and clone mammalian DNA sequences (Chapter 9). Investigators can order a BAC that contains any gene in the genome, flanked by 50–100 kb of genomic sequences on either side of the gene. Thus, a transgenic construct could be produced by inserting a transgene at the **transcriptional start site** of the endogenous gene in a BAC, either in addition to or replacing the endogenous gene. The transgene would then be located at the same location as the endogenous gene in relation to promoter elements. The limitation to using this technology is that recombinant DNA technology in BACs can be difficult due to the large size of the BAC compared to other expression vectors that are usually 3–10 kb in length.

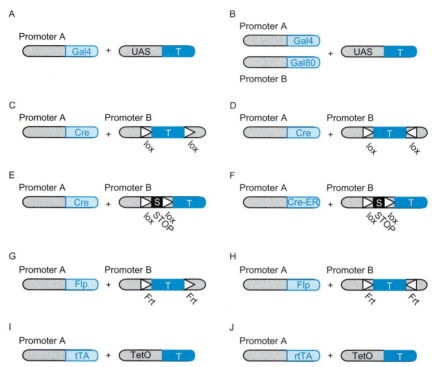

FIGURE 11.2 Binary expression systems. These systems use two or more constructs to spatially or temporally restrict transgene expression. **(A)** The Gal4/UAS system. Promoter A drives expression of the transcription factor Gal4. Gal4 binds to UAS and activates expression of Transgene T. Therefore, T expression is indirectly controlled by Promoter A. **(B)** The Gal4/Gal80/UAS system. As in **(A)**, Promoter A drives expression of Gal4, which binds to UAS and activates expression of Transgene T. Promoter B drives expression of Gal80, which inactivates Gal4. Therefore, T will be expressed in cells in which Promoter A is active but not in cells in which Promoter B is active. **(C)** The Cre/lox system. Promoter A drives the expression of Cre recombinase and Promoter B drives the expression of Transgene T. Cre recognizes lox sites and catalyzes a recombination event that removes DNA within the sites. Therefore, T will be expressed in cells in which Promoter B is active but not in cells in which Promoter A is active. **(D)** The Cre/lox system with inverted lox sites. When Cre recognizes two lox sites facing each other, Cre rearranges the DNA between the lox sites so that it becomes "flipped" in the reverse orientation. If a transgenic construct is produced in which the transgene is oriented in the opposite direction, a Cre construct can be used to flip the transgene in the correct orientation. Thus, Transgene T will be expressed in cells in which both promoters A and B are active. **(E)** The Cre/lox system used to remove a stop site. Cre is used to remove a stop codon that prevents the expression of Transgene T. Therefore, T is expressed in cells in which promoters A and B are active. **(F)** Promoter A expresses Cre recombinase fused to the estrogen receptor (ER). ER sequesters Cre in the cytoplasm until a scientist adds tamoxifen, which binds to ER and causes Cre to translocate to the nucleus, where it can remove a stop codon preventing expression of Transgene T. Therefore, T is expressed in cells in which Promoters A and B are active but only when tamoxifen is added. **(G)** The Flp/Frt system. Promoter A drives the expression of Flippase recombinase (Flp), and Promoter B drives the expression of Transgene T. Flp recognizes Frt sites and catalyzes a recombination event that removes DNA within the sites. Therefore, T will be expressed in cells in which Promoter B is active, but not in cells in which Promoter A is active. **(H)** The Flp/Frt system with inverted Frt sites. When Flp recognizes two Frt

Whether using small expression vectors or BACs, scientists make DNA constructs in DNA vectors that are circular, as described in Chapter 9. However, before injecting the transgene into cells to produce transgenic animals, the DNA is almost always made linear by using restriction enzymes to isolate the promoter/transgene sequence away from the vector. This helps the promoter/transgene region integrate into the genome as a linear fragment, as well as discards unnecessary coding sequences that may be in the vector, such as antibiotic resistance genes.

Transgenes typically integrate into the genome in random locations. Though there are certain regions of the genome that appear to be "hot-spots" of integration activity, scientists generally cannot control where the transgene integrates unless using specific gene-targeting methods described in Chapter 12. Because of this random integration, **position effects** can occur that influence the transgene's expression pattern. Depending on where the transgene lands in the genome, neighboring genes and regulatory sequences can alter the levels and pattern of expression. The number of transgene copies that integrate into the genome can also vary the levels and pattern of transgene expression. Therefore, transgenic organisms must be carefully evaluated to ensure that transgenes are expressed in the desired cell types. Just because a specific promoter is used to drive transgene expression, there is no guarantee that the transgene will be expressed in the desired cells.

BINARY TRANSGENIC SYSTEMS

The simplest method of making a transgenic animal is to use a single DNA construct, as just described, in which a promoter directly regulates the expression of a transgene. An alternate approach is a **binary expression system** that uses multiple constructs to further refine the spatial and temporal expression of a transgene (Figure 11.2 and Table 11.2). This is typically done by creating two different lines of transgenic animals, each line expressing one of the required constructs, and then mating the animals to produce offspring that express both constructs. These flexible expression systems have numerous advantages. The main advantage is that they can provide control over the timing and location

---

sites facing each other, Flp rearranges the DNA between the Frt sites so that it becomes "flipped" in the reverse orientation. If a transgenic construct is produced in which the transgene is oriented in the opposite direction, a Flp construct can be used to flip the transgene in the correct orientation. Thus, Transgene T will be expressed in cells in which both promoters A and B are active. **(I)** The Tet-off system. Promoter A drives the expression of the tetracycline transactivator (tTA) protein. tTA binds to the tetO promoter, activating expression of Transgene T. Therefore, expression of T is indirectly controlled by Promoter A. If a scientist applies a drug called doxycycline, tTA is reversibly inhibited. **(J)** The Tet-on system. Promoter A drives the expression of the reverse tetracycline transactivator (rtTA). rtTA is unable to bind to the tetO promoter unless doxycycline is present. Therefore, Promoter A indirectly activates expression of Transgene T but only when a scientist adds doxycycline.

TABLE 11.2 Promoters in Binary Transgenic Systems

Binary Transgenic System	Cell and Tissue Specificity Is Conferred by the Promoter(s) Regulating	The Gene of Interest Is Directly Regulated by the Promoter Called
Gal4/UAS	Gal4	UAS
Gal4/Gal80/UAS	Gal4 and Gal80	UAS
Cre/lox	Cre	Any promoter can be used.... Lox sites determine the location of DNA recombination
Flp/Frt	Flp	Any promoter can be used.... Frt sites determine the location of DNA recombination
tTA/tetO	tTA	tetO
rtTA/tetO	rtTA	tetO

of transgene expression. Another primary advantage is the ability to mix and match combinations of transgenes, saving time and effort in making new transgenic constructs.

The Gal4/UAS System

The Gal4/UAS system is a binary expression system primarily used in *Drosophila*, though it has also been applied to mice and zebrafish. In one construct, a cell-specific promoter drives the expression of the gene encoding Gal4, a transcription factor normally expressed in yeast. In a second construct, a transgene of interest is regulated by a promoter sequence called an upstream activation sequence (UAS). The Gal4 protein binds to the UAS sequence and highly expresses the transgene (Figure 11.2A). Thus, the transgene will only express in cells defined by the promoter regulating Gal4.

Why not simply place the transgene under the control of the promoter regulating Gal4? There are at least two reasons. First, the UAS expresses transgenes at much higher levels than endogenous promoters, so the Gal4/UAS constructs together allow for high levels of gene expression. Second, Gal4 lines now exist for hundreds of genes in the *Drosophila* genome. There are also many constructs in which a UAS drives a specific transgene. Therefore, the *Drosophila* scientific community has created an amazingly efficient system of creating and delivering transgenes to many different cell types (Figure 11.3).

Gal4/UAS can also be used in combination with another yeast protein, Gal80. The Gal80 protein binds to and inhibits the activity of Gal4. Thus,

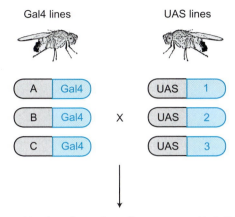

Any combination of genetic coding sequence (1, 2, 3, etc.)
driven by cell-specific promoter (A,B,C, etc.)

FIGURE 11.3 The utility of the Gal4/UAS system. Scientists have created hundreds of fly stocks in which endogenous promoters drive cell-specific expression of Gal4 transcription factors. In turn, any transgene of interest can be placed under the control of a UAS sequence. Therefore, a scientist can cross any combination of Gal4 line with a UAS line to drive transgene expression in specific cell types. This system has been used extensively in the fly as well as in *Xenopus* and zebrafish model organisms.

a second promoter element can drive expression of Gal80 to further restrict the cells expressing the transgene (Figure 11.2B). Furthermore, Gal80 can be made temperature sensitive: the Gal80^{ts} protein is active at 19° C but not 30° C. Therefore, it is possible to express a transgene in specific cells using the Gal4/UAS system and to regulate the timing of transgene expression by controlling the temperature, and thus the expression of Gal80^{ts}.

The Cre/Lox System

The **Cre/lox system** is a site-specific recombination system widely used in mice. The enzyme **Cre recombinase**, originally derived from the P1 bacteriophage, recognizes specific 34 base-pair DNA sequences called **lox** sites. Lox sites are added to a transgene so that they flank a sequence of DNA. This is referred to as a **"floxed"** sequence, a sequence *f*lanked by *lox* sites. Each lox site has an asymmetric sequence in the middle, providing the sequence with directionality. This directionality is important because Cre processes DNA depending on the orientation of the lox sites:

- Cre excises any DNA sequences between two lox sites oriented in the same direction (Figure 11.2C)
- Cre inverts any DNA sequences between two lox sites oriented in inverted directions (Figure 11.2D)

Therefore, the Cre/lox system allows for complex rearrangement of DNA in transgenic mice. Many transgenic lines now exist in which Cre recombinase is regulated by cell-specific promoters. In fact, the NIH is sponsoring a project (called GENSAT) that will create transgenic mice expressing Cre for each gene in the genome. The ability to engineer various combinations of lox sites and specific gene sequences provides unique experimental possibilities. For example, a transgenic mouse can selectively express a transgene with a "floxed stop" strategy: a "stop" codon is placed directly upstream from a transgene, preventing the transgene's expression. If two lox sites flank the "stop" codon, the transgene will only be expressed in cells that express Cre to remove the "stop" (Figure 11.2E).

Investigators can achieve temporal control of Cre/lox technology by fusing the gene encoding Cre recombinase with the gene encoding the **estrogen receptor** (ER). The estrogen receptor localizes to the cytoplasm, but translocates to the nucleus after binding to its ligand. The ER sequence fused to Cre is altered so it preferentially binds to the chemical **tamoxifen**, an estrogen antagonist, rather than endogenous estrogen present in an animal. In baseline conditions, Cre-ER fusion proteins are sequestered in the cytoplasm and unable to access lox sites in the DNA located in the nucleus. If a scientist injects tamoxifen into a mouse, it will bind to Cre-ER, causing it to translocate to the nucleus and recombine DNA (Figure 11.2F).

Another method for temporal and spatial control over Cre-mediated excision is the direct introduction of Cre transgenes through gene delivery methods as described in Chapter 10. Cre transgenes can be delivered to a floxed animal through electroporation or viral transduction, or to cells cultured from a floxed animal using *in vitro* methods of gene delivery.

The Flp/Frt System

The **Flp/Frt system** is a site-specific recombination system analogous to the Cre/lox system just described. The enzyme **Flippase recombinase** (Flp), originally derived from yeast, recognizes specific 34 base-pair DNA sequences called **Flippase recognition targets** (Frts), excising DNA between identical Frt sites (Figure 11.2G) and inverting DNA between inversely oriented Frt sites (Figure 11.2H). This system is used in both flies and mice.

The Tet-off/Tet-on System

The **Tet-off/Tet-on** system is a tool used in the fly and mouse to temporally control transgene expression. In the Tet-off system, a promoter drives the expression of the tetracycline transactivator (tTA) protein. This protein binds to another DNA sequence called the tetO operator, which regulates the expression of a transgene. Thus, in normal conditions, the transgene expression is regulated by the promoter that regulates tTA. However, a scientist can feed or inject a derivative of the antibiotic tetracycline called **doxycycline** into these transgenic mice. Doxycycline binds to tTA and renders it incapable of binding to tetO. Therefore, expression of the transgene can be reversibly inhibited with the addition of doxycycline (Figure 11.2I).

The Tet-on system works in the opposite way. A reversed tTA (rtTA) is only capable of binding to tetO when doxycyline is present. Therefore, the addition of doxycyline causes transcription of the transgene (Figure 11.2J). Like the Tet-off system, the effect of doxycycline is reversible.

These systems have gained popularity for their reversibility and temporal control, but also for the ability to control levels of transgene activation. Increasing the concentration of doxycycline in the Tet-off system gradually lowers and then shuts off transgene expression, while increasing concentrations of doxycycline in the Tet-on system gradually increases transgene expression.

MAKING TRANSGENIC ORGANISMS

After using recombinant DNA technology to produce a transgenic construct, an investigator must deliver the construct to recently fertilized eggs so that it can integrate into the genomes of the offspring. This section describes the process of creating transgenic animals in mice, flies, worms, and other animals.

Making Transgenic Mice

To make a transgenic mouse, a foreign piece of DNA is injected into the nucleus of a newly fertilized egg. In some of the new eggs, the transgene randomly incorporates into the mouse genome. After the egg develops into a fetus and eventually a postnatal offspring, the transgene may become stably expressed and can then be transferred to future offspring. In many institutions, individual investigators do not inject the transgenic construct into eggs themselves but submit their constructs to a specialized transgenic animal facility that has expertise in creating transgenic mice. As opposed to flies or worms, the harvesting and injection of recently fertilized eggs is a technically challenging process that most laboratories choose to outsource.

The transgenic construct is injected as a linear piece of DNA into the larger male **pronucleus** of a recently fertilized mouse egg (Figure 11.4). A pronucleus is the haploid nucleus of either the egg or sperm, both present in the fertilized egg prior to their fusion. This is the stage at which insertion of foreign DNA into the genome can occur. Because the embryo is at the one-cell stage, the incorporated gene is replicated and can end up in nearly all of the animal's cells in adulthood. In order for the transgene to be heritable, it must be incorporated into the germline (eggs or sperm).

After several embryos have been injected with the transgene, they are subsequently injected into a **pseudopregnant** mouse (Figure 11.5). The term *pseudopregnant* refers to the fact that the female is not actually pregnant but has been tricked into thinking she is pregnant by mating with a sterile male mouse. This is done to produce the hormones and physiological changes necessary

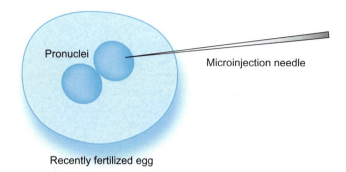

FIGURE 11.4 Injection of a transgenic construct in mice. To make a transgenic mouse, a scientist microinjects a transgenic construct into one of two pronuclei in a recently fertilized egg.

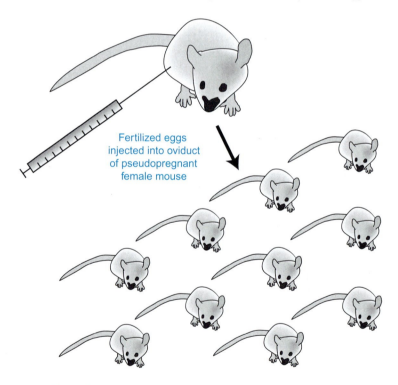

Some offspring will properly express the transgene, others will not

FIGURE 11.5 Injection of embryos to make transgenic mice. After injecting embryos with a transgenic construct, these embryos are subsequently injected into the oviduct of pseudopregnant female mice. Each female typically produces 6–10 offspring. However, only a fraction of the off-spring, usually around one-third of them, successfully express the transgene.

to see the embryo through to a viable fetus. The mouse eventually carries the embryos to term, and transgenic mice are born. Each transgenic facility uses different procedures, but often 2 to 4 pseudopregnant females are used to produce 20 to 30 transgenic offspring.

These transgenic offspring are called **founders**, as they represent the first generation of a new line of mice. However, not all offspring will have successfully incorporated the transgene, and those that do may not express the transgene in the desired cells and tissues. Therefore, once mice are born, the investigator must detect which mice have incorporated the transgene by genotyping the animals with a Southern blot and/or PCR (Chapter 9). Southern blots are required to fully characterize the number of transgene copies and verify the integrity of the transgene sequence. However, PCR of genomic tail DNA is a rapid way to reveal if the transgene is present and is the standard method for genotyping future offspring. Any mice that test negative for the transgene are discarded. Mice that test positive for the transgene are mated with wild-type mice so their offspring can be further investigated. These mice are systematically examined for the presence of the transgene and rigorously tested to ensure that the gene is expressed in the appropriate tissues. Because copy number and integration position can affect the expression pattern of a transgene, multiple founder lines are characterized. The total time necessary to produce and analyze the validity of a transgenic mouse is usually about 6–9 months, including the time necessary for breeding, gestation, development, and analysis of genotype and phenotype in the lab.

Making Transgenic Flies

The process of making a transgenic fly takes advantage of specialized, endogenous segments of DNA called **transposons** (also called **transposable elements**) that can move from one position to another within the genome. These mobile genetic sequences have only modest target site selectivity within the genome and can therefore insert themselves into random genomic locations. In **transposition**, an enzyme called **transposase** acts on specific DNA sequences at the end of a transposon, first disconnecting it from the flanking DNA and then inserting it into a new target DNA site.

The most commonly used transposons in *Drosophila* are called **P elements**. To make a transgenic fly, a DNA construct is first inserted between the two terminal sequences of the *Drosophila* P element (Figure 11.6). The terminal sequences enable the P element to integrate into *Drosophila* chromosomes when the P element transposase enzyme is also present. The investigator injects the transgenic construct into a very young fruit fly embryo along with a separate plasmid containing the gene encoding the transposase (Figure 11.7). If this technique is performed correctly, the injected gene enters the germ line as the result of a transposition event.

The ability to quickly produce transgenic flies, as well as the ability to express transgenes in specific cell types at specific times, have made *Drosophila*

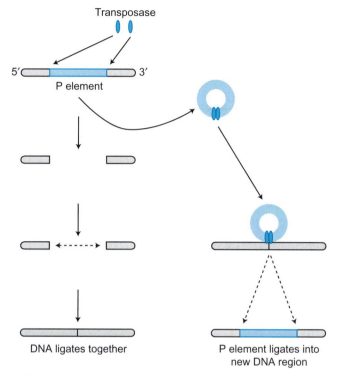

FIGURE 11.6 **P elements.** The transposase enzyme splices a P element out of its genetic environ-ment (the genome or a transgenic construct), inserting it into a new location in the genome.

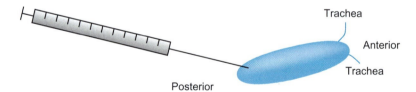

FIGURE 11.7 **Injection of a transgenic construct in flies.** To make transgenic flies, a scientist microinjects a transgenic construct into the posterior end of a fresh *Drosophila* embryo.

an outstanding and highly used model organism to study the nervous system (Box 11.1). In fact, scientists interested in mammalian behaviors and diseases continue to use *Drosophila* as a genetic model to gain knowledge about spe-cific genes and proteins that can be subsequently applied to future investigations using mammalian models.

Making Transgenic Worms

Transgenic techniques in *C. elegans* are conceptually similar to transgenic techniques in flies. An important difference is that worms are hermaphrodites

BOX 11.1 The Awesome Power of Fly Genetics

Scientists sometimes refer to "the awesome power of fly genetics" when describing the arsenal of genetic techniques available to use in *Drosophila*. Why is the fly considered to be such a tractable model organism? Aren't mice, worms, and other species also great model organisms for using transgenic technology?

The answer is that the fly seems to have not only a unique collection of genetic tools, but also a dedicated community of thousands of scientists who have collectively developed and validated these techniques over the past few decades. Consider the following advantages to using *Drosophila* as a model organism:

- It has a relatively short generation time (about 10–11 days at room temperature), so a scientist can create and study transgenic flies within weeks.
- It has a high fecundity: females can lay hundreds of eggs after fertilization.
- The housing and maintenance of *Drosophila* stocks require little space and equipment. Therefore, a lab can study multiple lines of transgenic or mutant flies at an extremely low cost.
- *Drosophila* can be used to study many different behaviors, including addiction, sleep, circadian rhythms, learning and memory, and sexual courtship, as well as behaviors that directly examine sensory and motor processes. Flies can also be used to model human diseases, such as Parkinson's disease, Huntington's disease, spinocerebellar ataxia, and Alzheimer's disease.
- Promoter and enhancer elements are well characterized in *Drosophila*, allowing scientists to target transgenes to very specific cell types.
- The *Drosophila* scientific community has, over many years, cultivated and advanced the Gal4/UAS system as a preeminent tool to drive many different transgenes in many different cell types. Using this binary delivery system, a scientist could deliver a specific transgene to a specific cell type without even having to make original targeting constructs, as hundreds of Gal4 and UAS lines are commonly available.
- Other genetic tricks, such as the addition of Gal80 constructs to the Gal4/UAS system, allow scientists to express transgenes at specific times throughout the fly life cycle.

These advantages offer unique experimental possibilities to neuroscientists. Of course, *Drosophila* also has relative disadvantages, such as limited cognitive abilities and a very different brain architecture compared with mammals. However, the combination of a large number of genetic tricks, a short breeding time, an inexpensive maintenance cost, and a dedicated research community all demonstrate why the study of the brain of a fruit fly has been so beneficial to the study of neuroscience.

and capable of carrying their own fertilized eggs. A transgene is injected into the gonad of a young adult hermaphrodite at the time the embryos are dividing. The gonad runs the length of the worm on the dorsal side, and, with practice, an investigator can become skilled at such injections (Figure 11.8). The transgene can either be carried as an extrachromosomal array (though worms will not necessarily carry the gene in the next generation), or if the worm is

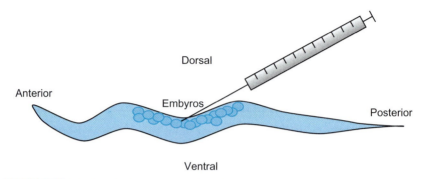

FIGURE 11.8 Injection of a transgenic construct in worms. To make transgenic worms, a scientist microinjects a transgenic construct into a worm's dorsal surface at the time embryos are dividing.

subjected to gamma or UV-irradiation, it can be integrated into a chromosome to make transgene expression more stable and less variable.

To ensure that the transgenic procedure was successful, investigators often use a co-injection marker that allows for easy identification of transgenic animals. For example, the investigator can co-inject the transgene with another construct encoding GFP. To avoid potential interference with the transgene of study, the co-injection marker can be controlled by a promoter that causes expression in a nonoverlapping, different cell-type. GFP can then serve as a positive control to verify that the transgenic manipulation was successful.

Transgenic techniques in *C. elegans* are good for studying gain-of-function (sufficiency), rescuing mutant lines, expressing reporter genes, and expressing RNAi constructs. Because there are many promoter elements that allow targeting to individual neurons, *C. elegans* is a powerful genetic model system used throughout neuroscience to study the genetic basis of development and behavior.

Making Other Transgenic Organisms

Although mice, flies, and worms are highly used genetic model organisms, in theory, a scientist can insert a transgene into any species by correctly injecting DNA into a recently fertilized egg. For example, different species of fish have stably expressed transgenes (including "GloFish," the first transgenic organism to be sold as pets). Cows, pigs, goats, and sheep have also expressed transgenes. A transgenic cow has been created that can produce human protein–enriched milk, a more nutritionally fortified product than regular bovine milk that can be provided to babies or the elderly with special nutritional needs. Other transgenic cows produce milk with less lactose or lower cholesterol. Products such as insulin, growth hormone, blood anticlotting factors, and so forth can also be produced in transgenic animals and collected through animal milk. Thus, transgenic technology is useful not only for studying the biology of organisms, but also for producing nutritional resources and biological reagents that can benefit the health of the human population.

CONCLUSION

This chapter has surveyed the amazing variety of transgenes useful for studying neurons and neural circuits, describing how these transgenes are stably inserted into the genomes for a variety of model organisms. Laboratories continually report new transgenes each year that add important tools to the genetic toolkit. For example, the *Channelrhodopsin-2* and *VSFP2* transgenes have only been reported within the past few years, and now multiple labs are expressing these genes in specific cell types to answer important questions about the nervous system. The next chapter discusses another method of studying genes and proteins, the process of disrupting their functional expression. One way to disrupt their expression is through genetic targeting, in which transgenes are delivered to exact locations in the genome to manipulate endogenous genes. These technologies serve as a compliment to the transgenic technologies mentioned in this chapter, and further expand the neuroscientist's toolkit of genetic techniques.

SUGGESTED READING AND REFERENCES

Books

Greenspan, R. J. (2004). *Fly Pushing: The Theory and Practice of Drosophila Genetics, 2nd ed.* Cold Spring Harbor Laboratory Press, Cold Spring Harbor, NY.

Pinkert, C. A. (2002). *Transgenic Animal Technology: A Laboratory Handbook*, 2nd ed. Academic Press, San Diego.

Review Articles

Barth, A. L. (2007). Visualizing circuits and systems using transgenic reporters of neural activity. *Curr Opin Neurobiol* **17**, 567–571.

Duffy, J. B. (2002). GAL4 system in Drosophila: a fly geneticist's Swiss army knife. *Genesis* **34**, 1–15.

Knopfel, T., Diez-Garcia, J. & Akemann, W. (2006). Optical probing of neuronal circuit dynamics: genetically encoded versus classical fluorescent sensors. *Trends Neurosci* **29**, 160–166.

Luo, L., Callaway, E. M. & Svoboda, K. (2008). Genetic dissection of neural circuits. *Neuron* **57**, 634–660.

Welsh, D. K. & Kay, S. A. (2005). Bioluminescence imaging in living organisms. *Curr Opin Biotechnol* **16**, 73–78.

Yamamoto, A., Hen, R. & Dauer, W. T. (2001). The ons and offs of inducible transgenic technology: a review. *Neurobiol Dis* **8**, 923–932.

Zhang, F., Aravanis, A. M., Adamantidis, A., de Lecea, L. & Deisseroth, K. (2007). Circuit-breakers: optical technologies for probing neural signals and systems. *Nat Rev Neurosci* **8**, 577–581.

Primary Research Articles—Interesting Examples from the Literature

Boyden, E. S., Zhang, F., Bamberg, E., Nagel, G. & Deisseroth, K. (2005). Millisecond-timescale, genetically targeted optical control of neural activity. *Nat Neurosci* **8**, 1263–1268.

Conti, B., Sanchez-Alavez, M., Winsky-Sommerer, R., Morale, M. C., Lucero, J., Brownell, S., Fabre, V., Huitron-Resendiz, S., Henriksen, S., Zorrilla, E. P., de Lecea, L. & Bartfai, T. (2006). Transgenic mice with a reduced core body temperature have an increased life span. *Science* **314**, 825–828.

Feng, G., Mellor, R. H., Bernstein, M., Keller-Peck, C., Nguyen, Q. T., Wallace, M., Nerbonne, J. M., Lichtman, J. W. & Sanes, J. R. (2000). Imaging neuronal subsets in transgenic mice expressing multiple spectral variants of GFP. *Neuron* **28**, 41–51.

Gong, S., Zheng, C., Doughty, M. L., Losos, K., Didkovsky, N., Schambra, U. B., Nowak, N. J., Joyner, A., Leblanc, G., Hatten, M. E. & Heintz, N. (2003). A gene expression atlas of the central nervous system based on bacterial artificial chromosomes. *Nature* **425**, 917–925.

Livet, J., Weissman, T. A., Kang, H., Draft, R. W., Lu, J., Bennis, R. A., Sanes, J. R. & Lichtman, J. W. (2007). Transgenic strategies for combinatorial expression of fluorescent proteins in the nervous system. *Nature* **450**, 56–62.

Marella, S., Fischler, W., Kong, P., Asgarian, S., Rueckert, E. & Scott, K. (2006). Imaging taste responses in the fly brain reveals a functional map of taste category and behavior. *Neuron* **49**, 285–295.

Protocols

Berkowitz, L. A., Knight, A. L., Caldwell, G. A., Caldwell, K. A. (2008). Generation of Stable Transgenic C. elegans Using Microinjection. JoVE, 18, http://www.jove.com/index/Details.stp?ID5833, doi: 10.3791/833.

Websites

Gensat Project: http://www.gensat.org/index.html

Human Genome ed. site: http://genome.wellcome.ac.uk/doc_wtd021044.html

Transgenic Fly Virtal Lab: http://www.hhmi.org/biointeractive/clocks/vlab.html

Chapter 12

Manipulating Endogenous Genes

After reading this chapter, you should be able to:

- Discuss methods that can be used for genetic loss-of-function experiments *in vitro* and *in vivo*
- Compare methods that disrupt genes or gene products, including genomic targeting, RNA interference (RNAi), and dominant negatives

Techniques Covered

- **Direct gene targeting:** knockouts, knockins, conditional knockouts (Cre/lox, FLP/FRT)
- **Disrupting gene products:** RNA interference (RNAi), morpholinos, dominant negatives

There are approximately 30,000 genes in the genome of humans and other mammals, such as rats and mice. At least two-thirds of these genes are expressed in the brain, with thousands of genes *exclusively* expressed in the nervous system. Therefore, it is reasonable to conclude that there are many hundreds, if not thousands, of genes whose main biological function seems to be regulating the development and function of the brain, spinal cord, and peripheral nerves. One goal of modern neuroscience research is to determine which genes are necessary and/or sufficient for a neurobiological process to occur, such as the development of an axon toward its postsynaptic target or the excitation of a population of neurons important for a specific behavior.

In Chapter 8, we surveyed methods of identifying and discovering which genes may be important for a biological process. One of these methods is an approach called **forward genetics,** in which scientists generate random mutations in many genes in order to discover which DNA sequences may be important for a certain phenotype. Often, the opposite approach is desirable; a scientist

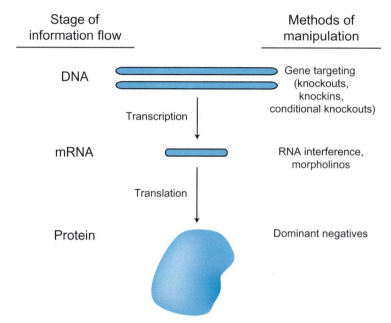

FIGURE 12.1 Methods of genetically perturbing endogenous genes and proteins. A scientist can manipulate gene and protein function at each stage of information processing. Gene-targeting techniques can be used to alter the genome, RNA interference and morpholino methods can be used to functionally block mRNA molecules, and other techniques, such as dominant negative constructs, can block protein function.

hypothesizes that a specific gene is necessary for a phenotype and wishes to produce an animal lacking the expression of that gene. This is called **reverse genetics** and is the subject of this chapter.

The purpose of this chapter is to describe methods used to deliberately perturb the expression of a specific gene. A scientist can deliberately disrupt the expression of a gene at different stages of the production of a protein (Figure 12.1). Using gene targeting techniques, scientists can remove or "knock out" a gene from the genome. Using RNA interference (RNAi) technology, it is possible to degrade mRNA transcripts, preventing the translation of the gene into protein and "knocking down" gene expression. Using dominant negative methods, gene function can be blocked by interfering with the protein product. Each of these methods provides scientists with powerful tools for perturbing the expression of genes in an animal's genome to determine the phenotypic effect on an animal's anatomy, physiology, and behavior.

GENETIC TARGETING

Gene targeting is the process of incorporating a DNA construct into a specific locus in an animal's genome. This process is distinct from making a standard transgenic animal, in which a DNA construct incorporates into the genome at a random location. In a genetic **knockout** animal, gene targeting is used to remove the coding region of a gene and replace it with a nonfunctional sequence. In a **knockin** animal, a genomic sequence is replaced with an alternate functional sequence, such as a different gene or different promoter. More complicated genetic strategies allow knockouts and knockins to occur in specific cell types at specific times. We examine the process of creating these animals in the following sections.

It should be noted that genetic targeting technology is almost exclusively used in the mouse. Most loss-of-function genetic mutations in flies and worms are discovered during forward genetic screens (Chapter 8). Furthermore, it is possible to use RNAi transgenes in invertebrates to knock down a specific gene of interest (see the following section). Knockout technology is not used in rats or other mammalian species because of the need to culture and inject embryonic stem (ES) cells during the targeting procedure. There are currently no reliable methods for culturing and maintaining ES cells from animals other than mice; thus, genetic targeting has almost exclusively been used to disrupt genes in the mouse.

Knockout

In a knockout mouse, the coding region of a gene is replaced with an alternate sequence that effectively deletes the gene from the genome. There are five major steps in the creation of a knockout mouse: (1) using recombinant DNA technology to create a targeting construct; (2) injecting this construct into ES cells and selecting cells that have properly incorporated the construct into the genome; (3) inserting successfully targeted ES cells into a blastocyst; (4) inserting the blastocyst into a female mouse; and (5) breeding the offspring of this mouse until a successful knockout is achieved. This lengthy process can take 1 to 2 years to produce a starting colony of knockout mice!

Creating a Targeting Construct

The first step in making a knockout mouse is to use recombinant DNA technology to produce an appropriate targeting construct that will replace the gene of interest. Replacing the target gene with the genetically engineered construct relies on the natural phenomenon of **homologous recombination**, in which regions with strong sequence similarity can exchange genetic material (Figure 12.2A). In nature, homologous recombination produces new combinations of DNA sequences during chromosomal crossover in meiosis, resulting in enhanced genetic variation in a population. In the lab, homologous recombination is used to replace endogenous genomic DNA sequences with targeting constructs containing homologous sequences of DNA (Figure 12.2B).

To use homologous recombination, a scientist designs a targeting construct with the replacement gene flanked by long segments of genomic DNA that are

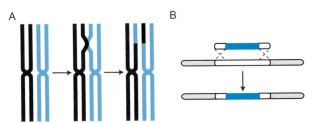

FIGURE 12.2 Homologous recombination. (A) Homologous recombination is a natural phenomenon in which genetic sequences with high sequence similarity can exchange genetic material, as occurs with homologous chromosomes during meiosis. **(B)** In the lab, a scientist can exploit homologous recombination to target a transgene to a specific locus in the genome. If the transgene is flanked by regions homologous with the locus, homologous recombination can allow the transgene to recombine into the genome.

FIGURE 12.3 A generic gene targeting construct. The replacement DNA sequence is flanked by two regions of DNA that are homologous to the genomic locus where the DNA is to be inserted. The neomycin resistance cassette (*Neo*) and thymidine kinase gene (*tk*) are also placed in the construct to select embryos that have properly incorporated the construct into the genome.

homologous to sequences in the endogenous mouse genome. These **homology arms** provide targeting specificity: the region between the two homologous segments replaces the target gene in the organism. Homologous end sequences are at least 1 kb (kilobase) in length, usually longer. A scientist can easily identify and choose homologous sequences surrounding a gene of interest by looking up their sequences on a mouse genome database.

The coding region in the middle of the construct contains a mutated version of the endogenous gene that no longer produces a functional protein product. This is often achieved by deleting the coding region within an exon. The targeting construct also contains a **selection cassette**, typically with two additional genes: a positive selection gene placed between the homology arms, and a negative selection gene placed outside the homology arms (Figure 12.3). The positive selection gene makes cells resistant to a drug, such as neomycin, puromycin, or hygromycin. Only cells that have incorporated the targeting construct into their genome will survive if a scientist exposes the cells to the drug. The negative selection gene, often thymidine kinase or diphtheria toxin A, can be used to kill cells that randomly integrated the targeted construct somewhere in the genome other than the desired locus. These genes allow for quality control while selecting properly targeted cells if and when homologous recombination takes place.

Injecting and Selecting ES Cells

After the scientist makes the targeting construct, the DNA is microinjected or electroporated into ES cells (Chapter 10). Because culturing and delivering

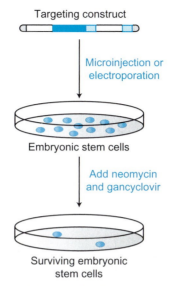

Targeting construct

Microinjection or
electroporation

Embryonic stem cells

Add neomycin
and gancyclovir

Surviving embryonic
stem cells

FIGURE 12.4 **Selecting embryonic stem cells.** The gene targeting construct is delivered into mouse ES cells by microinjection or electroporation. These are then exposed to the compounds neomycin and gancyclovir. Neomycin will kill cells that do not express the neomycin resistance gene, cells that did not incorporate the targeting construct into the genome. Gancyclovir is harmless to cells unless the *tk* gene is expressed, which can occur only if the targeting construct is incorporated into the wrong genomic location. ES cells that survive treatment with both compounds are further tested for expression of the construct and then injected into blastocysts.

DNA to ES cells is technically challenging, most scientists outsource this step to a specialized genetic engineering facility. Technicians inject dozens or even hundreds of cells to enhance the chances of finding a cell that has properly incorporated the targeting construct.

When the targeting construct is delivered to ES cells, three potential outcomes can occur: (1) it correctly recombines and integrates into the cell's genome; (2) it integrates with the cell's genome but in the wrong location; or (3) it fails to integrate into the genome at all. In order to select ES cells that have successfully recombined the targeting construct at the correct location, the technician uses the positive and negative selection genes in the targeting construct (Figure 12.4). For example, a selection cassette encoding the neomycin resistance gene (neo^R) can be used to ensure that the construct has integrated into the genome. After introducing the targeting construct, ES cells are exposed to neomycin, which kills all cells that do not express neo^R. Therefore, only ES cells that have incorporated the construct survive. Including the thymidine kinase (*tk*) gene in the selection cassette ensures that the targeting construct has inserted itself in the correct location. ES cells that have incorrectly incorporated the targeting construct into the genome will express the *tk* gene and die in the presence of a chemical called gancyclovir. Gancyclovir is harmless to normal cells, but it becomes highly toxic when phosphorylated by the *tk* protein product. Because the *tk* gene is outside the region of homology, if homologous recombination occurs, the gene will be lost. Therefore, neomycin and gancyclovir offer positive and negative selection measures, respectively, to ensure that the targeting construct has successfully inserted itself into the genome at a precise location. Statistically, homologous recombination is an extremely rare event, and only about 1% of the ES cells will survive these positive and negative selection measures.

Inserting ES Cells into a Blastocyst

After several positive ES cells are identified, they are injected into a blastocyst. The **blastocyst** is an early embryonic structure that follows the fertilization of an egg (Figure 12.5). There are two cell types within a blastocyst: the **trophoblast**, which forms the outer shell of cells around the blastocyst and later develops into the placenta, and the **inner cell mass**, a collection of embryonic stem cells that eventually give rise to all cells and tissues in the adult organism. ES stem cells that have survived the positive and negative selection process are carefully injected into the inner cell mass. The future tissues that develop from this blastocyst will be chimeric, with some tissues derived from the inner cell mass of the host blastocyst and some tissues derived from the injected ES cells.

Inserting a Blastocyst into a Female Mouse

The next step is to insert the blastocyst into the uterus of a **pseudopregnant** female mouse (Figure 12.6A). The pseudopregnant female is not actually pregnant, but has been tricked into believing she is pregnant by mating with a sterile male mouse. This is done so that the female's body produces the hormones and physiological changes necessary to carry the embryo through to a viable fetus. After injection, the mouse carries the embryos to term and gives birth to offspring.

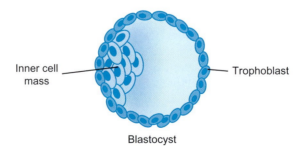

FIGURE 12.5 The blastocyst. A blastocyst is an early embryonic stage resembling a sphere of different cell types. The trophoblast forms the outer sphere of cells and gives rise to the placenta. The inner cell mass contains embryonic stem cells, which give rise to all cells and tissues in the adult organism. A scientist injects ES cells containing the targeting construct into the inner cell mass.

Breeding Offspring Mice

Once mice are born, it is necessary to determine which mice have incorporated the knockout construct. Often, the blastocyst is obtained from a white strain of mice, while the ES cells are derived from a brown strain (or vice versa). This allows the scientist to visually determine the success of the ES cell injection

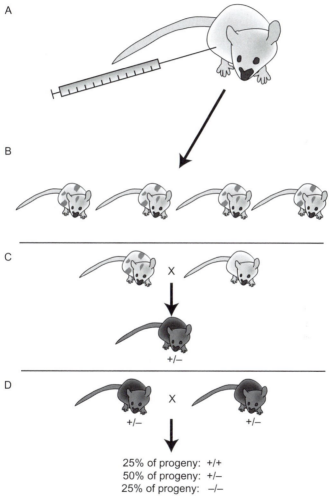

25% of progeny: +/+
50% of progeny: +/−
25% of progeny: −/−

FIGURE 12.6 Generating a genetically targeted mouse. (A) A scientist injects blastocysts into a pseudopregnant female mouse. **(B)** The mouse produces about 8 to 10 offspring, with tissues that developed from the endogenous ES cells residing within the blastocyst and the injected ES cells. The injected ES cells are typically removed from an animal with a brown coat color, allowing visualization when injected into mice with a white coat color. **(C)** Chimeric mice are bred with wild-type mice. Any offspring mice with a brown coat color stably express the targeting construct and are heterozygous for the endogenous gene. **(D)** In order to produce a full knockout, heterozygous mice are bred with each other.

into the blastocyst: if the offspring of the host mother are entirely white, then it is likely that no injected cells successfully combined with the blastocyst. If the offspring have both brown and white patches of fur, the knockout ES cells were successfully incorporated into the blastocyst (Figure 12.6B). Offspring mice will never be entirely brown because the blastocyst is composed of ES cells in the inner cell mass prior to injection of the knockout ES cells, and both cell types continuously divide to produce the animal's tissues. These mice are referred to as **chimeras** because they have a combination of tissues from the blastocyst-derived strain and the ES cell–derived strain. Some investigators use these chimeras to investigate the cell-autonomous role of the gene of interest by comparing the phenotypes of mutant cells to nonmutant cells. However, more steps must be taken to produce a full knockout.

The ultimate hope of the scientist is that the chimeric mice carry the knockout sequence in their **germ cells**, the cells that undergo meiosis to produce either sperm in males or eggs in females. Only genetic material in the germ cells will be passed on to future offspring. Therefore, some chimeras will be able to pass along the knockout sequence in the germline; those chimeras that do not carry the knockout sequence in germ cells will never pass along the knockout sequence. After visually identifying chimeric mice, the scientist breeds them with wild-type mice to produce nonchimeric heterozygous mice with one wild-type copy of the gene and one mutant copy (Figure 12.6C). Offspring will either be all white (no germline transmission in the chimera) or light brown.

The final step is to breed heterozygous mice with each other to produce full homozygous knockout animals (Figure 12.6D). This breeding scheme should result in Mendelian inheritance, with one-quarter homozygous wild-type offspring, one-quarter homozygous knockout offspring, and one-half heterozygous offspring. However, knocking out certain genes can lead to an **embryonic lethal** phenotype, in which a homozygous knockout mouse fails to survive gestation due to improper development. In these cases, scientists may still be able to examine the gene's role in early development. If the scientist is interested in studying the gene's role in postnatal animals, there are some alternative approaches. First, the scientist could compare homozygous wild-type animals with heterozygous animals to determine if having a single mutant version of the gene exerts any phenotypic effect. Alternatively, the scientist could employ a conditional knockout strategy in which the gene of interest is knocked out only in a particular cell type or at a particular time. Conditional knockouts are described below, as are other alternatives to genetic knockouts, including RNAi or dominant negative strategies to disrupt gene function in a postnatal animal.

In short, it takes *four* generations of animals to produce the first knockout mice of a colony: the pseudopregnant foster mothers, the chimeric offspring, the first heterozygous offspring of the chimeras, and the final homozygous knockout animals. Producing a knockout can require a full year or more because of the multiple generations of mice that must be bred and maintained.

Validating a Knockout Mouse

There are many methods that can be used to validate an animal's geno-type, and a thorough investigation uses multiple techniques (Figure 12.7). A **Southern blot** (Chapter 9) is a technique used to detect the presence of specific sequences of DNA from the genome. Samples from wild-type, heterozygous, and knockout mice can be compared to identify the presence or absence and the length of specific DNA sequences at the gene locus. Alternatively, a scientist may use a **PCR** (Chapter 9) genotyping strategy and design primers around the gene's locus. Well-designed PCR primers will produce different-sized PCR

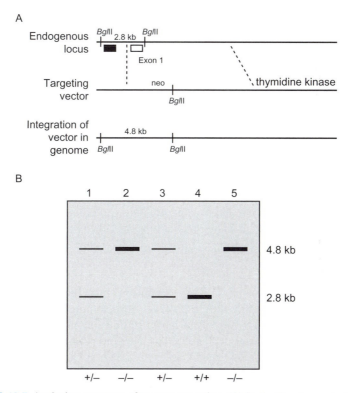

FIGURE 12.7 Analyzing genotype after gene targeting. (A) In the literature, scientists typically diagram a gene targeting procedure by depicting the endogenous locus, the targeting vector, and the new genomic sequence. In this example, scientists replaced the first exon of a gene with a new genetic sequence. Note the presence of the *Bgl*II restriction enzyme sites. In the endogenous locus, digestion of DNA with *Bgl*II results in a 2.8 kb fragment. In a mouse in which gene targeting occurred, the fragment will be 4.8 kb in length. The small black box depicted just beneath the endogenous locus represents a probe that can be used to analyze the size of the fragment. **(B)** In a Southern blot experiment, a scientist digests the DNA with a restriction enzyme—in this case *Bgl*II—and uses a probe to detect a specific genetic sequence. Using this method, it is possible to discriminate between fragments that are 2.8 and 4.8 kb in length, allowing the scientist to determine the genotype of individual mice.

products for wild-type, heterozygote, and knockout genomic DNA. Usually, a scientist performs a Southern blot to validate a knockout and then uses PCR for routine genotyping, as it is a simpler and quicker method.

PCR and Southern blotting techniques are the absolute gold standard in assessing the genotype of an animal. A scientist may also perform complementary tests, such as *in situ* **hybridization** and/or **immunohistochemistry** (Chapter 6) to ensure that there is no gene and protein expression. A **western blot** (Chapter 14) can also demonstrate the absence of a functional protein product.

Using Knockout Mice in Experiments

A knockout mouse is a powerful tool for investigating the necessity of a gene for biological processes, such as development or behavior. However, scientists must be careful to perform the appropriate control experiments to make valid conclusions when using this tool. Regardless of whether the scientist studies an anatomical, behavioral, or electrophysiological phenotype, knockout mice should be directly compared with their wild-type littermates. This reduces potentially confounding differences between mice of different genetic backgrounds. Some studies compare wild-type, heterozygous, and knockout littermates to demonstrate a dose-dependent role for the relative amount of a gene product, with a heterozygote phenotype intermediate between wild-type and knockout.

If no phenotype is immediately discernable, scientists must consider the possibility of a genetic compensatory mechanism. A knockout animal may compensate for the loss of a gene by altering the expression levels or pattern of another gene or adapting in some other way so that the resulting phenotype is still normal. Often, the phenotypic consequence of a genetic knockout is subtle, and more refined assays may be required to identify a difference between knockout and wild-type animals.

Finally, if a knockout mouse does exhibit a phenotype, the scientist must be careful about interpreting the gene's role. For example, if knocking out a specific gene causes an animal to lose its ability to respond to painful stimuli, it is incorrect to immediately conclude that the gene's function is to regulate pain. The only conclusion that can be reached is that the gene is *necessary* for the animal to report a painful stimulus. More rigorous tests must be performed to determine whether the gene is important for tactile sensation in general, for normal development of spinal cord neurons, for correct axonal targeting to the periphery, and so on. Full understanding of a gene's function comes from phenotypic analysis combined with investigations into the specific mechanisms of the gene's contributions to the phenotype.

Knockin

A knockin mouse is a genetically engineered mouse in which a genetic sequence is added at a specific locus in the genome. This differs from a standard transgenic mouse in that the DNA sequence is targeted to a particular location in the genome, rather than being randomly integrated. The procedure to make a knockin mouse uses the same homologous recombination strategy as the procedure just described for making a knockout mouse. However, a knockin mouse serves a very different purpose: to add functional genomic sequences, whether they are functional genes or promoter sequences. There are various reasons why a scientist may choose to make a knockin mouse:

- To replace a functional gene with a tagged version of the same gene. The resulting reporter fusion protein can then be used to visualize gene expression patterns in fixed or living tissue.
- To replace the gene of interest with a mutated version to mimic a human disease. For example, if a scientist hypothesizes that the cause of a human disorder is a specific mutation in a gene, the scientist could produce a mouse model of the disease that carries this mutation in the homologous mouse gene.
- To replace a regulatory region of DNA to investigate the role of a specific promoter or repressor element in the regulation of a gene product.

As with knockout mice, scientists should compare knockin animals with their wild-type littermates during experiments.

Conditional Knockout

It is often desirable to disrupt a gene only in a particular cell-type or at a particular time. For example, knocking out a gene in the entire organism could be lethal at the embryonic stage, but if the ablation is restricted to a specific tissue or cell-type or if the ablation occurs after the animal reaches adulthood, investigators may be able to avoid the lethality. In these scenarios, a scientist may choose to make a **conditional knockout mouse** in which a gene is removed from the genome in a precise spatial or temporal manner.

One of the most common methods of creating a conditional knockout is by using the **Cre/lox recombination system** (Figure 12.8; also described in Chapter 11). Homologous recombination is used to replace an endogenous target gene with a fully functional version of the gene flanked by a pair of short DNA sequences called **lox sites** (referred to as a "floxed" sequence). These lox sites do not disrupt the gene they flank, so the phenotype of floxed mice remains normal. Lox sites are specific 34 base pair sequences recognized by a protein called **Cre recombinase** (or simply Cre) that is not normally expressed in animals. Cre recognizes and recombines lox sites, excising any floxed genetic material. To produce conditional knockout mice, mice expressing a floxed gene are mated with transgenic mice expressing Cre under the control of a specific promoter. Cre catalyzes recombination between the lox sequences, excising the target gene from the genome. Many Cre lines now exist for distinct tissue and neural cell types. An alternative to the Cre/lox system

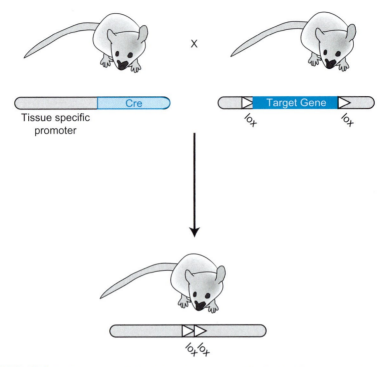

FIGURE 12.8 Producing conditional knockout mice with the Cre/lox system. A scientist makes a mouse in which a gene of interest is flanked by two lox sites. If this mouse is mated with another mouse expressing Cre recombinase, Cre can recombine the genetic material between the two lox sites and remove the gene from the genome.

is the **Flp/Frt recombination system.** This system is similar to Cre/lox, with the enzyme flippase recombinase (Flp) excising Flp recombinase target (Frt) sequences that flank the genomic region of interest.

To exert temporal control over a genetic knockout, such as knocking out a gene in adulthood, scientists may use inducible knockout systems. A popular method is to combine the Cre/lox recombination system with the **ER/tamoxifen system**. The estrogen receptor (ER) is a protein that remains in the cytoplasm unless it binds to its chemical ligand. Cre fused to the estrogen receptor (ER) remains in the cytoplasm, where it cannot act to recombine a floxed gene. The drug tamoxifen binds to the ER receptor, causing the fused Cre-ER protein to translocate into the nucleus to exert its effects and excise the floxed gene. After mating an inducible Cre line of mice with the floxed line of mice, investigators control the timing of the knockout through the timing of tamoxifen delivery.

Now that we have described gene targeting and methods of manipulating function at the genomic level, we will survey methods of disrupting a gene by interfering with its products—either the mRNA transcript or the protein product.

DISRUPTING GENE PRODUCTS

One of the limitations of using genetic targeting strategies and random mutagenesis screens to knock out functional gene expression is that these approaches can only be used in a few model organisms. However, scientists can also perturb endogenous gene function by interfering with a gene's mRNA transcript or its translated protein product (Figure 12.1). Transgenes encoding an RNAi or dominant negative construct can also be used to generate loss-of-function transgenic organisms. These techniques effectively inhibit gene function and are amenable to virtually all traditional, as well as nontraditional model organisms.

RNA Interference (RNAi)

RNA interference (RNAi) is the process of silencing gene expression through endogenous cellular mechanisms that degrade mRNA. Cells naturally use RNAi to regulate normal gene expression, as well as to degrade foreign mRNA transcripts from viruses. Scientists now use this natural process to selectively knock down gene expression and degrade mRNA for a gene of interest. Though knockdown efficiencies do not reach 100%, in good knockdown experiments, only a minimal amount of mRNA remains.

The process of RNAi begins when a piece of double-stranded RNA (dsRNA) comes in contact with an enzyme called Dicer (Figure 12.9). Dicer cleaves dsRNA into small duplex fragments, about 23 base pairs long, called small interfering RNA (siRNA). The siRNA-Dicer complex recruits additional cellular proteins to form an RNA-induced silencing complex (RISC) that uses only one strand of the siRNA duplex, the guide strand. RISC interacts with and degrades RNA strands that have a complementary nucleotide sequence to the siRNA guide strand. Thus, the RNAi process degrades mRNA before it is translated by a ribosome into a protein product.

In theory, there are three ways to make use of the RNAi pathway to disrupt a gene: (1) deliver a dsRNA molecule that is homologous to the mRNA transcript; (2) deliver synthetic siRNAs (19–23 bp) that will attract the RISC complex to bind to the homologous mRNA transcript; and (3) deliver short hairpin RNA (shRNA), a strand of RNA that contains complementary sequences at either end that cause the molecule to fold onto itself, producing a hairpin shape (Figure 12.10). Endogenous cellular RNAi machinery can then process shRNA into small dsRNA.

The model organism used partially determines what type of RNAi molecule to use, as different organisms respond with varied efficacy to these distinct methods. Though dsRNA works well in *Drosophila* and *C. elegans*, delivering dsRNA into mammalian cells generally causes a nonspecific interferon stress response, making experiments difficult to interpret. siRNA can knock down a gene of interest, but this is a transient effect, which may be suboptimal for

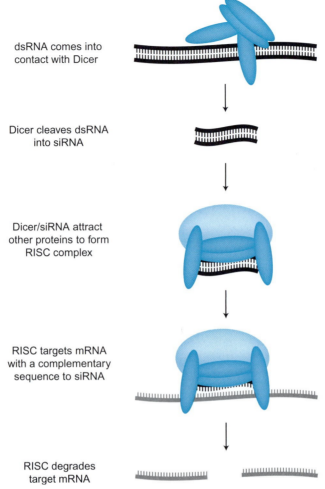

dsRNA comes into
contact with Dicer

Dicer cleaves dsRNA
into siRNA

Dicer/siRNA attract
other proteins to form
RISC complex

RISC targets mRNA
with a complementary
sequence to siRNA

RISC degrades
target mRNA

FIGURE 12.9 The concept behind RNAi technology. When double-stranded RNA (dsRNA) comes into contact with the Dicer enzyme, Dicer cleaves the RNA into small fragments, creating small interfering RNA (siRNA) molecules. The siRNA-Dicer complex recruits other proteins to form a RISC complex. RISC targets mRNA molecules with a complementary sequence to one of the two siRNA strands, degrading the mRNA and knocking down gene expression.

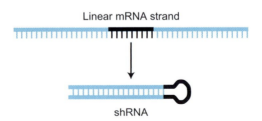

Linear mRNA strand

shRNA

FIGURE 12.10 Short hairpin RNA (shRNA). An shRNA molecule is composed of two complementary sequences separated by a small spacer. The molecule naturally folds into itself, resembling a hairpin shape.

certain experiments. For long-term, stable gene knockdown, scientists deliver shRNA into cells in a way that allows the shRNA to be stably integrated into the genome. This can be achieved by making a transgenic organism in which the shRNA sequence is placed under the control of a strong promoter.

There are two critical issues to consider in the design of a good RNAi experiment: the effectiveness of the RNA probe and the use of appropriate controls. An effective probe has the ability to knock down the expression of a specific mRNA strand without off-target effects. Usually, a scientist designs multiple RNAi probes against the gene of interest and compares their knockdown efficiency in cell culture before choosing one for an actual experiment. The scientist designs these probes to have sequences unique to the gene of interest and no other genes (this can be done using the BLAST program described in Chapter 8). Other, additional control probes are also necessary for a good RNAi experiment. Typical controls include a probe with a scrambled sequence, as well as a probe against a separate, nonessential gene to show the specificity of an effect. The most conclusive experiments are not only repeated multiple times, but demonstrate that multiple probes against different regions of the same gene result in the same knockdown phenotype. The gold standard of RNAi controls is a phenotypic rescue experiment, in which a version of the knocked down gene that is not targeted by the RNAi probe is delivered together with the RNAi probe to show that it can restore a normal phenotype. The rescuing gene is often an orthologue from another species with the same function but a slightly different genetic sequence.

Morpholinos

An alternative to RNAi is the use of **morpholinos** to block proper mRNA translation. Morpholinos are stable, synthetic 22–25 bp antisense oligonucleotide analogues designed to complement an RNA sequence. A morpholino and its complementary mRNA will hybridize to each other, preventing a ribosome from accessing the mRNA and translating it into a protein (Figure 12.11). Unlike RNAi, morpholinos do not degrade their target mRNA molecules. This is especially useful in organisms such as *Xenopus* frogs that do not seem to respond to an RNAi approach. Traditionally, morpholinos have been used in zebrafish, *Xenopus*, and chicks by injecting them into eggs and producing morphant embryos. Controls used in morpholino experiments are similar to those used for RNAi experiments.

Dominant Negatives

Dominant negative constructs encode protein variants that interfere with the normal function of the wild-type protein. Most dominant negatives inhibit normal protein function by competing for a substrate or binding partner without performing the protein's regular function.

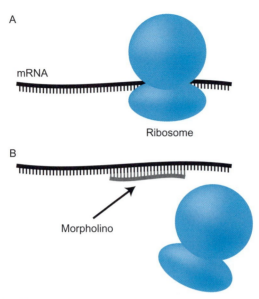

A

mRNA

Ribosome

B

Morpholino

FIGURE 12.11 **Morpholinos.** (**A**) During translation, a ribosome binds to an mRNA molecule and translates specific nucleotide sequences into chains of amino acids. (**B**) A morpholino functionally blocks the mRNA, preventing contact with a ribosome.

There are a variety of methods used to design dominant negative molecules. One common method is to engineer a mutation that incapacitates one functional domain of a protein while leaving other regions intact to interact as normal. Thus, dominant negative molecules will compete with the functional wild-type protein for endogenous binding partners. This approach is often used for transcription factors; the DNA-binding domain is left intact but the transcriptional regulatory domain is removed. Similarly, dominant-negative receptors may have intact extracellular ligand binding domains but may be missing the cytoplasmic intracellular signaling domain; they compete for the ligand but will not pass on the signal. Another common example is creating a dominant negative subunit of an oligomeric protein (Figure 12.12A). The dominant negative subunit can bind the other subunits, but prevents the proper function of the protein (Figure 12.12B).

CONCLUSION

This chapter has surveyed common methods of disrupting the function of a specific gene. Whether these methods target expression at the genomic, RNA, or protein level, they are all used to test the necessity of a gene/protein for a particular phenotype. Scientists can complement these experiments with methods that cause a genetic gain-of-function, resulting in their overexpression (Chapter 11).

A

Protein with four functional subunits

B

Expression of one dominant negative
subunit causes dysfunction of entire
oligomeric protein

FIGURE 12.12 **Dominant negatives.** (**A**) In this example, a functional protein is composed of four identical subunits. (**B**) A transgenic construct produces a slightly mutated subunit that is still able to form an oligomeric protein but inhibits the protein's function.

SUGGESTED READING AND REFERENCES

Books

Appasani, K. (Ed.) (2005). *RNA Interference Technology: From Basic Science to Drug Development*. Cambridge University Press, Cambridge.

Schepers, U. (2005). *RNA Interference in Practice: Principles, Basics, and Methods for Gene Silencing in C. elegans, Drosophila, and Mammals*. Wiley-VCH, Wenheim, Germany.

Review Articles

Herskowitz, I. (1987). Functional inactivation of genes by dominant negative mutations. *Nature* **329**, 219–222.

Lehner, B., Fraser, A. G. & Sanderson, C. M. (2004). Technique review: how to use RNA interference. *Brief Funct Genomic Proteomic* **3**, 68–83.

Miller, V. M., Paulson, H. L. & Gonzalez-Alegre, P. (2005). RNA interference in neuroscience: progress and challenges. *Cell Mol Neurobiol* **25**, 1195–1207.

Sandy, P., Ventura, A. & Jacks, T. (2005). Mammalian RNAi: a practical guide. *Biotechniques* **39**, 215–224.

Primary Research Articles—Interesting Examples from the Literature

Doetschman, T., Gregg, R. G., Maeda, N., Hooper, M. L., Melton, D. W., Thompson, S. & Smithies, O. (1987). Targetted correction of a mutant HPRT gene in mouse embryonic stem cells. *Nature* **330**, 576–578.

Fire, A., Xu, S., Montgomery, M. K., Kostas, S. A., Driver, S. E. & Mello, C. C. (1998). Potent and specific genetic interference by double-stranded RNA in Caenorhabditis elegans. *Nature* **391**, 806–811.

Hayashi, M. L., Rao, B. S., Seo, J. S., Choi, H. S., Dolan, B. M., Choi, S. Y., Chattarji, S. & Tonegawa, S. (2007). Inhibition of p21-activated kinase rescues symptoms of fragile X syndrome in mice. *Proc Natl Acad Sci USA* **104**, 11489–11494.

Jessberger, S., Aigner, S., Clemenson, G. D., Toni, N., Lie, D. C., Karalay, O., Overall, R., Kempermann, G. & Gage, F. H. (2008). Cdk5 Regulates Accurate Maturation of Newborn Granule Cells in the Adult Hippocampus. *PLoS Biol* **6**, e272.

Mezghrani, A., Monteil, A., Watschinger, K., Sinnegger-Brauns, M. J., BarrËre, C., Bourinet, E., Nargeot, J., Striessnig, J. & Lory, P. (2008). A destructive interaction mechanism accounts for dominant-negative effects of misfolded mutants of voltage-gated calcium channels. *J Neurosci* **28**, 4501–4511.

Papaioannou, V. E., Gardner, R. L., McBurney, M. W., Babinet, C. & Evans, M. J. (1978). Participation of cultured teratocarcinoma cells in mouse embryogenesis. *J Embryol Exp Morphol* **44**, 93–104.

Ralph, G. S., Radcliffe, P. A., Day, D. M., Carthy, J. M., Leroux, M. A., Lee, D. C., Wong, L. F., Bilsland, L. G., Greensmith, L., Kingsman, S. M., Mitrophanous, K. A., Mazarakis, N. D. & Azzouz, M. (2005). Silencing mutant SOD1 using RNAi protects against neurodegeneration and extends survival in an ALS model. *Nat Med* **11**, 429–433.

Raoul, C., Abbas-Terki, T., Bensadoun, J. C., Guillot, S., Haase, G., Szulc, J., Henderson, C. E. & Aebischer, P. (2005). Lentiviral-mediated silencing of SOD1 through RNA interference retards disease onset and progression in a mouse model of ALS. *Nat Med* **11**, 423–428.

Thomas, K. R. & Capecchi, M. R. (1987). Site-directed mutagenesis by gene targeting in mouse embryo-derived stem cells. *Cell* **51**, 503–512.

Protocols

Birmingham, A., Anderson, E., Sullivan, K., Reynolds, A., Boese, Q., Leake, D., Karpilow, J. & Khvorova, A. (2007). A protocol for designing siRNAs with high functionality and specificity. *Nat Protoc* **2**, 2068–2078.

Dann, C. T. & Garbers, D. L. (2008). Production of knockdown rats by lentiviral transduction of embryos with short hairpin RNA transgenes. *Methods Mol Biol* **450**, 193–209.

Kappel, S., Matthess, Y., Kaufmann, M. & Strebhardt, K. (2007). Silencing of mammalian genes by tetracycline-inducible shRNA expression. *Nat Protoc* **2**, 3257–3269.

Morozov, A. (2008). Conditional gene expression and targeting in neuroscience research. *Curr Protoc Neurosci*, Chapter 4, Unit 4.31.

Mortensen, R. (2003). Production of a heterozygous mutant cell line by homologous recombination (single knockout). *Curr Protoc Neurosci*, Chapter 4, Unit 4.30.

Mortensen, R. (2007). Overview of gene targeting by homologous recombination. *Curr Protoc Neurosci*, Chapter 4, Unit 4.29.

Websites

Neuromice http://www.neuromice.org/

Nagy, A. creXmic: Database of Cre Transgenic Lines http://nagy.mshri.on.ca/cre/

Ambion RNA Interference Resource http://www.ambion.com/RNAi/

Nature Reviews RNAi Collection http://www.nature.com/focus/rnai/

The RNAi Consortium, Broad Institute http://www.broad.mit.edu/rnai/trc

Chapter 13

Cell Culture Techniques

After reading this chapter, you should be able to:

- Explain the advantages and disadvantages of using *in vitro* culture techniques
- Compare types of cells used to examine nervous system function *in vitro*
- Describe techniques for manipulating *in vitro* cultures

Techniques Covered

- **Tools and reagents used in culture:** equipment, media
- **Types of cultured cells:** immortalized cell lines, primary cell and tissue culture, stem cells
- **Manipulating cells in culture:** transfection, coculture, pharmacology, antibody interference

Cell culture is the process by which a scientist grows and maintains cells under carefully controlled conditions outside of a living animal. There are many reasons why this approach is extremely desirable. Examining the nervous system *in vitro* ("within glass") allows scientists to simplify the cellular environment, providing greater control over experimental manipulations and reducing potentially confounding interactions with other biological systems. *In vitro* tools and techniques make experiments possible that would otherwise be difficult or impossible to perform (or interpret) in intact organisms, such as performing multiple assays in parallel with the exact same number of cells. In addition to reducing the complexity of the experimental preparation, *in vitro* experiments tend to be faster, less expensive, and require fewer animals than experiments performed *in vivo*.

However, cell culture experiments also raise interesting questions: Does a cell in a culture flask truly behave like it would inside the brain or body? What does it mean to culture neurons, cells notorious for their intercellular communication and neural networks, in relatively isolated conditions in which they do not form synaptic connections? Are immortalized cell lines, cells that continue to divide indefinitely, actually good models for cells that degenerate and perish in the brain? Because the goal of most *in vitro* experiments is to generate a hypothesis and conclusion about what occurs *in vivo*, it is important to design experiments carefully and extrapolate results cautiously.

This chapter surveys the cell culture approach to studying the brain. We start with a quick summary of the equipment and reagents necessary to maintain cells outside their endogenous environments. Then we describe three categories of cells used in culture experiments: immortalized cell lines, primary cell cultures, and a special class of cell that has received much attention over the past decade, stem cells. Finally, we talk about some of the methods scientists can use to perturb cells in cultured environments.

CELL CULTURE EQUIPMENT AND REAGENTS

Specialized equipment and reagents are necessary to provide cultured cells with an environment that can support their continued growth and health outside a living organism. Most of these supplies are used to artificially mimic the endogenous, *in vivo,* cellular environment. Taking a cell out of the brain and expecting it to survive in a cell culture dish without providing basic, life-sustaining factors would be like catching a fish and expecting it to stay alive in a cage: unless a scientist correctly provides the elements necessary to sustain life, the cell will quickly die. Thus, most equipment and reagents in cell culture labs continually supply cells with oxygen, nutrients, growth factors, and other elements necessary to keep cells alive.

Other tools and reagents prevent contamination. Cells *in vivo* have the benefit of an active immune system to prevent contamination from bacteria, fungi, and other microorganisms. Cells in the brain have the added benefit of the blood-brain barrier to further prevent contamination. In culture conditions, cells are incredibly vulnerable, and scientists must take great care in avoiding contamination, especially since microorganisms are ubiquitous in the environment and can easily penetrate cell culture plates.

Equipment

Although individual laboratories study different cell lines and ask different scientific questions, cell culture rooms tend to contain the same fundamental pieces of equipment. This equipment includes the following:

- **Biosafety Hood.** A biosafety hood or laminar flow cabinet is used to prevent contamination by microorganisms (Figure 13.1). When not in use, these chambers are often illuminated with UV light that helps sterilize exposed surfaces. Just before use and throughout experiments, scientists spray all surfaces with 70% ethanol to provide further decontamination. During experiments, scientists always keep reagents and cell culture flasks/plates

covered until they are ready for use; when exposed, a scientist should never pass their gloved hands or other equipment over open bottles or, even worse, the cells themselves.

- **Cell Incubator.** Cell incubators house and store culture flasks/plates. These incubators maintain an appropriate temperature, humidity, and gas concentration to mimic endogenous conditions. They are usually set at 37° C with 5% CO_2 levels. Although 5% CO_2 is not the level experienced in the body *in vivo*, it maintains the pH of buffers in the growth media at proper physiological levels.

- **Treated cell culture flasks/plates.** Tissue culture flasks and plates come in many varieties depending on the needs of the investigator. Flasks typically range in sizes of $12.5\,cm^2$ to large, multitiered chambers with surface areas of $1800\,cm^2$. Multiwelled plates are often used for actual tissue culture experiments, as these wells can hold an exact numbers of cells, and it is easy to keep track of various experimental conditions across wells. Plates are typically sold with 6, 12, 24, 48, 96, or 384 wells. Many cells, including neurons, need to adhere to a substrate to grow. Therefore, tissue culture dishes or glass coverslips are often coated with attractive amino acids, such as lysine or ornithine, or extracellular matrix components like collagen and laminin.

- **Refrigerator.** A refrigerator (properly referred to as a 4° C incubator) maintains cell culture media and other reagents when not in use. The growth factors and antibiotics in culture media degrade over time, but can last for weeks if stored at 4° C.

- **Water bath.** A water bath is often set at 37° C and is used to quickly warm cell culture media and other reagents stored at 4° C just prior to being added to cell culture flasks/plates. If a scientist does not warm media before adding it to cells, the cells could be shocked by the extreme cold temperatures and die.

- **Microscope.** Microscopes are used in most tissue culture rooms for routine observation of cell culture flasks/plates to inspect the health and confluence of cells. **Confluence** refers to the percentage of the surface of the bottom of the plate covered by cells. Most cell lines should never become 100% confluent, as cells in dense populations tend to inhibit each other's growth, as well as quickly drain nutrients from the culture media.

Culture Media

Growth media is critical to cell culture experiments, supplying nutrients (amino acids and vitamins) and a source of energy (glucose) for cells. Growth media can vary in pH, nutrient concentration, and the presence of growth factors or other biologically relevant components. To survive, cells must be bathed in an isotonic fluid that has the same concentration of solute molecules as inside the cell. The media is buffered to maintain a compatible pH (usually 7.4, though there are some cell-specific variations). In the 5% CO_2 environment of culture incubators, bicarbonate buffer maintains the physiological pH as well as provides nutritional benefits to the cells. HEPES (4-(2-hydroxyethyl)-1-piperazineethanesulfonic acid) is added to the culture media for extra buffering capacity when the cells in a culture experiment require extended periods of manipulation outside a CO_2 incubator.

Serum is often added to culture media for its ability to promote survival through undefined mixtures of growth factors, hormones, and proteins, like PDGF (platelet-derived growth factor), insulin, and transferrin. For stricter control over the cellular environment, investigators use serum-free, chemically defined supplements, such as N2 or B27, that contain known formulations of survival factors. A typical growth medium in neuronal cultures is NeuroBasal, which provides optimized amino acids and nutrients to cultured neurons, supplemented with N2 or B27.

Adding or removing specific ingredients to or from the media can influence cellular behavior. For example, to maintain a neural progenitor pool that continues to divide in culture, scientists add growth factors such as epidermal growth factor (EGF) and fibroblast growth factor (FGF) to prevent differentiation. Specific recipes for culture media are tailored to provide optimal conditions for promoting the health and proper physiological function of the cultured cells and maintaining cellular behavior.

Now that we have described some of the equipment and reagents necessary for cell culture, we survey the different categories of cells commonly used in cell culture experiments.

IMMORTALIZED CELL LINES

Immortalized cell lines are cells that have been manipulated to proliferate indefinitely and can thus be cultured for long periods of time (Table 13.1).

TABLE 13.1 Commonly Used Immortalized Cell Lines

	Cell Type and Origin	Comments
3T3	Mouse embryonic fibroblast	Robust and easy to handle; contact inhibited; stops growing at very high densities
HeLa	Human epithelial cell	From cervical cancer in a human patient named Henrietta Lacks; may contaminate other cultured cell lines; able to grow in suspension
COS	Monkey kidney	Efficiently transfected; commonly used as an expression system for high-level, short-term expression of proteins
293/293T/ HEK-293T	Human embryonic kidney	Easy to transfect and manipulate; commonly used as an expression system to study signaling and recombinant proteins
MDCK	Dog kidney epithelial cell	Polarized with distinct apical and basal sides, used in studying trafficking
CHO	Chinese hamster ovary	Useful for stable gene expression and high protein yields for biochemical assays; commonly used as an expression system for studying cell signaling and recombinant proteins
S2	*Drosophila* macrophage-like cells	Well-characterized *Drosophila* cell line; highly susceptible to RNAi treatment
PC12	Rat pheochromocytoma chromaffin cell	Neuron-like, derived from a neuroendocrine adrenal tumor; can differentiate into a neuron-like cell in the presence of NGF
Neuro-2a/ N2a	Mouse neuroblastoma	Model system for studying pathways involved in neuronal differentiation; can be driven to differentiate by cannabinoid and serotonin receptor stimulation
SH-SY5Y	Human neuroblastoma, cloned from bone marrow	Dopamine beta hydroxylase activity, acetylcholinergic, glutamatergic, adenosinergic; grow as clusters of neuroblast-type cells with short, fine neurites

Immortalized cell lines are derived from a variety of sources that have chromosomal abnormalities or mutations that permit them to continually divide, such as tumors. Because immortalized cells continuously divide, they eventually fill up the dish or flask in which they are growing. By **passaging** (also known as **splitting**), scientists transfer a fraction of the multiplying cells into new dishes to provide space for continuing proliferation.

There are many advantages to using immortalized cell lines. Because there are standard lines used by many different labs, immortalized cells are fairly well characterized. They are, at least theoretically, homogeneous, genetically identical populations, which aids in providing consistent and reproducible results. Immortalized cells tend to be easier to culture than cells used in primary cultures in that they grow more robustly and do not require extraction from a living animal. Also, because they grow quickly and continuously, it is possible to extract large amounts of proteins for biochemical assays (Chapter 14). It is also possible to create cell lines that continuously express a gene of interest, such as a fluorescently tagged or mutant version of a protein.

The major disadvantage to using immortalized cells is that these cells cannot be considered "normal," in that they divide indefinitely and sometimes express unique gene patterns not found in any cell type *in vivo*. Therefore, they might not have the relevant attributes or functions of relatively normal cells. Also, after periods of continuous growth, cell characteristics can change and become even more different from those of a "normal" cell. Thus, it is important to periodically validate the characteristics of cultured cells and not use cells that have been passaged too many times.

Immortalized cell lines of neuronal origin can be used to study properties unique to neurons. Scientists have used neuronal cell lines to investigate processes that occur during differentiation in neurons, such as axon selection, guidance, and growth. However, most neuronal immortalized cell culture models are derived from tumors and are sometimes genomically abnormal. One popular neuronal cell line is a rat pheochromocytoma cell line derived from an adrenal gland tumor called PC12. The addition of nerve growth factor (NGF) causes PC12 cells to reversibly differentiate with a neuronal phenotype (Figure 13.2). These cells can synthesize catecholamines, dopamine, and norepinephrine, and express tyrosine hydroxylase (TH) and choline acetylase (ChAT), enzymes involved in the production of neurotransmitters. PC12 cells have been used to study molecular phenomena associated with neuronal differentiation and have even been used in experiments to replace dopaminergic neurons in an animal model of Parkinson's disease. Neuroblastoma cell lines, like mouse Neuro2A, can express TH, ChAT, and acetylcholinesterase (AChE) and have been used in electrophysiology and development studies.

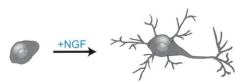

FIGURE 13.2 **Adding NGF to PC12 cells causes them to differentiate into neuron-like cells.** Differentiated PC12 cells grow neurites and synthesize neurotransmitters, such as dopamine and norepinephrine.

Although immortalized cell lines of neural origin can be useful for some experiments, it is advantageous, when possible, to use primary cultured cells— cells extracted from living animals.

PRIMARY CELL AND TISSUE CULTURE

Primary cell and tissue culture uses tissue removed directly from a living animal rather than immortalized cells that continuously divide in a dish. Primary tissue culture allows scientists to directly investigate cells of interest, probing the functions of specific neuronal or glial subtypes and the cellular and molecular dynamics specific to certain cell types. Examining neural tissue *in vitro* is better for recording electrophysiological properties, visualizing dynamic structure and function, and using pharmacology to manipulate function. By examining tissue cultured from genetically manipulated organisms, scientists can observe the cellular and molecular effects of the genetic modification. In addition, primary tissue culture allows investigators to more confidently extrapolate results directly to the intact nervous system.

However, there are some disadvantages to using primary cultures (Table 13.2). For example, primary cultures have a limited lifetime, unlike immortalized cell lines. The age of the animal source influences the health and robustness of the cell culture: tissue from younger, embryonic or early postnatal animals survives better and tends to be healthier than tissue from older animals. Also, a population of primary cells will always be more heterogeneous than a culture of immortalized cells, no matter how careful the scientist was in extracting and purifying the cells of interest. There are three main categories of primary tissue culture: dissociated cultures, explants, and slice cultures.

TABLE 13.2 Immortalized Cell Lines versus Primary Cell Culture

	Advantages	Disadvantages
Immortalized Cell Lines	• Easier to use (grow, transfect, etc.) • Homogeneous • Fairly well characterized • Can create stable cell lines expressing gene of interest	• May not have the same properties as neurons or primary cell type of interest
Primary Cell Culture	• Relevant cell type, physiology, and circuitry	• Heterogeneous populations with high variability

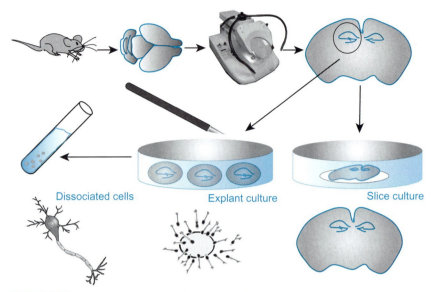

FIGURE 13.3 **Basic steps involved in primary tissue culture**. After a brain is removed from an animal, it can be sectioned using a vibratome (Chapter 6) and cultured as slices, more finely dissected into explants, or enzymatically dissociated into single cells.

Dissociated Cell Cultures

In **dissociated cultures**, neural tissue is separated into individual cells that are then grown on two-dimensional coated glass coverslips or within three-dimensional substrates (Figure 13.3, left). After extracting an animal's brain, specific regions can be micro-dissected to isolate a specific neuronal subtype (e.g., cortical, hippocampal). This region is then mechanically or enzymatically digested within a liquid suspension to separate individual cells from each other. The neurons are then removed from the suspension and plated onto a substrate on which the cells can attach and grow.

Neurons dissociated from different regions of the brain retain their initial identities. The morphological, molecular, and physiological properties of cell populations present in culture correspond closely to the characteristics of the cell population present in the region of origin in a living organism. With the proper growth factors and care, it is possible to maintain a dissociated culture for weeks, during which time cells acquire properties of mature neurons: they develop characteristic axons and dendrites, form synapses with one another, and express receptors and ion channels specific to their cell types, even producing spontaneous electrical activity.

Because a dissociated culture is a sparser collection of cells in a controlled environment, individual neurons can be probed and examined, making it easier to observe their cellular processes and subcellular properties. Scientists can also perturb this system to better understand the molecular components regulating the cells themselves. Dissociated neuronal cell cultures have been used to study neurite outgrowth, synapse formation, and electrophysiological properties.

The ability to probe individual neurons, however, comes at the expense of losing the organization and connectivity critical to *in vivo* functions. Another limitation of dissociated cultures is the small quantity of material produced, which can make biochemical analyses more difficult. Also, most primary cell cultures are not homogeneous. Neuronal cultures are often mixtures of both glia and neurons that respond to different neurotransmitters, so identifying an individual population of cells can be difficult. To minimize heterogeneity, investigators usually attempt to dissect regions as precisely as possible in order to maximize the presence of the desired cell type in culture.

Various methods have been developed to purify specific cell populations. In a technique call **immunopanning**, a scientist coats the bottom of a plate with antibodies that recognize cell-surface markers on the outside of specific cells. When heterogenous populations of cells are added to the plate, the scientist can purify the cells of interest by allowing the cells to bind to the bottom of the plate and then wash off the undesired, unbound cells. This technique has been used to culture oligodendrocyte precursors as well as retinal ganglion cells and corticospinal motor neurons.

Slice Cultures

Slice cultures maintain the structure and organization of the brain in a relatively thick (250–400 µm) section of brain tissue cut using a vibratome (Chapter 6). Slices provide greater access to and visibility of deep subcortical structures, like the hippocampus and thalamus, which are difficult to access *in vivo* (Figure 13.3, right). These slices can either be **acute cultures** that are used immediately or they can be **organotypic slice cultures** that are cultured over longer periods of time. Acute slices are typically used for short-term electrophysiology experiments, while organotypic cultures are used to observe structural and morphological changes over an extended period of time (days to weeks), such as neuronal migration, axon outgrowth, or synapse formation.

Slices require an air/liquid interface to properly regulate gas exchange throughout the slice. Different protocols for culturing slices have varied success in preserving the structure and development of the tissue to parallel that seen *in vivo*.

Explant Cultures

Dissociated cultures allow the examination of isolated neurons, while slice cultures allow the examination of neurons connected in an organized structure. **Explant cultures** are intermediate, involving intact fragments of tissue (Figure 13.3, center). While they do not necessarily preserve the precise organization and orientation of the endogenous nervous system, explants contain the same mixture of cell types. The cells form more coherent groups than in dissociated cultures, though the groups are generally not as large as in slice cultures. Unlike slices, which require an air interface for proper oxygenation, explant cultures can be submerged in the bath media. Explants are often used in coculture assays and for studies on neurite outgrowth and neuronal migration.

STEM CELL CULTURES

Stem cells are **pluripotent** cells with the capacity to generate any type of cell (e.g., neuron, muscle, blood) and an unlimited ability to renew itself. They are like immortalized cell lines in that they should theoretically be able to be propagated in culture continuously, but they are also primary cells derived directly from a living organism. Stem cells are characterized through gene and protein marker expression as well as function.

Stem cells are classified by their source (embryonic stem cells, adult stem cells, or induced pluripotent stem (iPS) cells), as well as the tissue they typically generate (neural stem cells, hematopoietic stem cells, skin stem cells, etc.). Stem cells defined by the tissue they generate are **multipotent**—able to give rise to all types of cells found in the tissue and able to continuously self-renew. For example, neural stem cells can give rise to all three neural lineages—neurons, astrocytes, and oligodendrocytes—as well as additional neural stem cells. Stem cells are distinct from **progenitor cells**, which have a more limited capacity for self-renewal and may be **unipotent**, giving rise to a single cell type.

Stem cell culture is a form of dissociated primary culture. Stem cells must be cultured in the presence of growth factors (epidermal growth factor (EGF) and basic fibroblast growth factor (bFGF)) in specially formulated media that preserves their multipotency and ability to self-renew. However, because scientists do not yet know all of the environmental components required to recreate *in vivo* conditions for preserving stem cell function, *in vitro* studies may not completely capture the effect of endogenous environmental influences.

Stem cell culture has a variety of uses in neuroscience. Embryonic stem (ES) cells are examined for their ability to differentiate into specific neuronal subtypes. Neural stem cells (NSCs) can be cultured to study the basic biology of development and aging. Induced pluripotent stem (iPS) cells can be used to generate clones of cells from patients with neurological disorders to characterize cellular and molecular changes in diseased neurons. While an extensive history and theory of stem cells already exists, basic stem cell biology is still a rapidly evolving field. Here we focus on specific applications of stem cell culture in neuroscience and describe common *in vitro* techniques used to identify stem cells.

Embryonic Stem Cells

Pluripotent **embryonic stem (ES) cells** can give rise to all tissues in an organism. Derived from the inner cell mass of a blastocyst embryo, culturing ES cells *in vitro* essentially traps the cells in a pluripotent state by growing them in the presence of factors that prevent the cells from differentiating. Investigators can reconstruct the environment of ES cells in a dish so that culture conditions contain the specific molecules and mitogens that specify the formation of a neuron in the developing embryo.

Using special culture conditions, ES cells are first induced to become general **neural progenitors**, precursors that are committed to a neural fate. Once they have become neural progenitors, the specific molecules known to act during normal neuronal development are added to the culture to direct differentiation of specific neuron subtypes. For example, to make motor neurons, cells are exposed to high concentrations of sonic hedgehog (SHH) and retinoic acid, morphogens that pattern spinal cord motor neurons during development. To make midbrain dopaminergic neurons, cells are exposed to SHH and fibroblast growth factor 8 (FGF8), morphogens that specify dopaminergic fate during development.

Investigators monitor the progression of differentiation by examining the culture for activation of transcription factors relevant to development and the appearance of markers known to promote differentiation of specific neuronal subtypes. For example, when trying to induce midbrain dopaminergic neurons, scientists look for midbrain-specific transcription factors (Pitx3, En1, Lmx1b, Nurr1) and later for markers of mature dopamine neurons (TH, DAT, Girk2).

This process, from the initial establishment of the ES cells to their differentiation into specific neuronal subtypes, can take months. Also, despite attempts to create pure populations of a specific neuronal subtype, differentiated ES cell populations are still heterogeneous, with many other cells mixed in. However, the ability to differentiate ES cells into specific neuronal cell types is useful for investigating therapies for a number of neurodegenerative disorders such as Parkinson's and Alzheimer's disease.

Neural Stem Cells

Multipotent **neural stem cells (NSCs)** exhibit self-renewal properties and the ability to differentiate into all neural subtypes. They can be extracted from regions of embryonic or adult brains, where they normally divide and give rise to neurons or glia. The embryonic brain contains many NSCs, while adult brains have far fewer NSCs. Furthermore, the ability of adult NSCs to produce neurons decreases with age.

Investigators typically culture adult NSCs from the subgranular zone of the dentate gyrus in the hippocampus and the subventricular zone in the wall of the lateral ventricles. After dissection, the region containing NSCs is dissociated and can be moderately purified through centrifugation. NSCs can also be grown and passaged in special media that promote NSC survival. By removing growth factors from the media, NSCs can be differentiated into neurons, astrocytes, and oligodendrocytes.

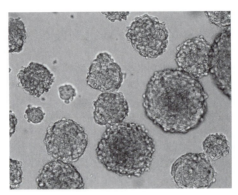

FIGURE 13.4 **Neurospheres**. Dividing neural progenitor and stem cells form spheres in culture. (Courtesy of Victoria Rafalski).

It is difficult to distinguish neural progenitors from NSCs, and the terms often get confused. Both types of cells proliferate and express common sets of molecular markers (e.g., nestin, Pax6). Thus, both molecular and functional assays must be used to determine the identity of cultured cells. Two defining features of NSC cultures that can be tested to confirm cellular identity are (1) the ability to give rise to all three types of the neural lineage (multipotency) and (2) the ability to propagate more NSCs (self-renewal).

Self-renewal is tested through the use of primary **neurosphere** and **secondary neurosphere assays**. When NSCs are cultured in nonadherent conditions in the presence of the growth factors EGF and bFGF, they give rise to **neurospheres**, balls of dividing cells (Figure 13.4). Neurospheres form as a general result of dividing neural precursors, including progenitor cells that do not continuously self-renew. So, while this assay is frequently used, it does not definitively distinguish between stem and progenitor cells.

The secondary neurosphere assay involves culturing cells from established neurospheres to see if the cells generated by the first neurosphere are able to continue proliferating, an indicator of self-renewal. These assays can be difficult to quantify, as cells and spheres can each fuse, so a sphere may not necessarily form from a single NSC. Also, neurospheres are species-specific; rat and human NSCs do not form neurospheres as often as mouse NSCs.

Induced Pluripotent Stem Cells

Induced pluripotent stem (iPS) cells are differentiated cells that scientists have reverted back to a stem cell state. A variety of strategies have been tested to do this. One common strategy is to use viruses to introduce four specific transcription factors into somatic cells, such as fibroblasts. The presence of these transcription factors seems to allow these cells to exhibit stem cell behaviors. Other strategies use a combination of viral delivery and chemical manipulations to coax somatic cells into a pluripotent state.

Once iPS cells are generated, they are tested to see whether they possess characteristics indicative of "stemness" by comparing them to ES cells in various ways. Scientists typically assess transcription factor profiles, examining the presence of markers like Oct4, Nanog, Sox2, AP, SSEA4, and TRA-1-80, using immunohistochemistry (Chapter 6) and reverse transcription polymerase chain reaction (RT-PCR) techniques (Chapter 9). The potential for these iPS cells to differentiate is assessed by driving cells into specific lineages using defined protocols. The gold standard of determining whether an iPS cell is like a stem cell is to examine whether these cells retain the ability to generate chimeric mice (Chapter 12).

In neuroscience, the ability to reprogram fully differentiated cells into an ES-like state opens up the possibility of modeling human disease *in vitro*, directly from patient-derived cells. This could reveal the cellular and molecular pathogenesis of diseases such as Parkinson's and autism spectrum disorders. Induced pluripotent stem cells may also be potentially used for therapeutic purposes, allowing somatic cells derived from a patient to take the form of a cell type depleted as the result of neurodegeneration, such as dopaminergic cells in the substantia nigra in Parkinson's disease.

MANIPULATING CELLS IN CULTURE

One of the major advantages of studying the nervous system *in vitro* is the ability to control the cellular environment. Thus, investigators can observe cellular behavior under a variety of conditions, which is important for probing the functions of both intracellular and extracellular factors. The roles of particular genes, cell types, or proteins in particular behaviors can be characterized by selectively manipulating the genes, cell types, or proteins exposed to the culture system.

Transfection

Transfection is the process of delivering DNA to cells using nonviral methods. One of the major benefits of cell culture systems, especially dissociated cultures, is that they are very amenable to cell transfection. Therefore, it is usually easy to introduce recombinant DNA molecules into cultured neurons to mark specific subcellular organelles or perturb function. Transfection is discussed in much greater detail in Chapter 10.

Coculture Systems

Coculturing allows a variety of cell types to be cultured together to examine the effect of one culture system on another (Figure 13.5). This is useful when examining the effect of one type of tissue on another, one region of the brain on another, or how a particular secreted molecule leads to changes in neural

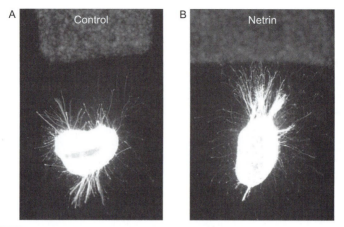

FIGURE 13.5 Co-culturing midbrain explants with COS cell aggregates demonstrates the attractive effect of the guidance cue netrin-1. COS cells (top) transfected with a control plasmid (A) do not influence the direction of axon outgrowth. COS cells transfected with netrin-1 (B) attract axons, (Courtesy of Dr. Jie Li and Dr. Mary Hynes.)

development or physiology. For example, cocultures of different regions of spinal cord explants initially revealed differing effects on the ability to attract or repel neurite outgrowth. Biochemical purification from explants in coculture experiments led to the identification of specific molecules that could then be introduced into immortalized cell lines to express and secrete these molecular guidance cues. Transfected cell lines could then be cocultured with spinal cord explant cultures to examine the neurite outgrowth response of the spinal cord neurons to specific guidance cues. Such approaches have deciphered a large variety of chemoatttactants and chemorepellants that have since been validated *in vivo*.

Pharmacology

Drugs and chemicals that can activate or inhibit specific proteins, channels, or receptors can be used to investigate specific pathways *in vivo* and *in vitro* (Chapter 3). In cell culture experiments, a scientist can include pharmacological agents in the culture media to examine the cellular and biochemical effects. See Chapter 3 for a more detailed description of pharmacological techniques.

Antibody Interference

Antibodies are molecules that recognize specific antigens, usually parts of a protein, and bind those antigens with high affinity (Chapter 14). In nature, these antibodies function as part of the immune system. In the lab, antibodies can be used in many different assays to study protein expression and binding partners. If the antigen of an antibody is accessible in living cells, it is also sometimes possible to block protein function by applying antibodies against the protein of interest. These antibodies bind the protein at a site critical for its normal function, thus inhibiting function *in vitro* or *in vivo*.

CONCLUSION

The ability to grow cells and tissue outside of a living organism allows scientists much greater access and control over the cellular environment compared with *in vivo* conditions. Therefore, scientists can investigate the cellular and molecular mechanisms of nervous system function with great detail. The tools and techniques of *in vitro* culture techniques can provide many informative results, but ultimately this knowledge must be tested in the intact nervous system to fully understand the subtleties of a fully connected, interactive network.

SUGGESTED READING AND REFERENCES

Books
Banker, G., & Goslin, K. (1998). *Culturing Nerve Cells*, 2nd ed. MIT Press, Cambridge, MA.

Butler, M. (2004). *Animal Cell Culture and Technology*, 2nd ed. BIOS Scientific Publishers, London.

Freshney, R. I. (2005). *Culture of Animal Cells: A Manual of Basic Techniques*, 5th ed. Wiley-Liss, Hoboken, NJ.

Martin, B. M. (1994). *Tissue Culture Techniques: An Introduction*. Birkhäuser, Boston.

Review Articles
Blau, H. M., Brazelton, T. R. & Weimann, J. M. (2001). The evolving concept of a stem cell: entity or function? *Cell* **105**, 829–841.

Jaenisch, R. & Young, R. (2008). Stem cells, the molecular circuitry of pluripotency and nuclear reprogramming. *Cell* **132**, 567–582.

Seaberg, R. M. & van der Kooy, D. (2003). Stem and progenitor cells: the premature desertion of rigorous definitions. *Trends Neurosci* **26**, 125–131.

Primary Research Articles — Interesting Examples from the Literature
Barberi, T., Klivenyi, P., Calingasan, N. Y., Lee, H., Kawamata, H., Loonam, K., Perrier, A. L., Bruses, J., Rubio, M. E., Topf, N., Tabar, V., Harrison, N. L., Beal, M. F., Moore, M. A. & Studer, L. (2003). Neural subtype specification of fertilization and nuclear transfer embryonic stem cells and application in parkinsonian mice. *Nat Biotechnol* **21**, 1200–1207.

Chen, Y., Stevens, B., Chang, J., Milbrandt, J., Barres, B. A. & Hell, J. W. (2008). NS21: redefined and modified supplement B27 for neuronal cultures. *J Neurosci Methods* **171**, 239–247.

Dotti, C. G. & Banker, G. A. (1987). Experimentally induced alteration in the polarity of developing neurons. *Nature* **330**, 254–256.

Dotti, C. G., Sullivan, C. A. & Banker, G. A. (1988). The establishment of polarity by hippocampal neurons in culture. *J Neurosci* **8**, 1454–1468.

Dugas, J. C., Mandemakers, W., Rogers, M., Ibrahim, A., Daneman, R. & Barres, B. A. (2008). A novel purification method for CNS projection neurons leads to the identification of brain vascular cells as a source of trophic support for corticospinal motor neurons. *J Neurosci* **28**, 8294–8305.

Ge, S., Yang, C. H., Hsu, K. S., Ming, G. L. & Song, H. (2007). A critical period for enhanced synaptic plasticity in newly generated neurons of the adult brain. *Neuron* **54**, 559–566.

Pan, P. Y., Tian, J. H. & Sheng, Z. H. (2009). Snapin facilitates the synchronization of synaptic vesicle fusion. *Neuron* **61**, 412–424.

Raineteau, O., Rietschin, L., Gradwohl, G., Guillemot, F. & Gahwiler, B. H. (2004). Neurogenesis in hippocampal slice cultures. *Mol Cell Neurosci* **26**, 241–250.

Soldner, F., Hockemeyer, D., Beard, C., Gao, Q., Bell, G. W., Cook, E. G., Hargus, G., Blak, A., Cooper, O., Mitalipova, M., Isacson, O. & Jaenisch, R. (2009). Parkinson's disease patient-derived induced pluripotent stem cells free of viral reprogramming factors. *Cell* **136**, 964–977.

Takahashi, K. & Yamanaka, S. (2006). Induction of pluripotent stem cells from mouse embryonic and adult fibroblast cultures by defined factors. *Cell* **126**, 663–676.

Tessier-Lavigne, M., Placzek, M., Lumsden, A. G., Dodd, J. & Jessell, T. M. (1988). Chemotropic guidance of developing axons in the mammalian central nervous system. *Nature* **336**, 775–778.

Watanabe, K., Kamiya, D., Nishiyama, A., Katayama, T., Nozaki, S., Kawasaki, H., Watanabe, Y., Mizuseki, K. & Sasai, Y. (2005). Directed differentiation of telencephalic precursors from embryonic stem cells. *Nat Neurosci* **8**, 288–296.

Winn, S. R., Tresco, P. A., Zielinski, B., Greene, L. A., Jaeger, C. B. & Aebischer, P. (1991). Behavioral recovery following intrastriatal implantation of microencapsulated PC12 cells. *Exp Neurol* **113**, 322–329.

Protocols

Current Protocols in Neuroscience. Chapter 3: Cellular and Developmental Neuroscience (2006). John Wiley & Sons, Inc: Hoboken, NJ.

Loring, J. F., Wesselschmidt, R. L., & Schwartz, P. H. (eds.) (2007). *Human Stem Cell Manual: A Laboratory Guide*. Academic Press, Amsterdam.

Poindron, P., Piguet, P., & Förster, E. (eds.) (2005). *New Methods for Culturing Cells from Nervous Tissues*, Karger, Basel.

Zigova, T., Sanberg, P. R., & Sanchez-Ramos, J. R. (eds.) (2002). *Neural Stem Cells: Methods and Protocols*. Humana Press, Totowa, NJ.

Websites

American Type Culture Collection: http://www.atcc.org

European Collection of Cell Cultures: http://www.hpacultures.org.uk/aboutus/ecacc.jsp

Chapter 14

Biochemical Assays and Intracellular Signaling

After reading this chapter, you should be able to:

- Discuss the basics of intracellular signaling and why studying proteins is important to neuroscience
- Explain how antibodies are designed, produced, and used to study proteins
- Describe biochemical assays used to investigate intracellular signaling

Techniques Covered

- **Antibody production:** making monoclonal and polyclonal antibodies
- **Techniques used to purify proteins:** affinity chromatography, immunoprecipitation assay
- **Techniques used to measure the amount and location of proteins within cells and tissue:** western blot (immunoblot), enzyme-linked immunosorbant assay (ELISA), radioimmunoassay (RIA), immunohistochemistry (IHC), immunoelectron microscopy, reporter proteins
- **Techniques used to investigate protein-protein interactions:** coimmunoprecipitation, protein affinity chromatography, yeast two-hybrid assay
- **Techniques used to investigate posttranslational modifications:** posttranslational modification assays (kinase assay), posttranslational modification-specific antibodies
- **Techniques used to investigate protein-DNA interactions:** electromobility shift assay (EMSA), chromatin immunoprecipitation assay (ChIP), ChIP on chip assay, luciferase assay

Proteins are molecular machines responsible for virtually all of the structural and functional properties of cells. Structural proteins maintain a cell's shape, size, and durability. Transmembrane proteins serve as receptors for hormones and neurotransmitters, allowing the cell to receive information from the extracellular environment. Other proteins act as downstream messengers from cell surface receptors and mediate the cellular response to an extracellular cue. Many proteins serve as enzymes that facilitate the metabolic and physiological needs of the cell. Finally, proteins regulate the transcription and expression of genes that code for other proteins to further affect a cell's physiology. Because of their importance to neurons and neural systems, anyone interested in how the brain works should appreciate the vital roles that proteins play in cells, as well as understand the techniques used by scientists for their investigation.

Proteins interact with each other and other biomolecules much like neurons interact in functional circuits. Instead of acting in isolation, they communicate with upstream signaling molecules and subsequently affect downstream proteins and biochemical substrates. **Intracellular signaling** (or signal transduction) is a term used to describe the biochemical pathways that affect a cell's physiology and metabolism. Often, these pathways begin with the activation of a cell surface receptor, which leads to a series of biochemical events that culminate in a change in gene expression in the nucleus. Just as an organism senses a change in the environment and uses neural circuits to produce an appropriate response, an individual neuron senses a change in its environment and uses intracellular signaling pathways to change its transcriptional, physiological, and metabolic properties (Figure 14.1).

The purpose of this chapter is to survey the rationale and methods used to study the important roles that proteins play within neurons. We will begin with a general description of intracellular signaling events and why they are useful for understanding the properties of neurons. Next, we will describe two fundamental methods used to study proteins: producing and using antibodies, and purifying proteins of interest. The bulk of the chapter then surveys a variety of methods that answer useful questions about intracellular signaling events: (1) whether a protein is expressed in a particular population of cells and how much is present; (2) whether a protein interacts with another protein; (3) what types of posttranslational modifications can occur for a protein and which proteins cause these modifications; and (4) whether a protein is capable of interacting with DNA and regulating gene expression.

INTRODUCTION TO SIGNAL TRANSDUCTION AND INTRACELLULAR SIGNALING

The study of signal transduction pathways is fundamental to several subfields of neuroscience, including neural development, synapse formation, learning and memory, and homeostatic mechanisms. For example, many researchers have investigated the signaling pathways responsible for axon guidance and synapse formation. Extracellular guidance molecules bind to membrane receptors, signaling which direction the neuron should grow. Intracellular signaling cascades then provide the metabolic and structural changes necessary for the axon to grow out in the correct direction. When an axon reaches its postsynaptic

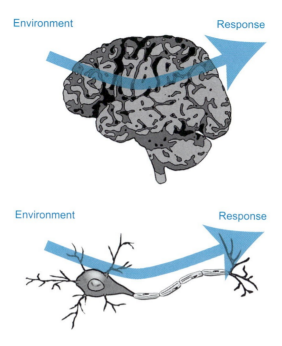

FIGURE 14.1 A metaphor for intracellular signaling. Just as the brain senses changes in the environment and uses neural circuitry to judge and execute an appropriate behavioral response, a neuron senses changes in its extracellular environment and uses intracellular signaling to produce changes in metabolism and gene expression.

target, other extracellular cues shut off this intracellular signaling and new cascades develop to strengthen synaptic connections. An understanding of the biochemical events within a neuron is necessary to truly understand the nature of axon guidance, as well as other vital neuronal functions.

An example of a well-known cell signaling pathway is the JAK-STAT pathway. Various extracellular proteins such as **cytokines** and **growth factors** bind to specific transmembrane receptors (Figure 14.2A). Some of these receptors bind to proteins called Janus Kinases (JAKs). **Protein kinases** are proteins that, when activated, have the ability to add a phosphate group to other proteins. When a cytokine binds to its receptor, JAKs transfer a phosphate group to the receptor (Figure 14.2B). This phosphate group causes a conformational change in the three-dimensional structure of the receptor that allows other proteins to bind called Signal Transducers and Activators of Transcription (STATs). When STAT proteins bind to the receptor, JAKs add an additional phosphate group to the STAT proteins, causing them to dissociate from the receptor and bind to each other (Figure 14.2C). These new STAT dimers translocate to the nucleus and activate specific target genes (Figure 14.2D).

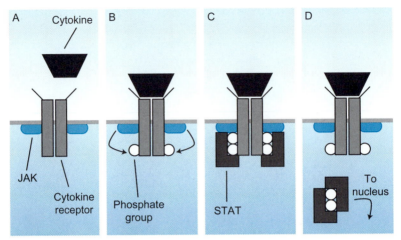

FIGURE 14.2 **The JAK-STAT signaling pathway. (A)** Some receptors for cytokines and growth factors interact with Janus Kinase (JAK) proteins. **(B)** When a cytokine binds to the receptor, JAKs add a phosphate group to the intracellular domain of the receptor. **(C)** These phosphate groups attract STAT proteins to the receptor. JAKs add a phosphate group to STAT proteins, causing them to **(D)** dissociate from the receptor and bind to each other. These STAT dimers then translocate to the nucleus, where they can affect gene transcription.

The JAK-STAT pathway is one of several pathways that play a prominent role in the development and functional physiology of the nervous system. Several elements of this pathway are consistent among other signaling pathways: an extracellular cue binds to a membrane protein, causing an intracellular signaling cascade that eventually affects the expression of genes. These genes code for other proteins or genetic regulatory sequences that play a role in the cell adapting to its environment.

FUNDAMENTAL TOOLS FOR STUDYING PROTEINS

This chapter describes many techniques that can be used to answer specific questions about intracellular signaling. Many of these techniques routinely make use of two fundamental biochemical tools that are essential for studying proteins: making and using antibodies, and purifying proteins of interest. Therefore, before discussing any specific assays, we first examine these fundamental methods.

Making and Using Antibodies

Antibodies are proteins used by an organism's immune system to detect foreign proteins, called **antigens**. An antibody recognizes a specific region of an antigen, the **epitope**. Because they bind specific proteins with high affinity,

A Monoclonal antibodies   B Polyclonal antibodies

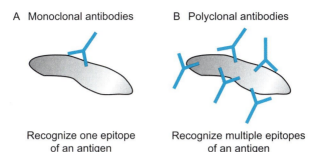

Recognize one epitope   Recognize multiple epitopes
of an antigen   of an antigen

FIGURE 14.3 Monoclonal versus polyclonal antibodies. A monoclonal antibody recognizes a single epitope of an antigen, while a polyclonal antibody is a mixture of several antibodies that recognize multiple epitopes of an antigen.

antibodies are useful for multiple purposes. For example, a labeled antibody can bind and indicate the presence of a protein in a sample. Alternatively, an antibody can be attached to a column or plate in order to collect and purify a protein of interest.

Antibodies for research are produced either in specialized cell lines or live animals. **Monoclonal antibodies** are collections of antibodies that recognize the same epitope. They are derived from single cell lines, usually cells obtained from mice (Figure 14.3A). **Polyclonal antibodies** are a collection of antibodies that may not be completely homogeneous, recognizing distinct regions, or epitopes, of the same antigen (Figure 14.3B). These antibodies are commonly produced in rats, rabbits, chickens, goats, sheep, and donkeys. The procedures for producing monoclonal and polyclonal antibodies are described in Figures 14.4 and 14.5, respectively.

A great variety of antibodies are commercially available and range in cost from $50 to as much as $1000 ($100–200 is common). Individual labs may produce their own antibodies, but usually this work is contracted to an antibody-production service that produces custom antibodies on demand. If there is no good antibody available, scientists can also produce tagged versions of the protein of interest using recombinant DNA techniques. They can then use antibodies that recognize these common tags (Box 14.1).

Several biochemical techniques depend on the use of antibodies (Table 14.1). A specific batch of antibody may be useful for one, several, or (rarely) all of these techniques, depending on the particular binding affinity of the antibody-protein complex. For example, an antibody that works well for immunohistochemistry (IHC) may work very poorly for chromatin immunoprecipitation (ChIP). Even within IHC, the antibody may work much better for labeling proteins in cell culture than labeling proteins in tissue sections. Therefore, the only way to know if an antibody is useful for a particular technique is to try it. Often, commercial manufacturers indicate which biochemical techniques an antibody can be used for based on internal testing.

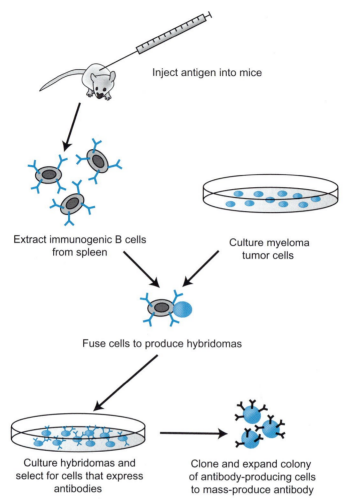

Inject antigen into mice

Extract immunogenic B cells
from spleen

Culture myeloma
tumor cells

Fuse cells to produce hybridomas

Culture hybridomas and
select for cells that express
antibodies

Clone and expand colony
of antibody-producing cells
to mass-produce antibody

FIGURE 14.4 **Producing a monoclonal antibody.** A scientist injects an antigen into an animal, usually a mouse. The mouse's immune system naturally produces antibodies as a response. After several days, the scientist extracts immunogenic B cells that produce antibodies from the spleen. These cells are fused with tumor cells that divide indefinitely. After culturing these cells and selecting for hybrids that continuously produce an antibody of interest, a scientist can mass-produce a monoclonal antibody.

Purifying Proteins

One of the most fundamental tools in the study of proteins is the ability to purify a protein of interest. Purifying proteins allows an investigator to generate antibodies, determine a protein's sequence, and determine a protein's binding partners. Protein purification is usually accomplished in either of two ways: chromatography or immunoprecipitation (IP).

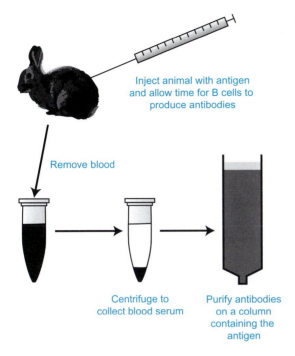

FIGURE 14.5 Producing a polyclonal antibody. A scientist injects an antigen into an animal, often multiple times over the course of weeks. The animal's immune system produces many antibodies as a response. The scientist then collects blood, separates the serum from the red blood cells, and purifies the serum using affinity chromatography.

Inject animal with antigen and allow time for B cells to produce antibodies

Remove blood

Centrifuge to collect blood serum

Purify antibodies on a column containing the antigen

BOX 14.1 Creating and Using Tagged Proteins

Most of the techniques used to detect specific proteins rely on antibodies. However, antibodies do not exist for all native proteins, producing good antibodies can sometimes be difficult and expensive, and the antibodies are not guaranteed to be effective for all techniques. Therefore, many scientists use standard genetic engineering techniques (Chapter 9) to add short amino acid tags to a protein of interest. The genetic sequence for the protein of interest can be inserted into a plasmid vector containing the sequence for one of these tags to create a fusion protein. The scientist can then use antibodies against the tag to examine or purify the tagged protein. Common tags include c-Myc, His, FLAG, GST, and HA. FLAG and GST tags are often used for western blot and IP assays, while c-Myc and HA tags are often used in immunohistochemistry.

Chromatography

Chromatography allows a mixture of proteins (and other biomolecules) to be separated on a column. In this process, a scientist runs a solution through a column containing a porous solid matrix (Figure 14.6). The movement of the different proteins through the column depends on their interaction with the matrix, and therefore proteins can be differentially collected as they flow out the bottom.

TABLE 14.1 Techniques that Depend on Protein-Specific Antibodies

Technique	Role of Antibody in Technique
Western blot	Antibody binds to a protein that has been run on a polyacrylamide gel to mark the presence, approximate size, and relative quantity of the protein in a sample.
Enyzme-linked immunosorbent assay (ELISA)	Antibody binds to a protein to mark the presence and measure the quantity of the protein in a sample.
Radioimmunoassay	Antibody binds to radioactive proteins of known concentration. A sample containing nonradioactive proteins is added to compete with the radioactive proteins, allowing a scientist to determine the protein concentration in a sample.
Immunohistochemistry	Antibody binds to a protein to show spatial expression in tissue or cells.
Immunoprecipitation	Antibody is bound to small beads. A sample is mixed with the beads, so the antibody binds and purifies a protein. The protein sample is then eluted off the beads.
Coimmunoprecipitation	Antibody is bound to small beads. A sample is mixed with the beads, so the antibody binds and purifies a protein. After the protein is eluted, the investigator uses a separate antibody in conjunction with a western blot or ELISA experiment to detect the presence of any secondary proteins that may interact with the immunoprecipitated protein.
Chromatin immunoprecipitation (ChIP)	Antibody is bound to small beads in a column. A sample is run through the column, so the antibody binds and purifies a protein. After the protein is eluted, the investigator uses PCR to determine the presence of any bound DNA sequences.
Antibody interference	Antibody binds to a protein, preventing the proper function of the protein.

Proteins can be separated on the basis of many different properties. For example, in **gel-filtration chromatography**, proteins descend through a column based on their size. In **ion-exchange chromatography**, proteins are separated based on their charge. A more specific way to purify a protein, **affinity chromatography**, uses small molecules fixed to the porous matrix to bind to proteins that may have a high affinity, such as a substrate binding to an enzyme.

Immunoprecipitation (IP)

Immunoprecipitation uses antibodies to purify a protein of interest (Figure 14.7). Specific antibodies are first bound to a mixture of beads. When a solution is mixed with these antibody-conjugated beads, the antibody will bind to

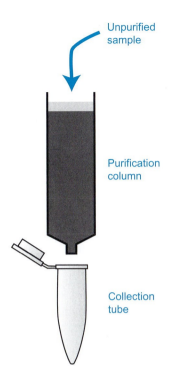

Unpurified sample

Purification column

Collection tube

FIGURE 14.6 **Chromatography.** A biological sample is run through a column composed of a porous membrane. Different columns have different properties and separate proteins based on their size, charge, binding partners, etc.

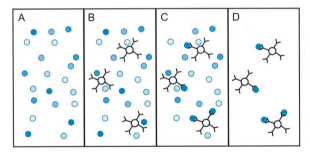

FIGURE 14.7 **Immunoprecipitation.** (A) Before an immunoprecipitation procedure begins, a protein of interest exists in a heterogenous sample with many other proteins. (B) A scientist adds antibodies that recognize the protein to small beads and then adds the beads to the sample. (C) The antibodies bind to the proteins, causing them to precipitate out of solution and adhere to the beads. (D) The beads are removed from the sample, purifying the protein of interest. The proteins can be removed from the beads by changing the salinity or pH of the solution.

its target, precipitating the protein out of solution. After the initial solution is washed away, a scientist can elute the protein off the beads by removing the antibodies, usually by changing the salt concentration or pH of the solution. This elution product can be further purified, although this can also reduce the yield of the final product.

The use of antibodies to bind proteins and the ability to purify proteins based on their binding partners, charge, size, and other properties are common tools used in many cell signaling assays, which we survey now.

INVESTIGATING PROTEIN EXPRESSION

A fundamental question in any signaling experiment is whether a particular protein is expressed in a population of cells and, if so, how much of the protein is present. There are a variety of techniques a scientist may use to answer these questions, some more quantitative and others with better spatial resolutions. A quick comparison of the strengths and limitations of each technique is presented in Table 14.2.

Western Blot (WB)

The **western blot** (also known as an immunoblot) is the most commonly used method to measure the expression of a protein in a cell signaling experiment (Figure 14.8). First, scientists extract proteins from freshly dissected brain tissue or cultured cells using a simple procedure that lyses the cells and removes DNA contaminants. The sample is then run using what is known as **sodium dodecyl sulfate (SDS) polyacrylamide gel electrophoresis (PAGE)**, or simply SDS-PAGE. SDS is a negatively charged detergent that binds to hydrophobic regions of the protein, allowing it to become soluble. The detergent also causes the protein to take on a net negative charge, regardless of its amino acid

TABLE 14.2 Methods of Measuring Protein Expression

Technique	Description
Western blot	Allows for the detection of a protein in a sample, with measurements of relative concentration between samples.
Enyzme-linked immunosorbent assay (ELISA)	Allows for measurement of protein concentration in a sample. More sensitive than a western blot.
Radioimmunoassay	Precise measurement of protein concentration in samples in which the protein may be very dilute.
Immunohistochemistry	Measures the spatial distribution of the expression of a protein in cells and tissues. Not very quantitative.
Immunoelectron microscopy	Measures the spatial distribution of the expression of a protein within cells at a very high resolution.
Reporter proteins	Measures the spatial distribution of the expression of a protein without the need to use IHC. Can be performed in live cells or tissues.

composition. Therefore, the protein will migrate toward a positive electrode when voltage is applied during electrophoresis.

The scientist loads the protein sample onto a polyacrylamide gel (Figure 14.8A). In the presence of an electric field, the gel acts as a physical filter so that larger proteins travel relatively slowly and smaller proteins travel relatively quickly. This allows the investigator to separate a heterogeneous mixture of proteins into a spectrum of proteins with different molecular weights.

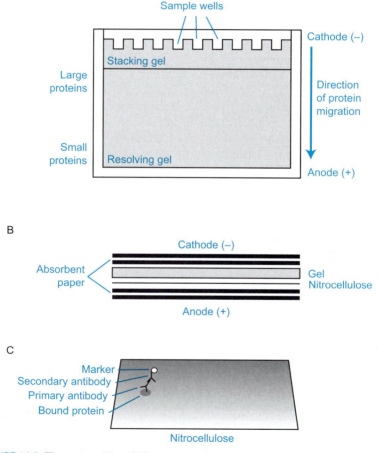

FIGURE 14.8 The western blot. (A) Proteins samples are run on an SDS-PAGE gel, which separates proteins on the basis of size. **(B)** The proteins within the gel are then transferred to a thin nitrocellulose membrane. The gel and membrane are stacked and an electric field causes the proteins to migrate from the gel to the membrane. **(C)** To detect a specific protein of interest, a scientist incubates the nitrocellulose membrane in a primary antibody and then a secondary antibody that recognizes the primary antibody. The secondary antibody contains a marker that can be used to visualize the protein on the membrane.

At this stage of the WB process, the investigator can visualize the protein bands on the polyacrylamide gel using a stain such as **Coomassie blue**. This dye will stain all of the proteins in each lane of the gel. The result can appear like a smear, with some prominent bands representing highly expressed proteins in the sample. A scientist may perform this step to demonstrate that the amount of protein in each lane is the same.

After the protein sample is run on a gel, the final step in a WB procedure is to identify the presence of a specific protein. Because the gel is relatively thick and extremely delicate, a scientist first transfers the proteins onto a thin, nitrocellulose membrane in the "blotting" process (Figure 14.8B). This process uses current to move the proteins from the gel onto the membrane. Once the proteins have transferred to the membrane, the investigator can detect the presence of a particular protein by incubating the membrane in a solution with an antibody specific to the protein (Figure 14.8C). This antibody (or a secondary antibody) is coupled to a radioactive isotope, an easily detectable enzyme, or a fluorescent dye. The labeled antibody marks the presence of the protein as a discrete band within a lane of the gel. The scientist can compare the location of the protein on the blot to a standardized protein ladder that highlights specific molecular weights in kD (kilodaltons).

Thus, the final data in a western blot experiment is a depiction of the band of protein, along with its molecular weight and necessary controls (Figure 14.9). The most important control in a WB is a "loading control," a measurement of a separate protein in the sample to ensure that each lane contains the same quantity of total protein. For example, an antibody to β-actin, a protein expressed in all cells, is often used to ensure that each lane contains the same total amount of protein, allowing a scientist to compare the intensity of bands in different samples to determine relative protein amounts. When characterizing a new antibody, a scientist should perform a negative control with a sample in which the specific protein is absent, as well as a positive control with a sample in which the specific protein is definitely present. This will allow the scientist to verify that the antibody is specific and capable of indicating the presence of the protein of interest.

An investigator may wish to know not only which cells express a protein of interest, but also where, within those cells, that protein is localized. In this case, the investigator can use a process called **cell fractionation** to divide a sample into its cellular components before a WB experiment begins (Figure 14.10). Individual organelles within a cell vary in their weight and can be pulled out of solution through a series of centrifugation steps. Tissue is first dissociated into a liquid homogenate and then centrifuged through a series of spins, each increasing in time and force. After each step, the pellet containing specific cellular components is collected. Finally, the scientist can run each sample in a separate lane of a WB experiment to determine which fraction contains a particular protein.

Enzyme-Linked Immunosorbent Assay (ELISA)

An **enzyme-linked immunosorbent assay** is an alternative to a WB, used to measure the expression of a protein, especially a protein that may be expressed

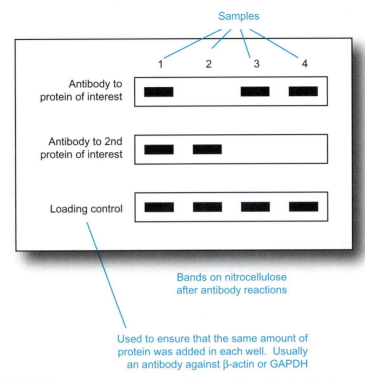

FIGURE 14.9 Typical data from a western blot experiment. The black bands represent proteins visualized on the nitrocellulose membrane.

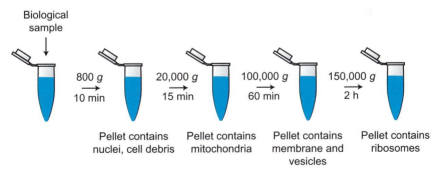

FIGURE 14.10 Cell fractionation. A biological sample can be centrifuged at successively higher speeds to collect specific organelles within cells.

in extremely low quantities. A traditional western blot requires milligrams of protein while an ELISA is able to quantify nanograms.

A scientist binds an antibody to the bottom surface of a plate (Figure 14.11). A sample is added to the plate, and any protein present that can bind to the antibody becomes adhered to the surface. The sample is washed away, but the

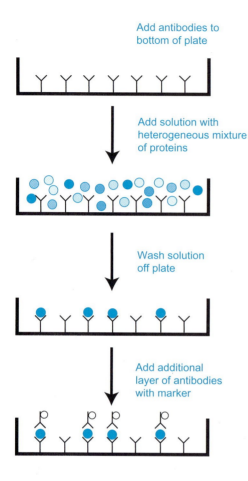

Add antibodies to
bottom of plate

Add solution with
heterogeneous mixture
of proteins

Wash solution
off plate

Add additional
layer of antibodies
with marker

FIGURE 14.11 ELISA. A scientist coats the bottom of a plate with antibodies specific to a protein of interest and then adds a biological sample. The proteins bind to the antibodies, and the sample is washed off the plate. Other antibodies conjugated to a marker are added to the plate, allowing a scientist to determine and quantify the amount of protein in the sample.

bound protein remains. A secondary detecting antibody that is linked to an enzyme is then applied to the plate, such that the protein of interest is "sandwiched" between both antibodies. A chemical substrate is added to the sample that is converted by the enzyme into a chromogenic or fluorescent signal. Finally, the scientist uses a specialized device to measure the signal and determine the quantity of protein present.

Radioimmunoassay (RIA)

A radioimmunoassay is a highly sensitive method of measuring very low concentrations of proteins. Samples are usually derived from blood, extracellular fluid, or cerebrospinal fluid.

To perform a RIA reaction (Figure 14.12A), a scientist first labels a known quantity of protein with a radioisotope, usually an isotope of iodine (^{125}I or ^{131}I), as iodine readily attaches to the amino acid tyrosine. The radiolabeled protein

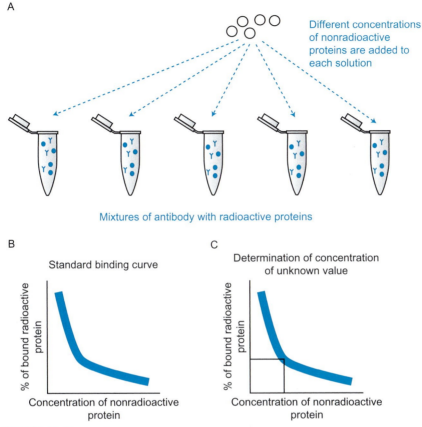

A

Different concentrations of nonradioactive proteins are added to each solution

Mixtures of antibody with radioactive proteins

B

Standard binding curve

% of bound radioactive protein

Concentration of nonradioactive protein

C

Determination of concentration of unknown value

% of bound radioactive protein

Concentration of nonradioactive protein

FIGURE 14.12 A radioimmunoassay. (A) A scientist creates a mixture of radioactive proteins and antibodies that recognize these proteins. This mixture is divided into several reaction tubes. To each tube, the scientist then adds different known concentrations of the nonradioactive protein. The nonradioactive and radioactive proteins compete to bind with the antibodies. Finally, the scientist collects the antibodies and bound proteins using a secondary antibody. **(B)** By measuring the radioactivity of each sample, a scientist can produce a standard binding curve, comparing the amount of radioactive protein bound to an antibody with the concentration of nonradioactive protein added to the sample. **(C)** This binding curve can be used to determine the concentration of a protein from an experimental sample.

is then mixed with an antibody specific for that protein. This antibody-protein mixture is divided into several samples. To the first sample, the scientist adds a known amount of unlabeled, nonradioactive protein. This protein competes for antibody binding sites with the radioactive form of the protein. In the subsequent samples, the scientist continually increases the concentration of nonradioactive protein, increasing the amount of displaced radioactive protein. Finally, the scientist removes all proteins bound to antibodies from the samples using a secondary antibody so that only unbound proteins remain. The radioactivity of each sample is measured. From these data, a scientist can create a standard binding curve reflecting the ratio of bound to unbound protein versus the concentration of nonradioactive protein (Figure 14.12B).

In order to measure the concentration of a protein from a biological sample, the scientist runs this unknown quantity of protein in parallel with the known preceding quantities. Once the scientist identifies the ratio of bound to unbound protein, it is possible to use the standard binding curve to deduce the concentration of the unknown sample (Figure 14.12C).

A radioimmunoassay is sensitive and can measure extremely small protein concentrations, such as the concentration of neuropeptides in the extracellular fluid of brain tissue. However, this technique requires specialized equipment and is relatively expensive. It also requires safety precautions due to the use of radioactivity.

Immunohistochemistry (IHC)

Immunohistochemistry was previously described in Chapter 6 as a method to visualize the expression of proteins in cells or brain sections. This technique can be used to determine the spatial expression pattern of a protein, a relative advantage over the WB, ELISA, and RIA techniques described previously. IHC is not a very good technique for quantifying the amount of protein present in a sample, though it is possible to compare the relative signal from two different samples to show a relative difference in expression.

Immunoelectron Microscopy (IEM)

Immunoelectron microscopy is an extension of IHC used in combination with electron microscopy. This enhances the resolution of spatial data so that protein expression can be visualized within subcellular structures. IEM can be used either as an alternative to cell-fractionation followed by western blot or in combination with those techniques to complement and strengthen the evidence that a protein is expressed in a particular region of a cell, such as the synapse.

Reporter Proteins

Instead of using immunohistochemistry, a scientist can use a **reporter protein**, such as GFP, to genetically visualize protein expression (Chapter 6). There are at least two options: (1) the scientist can place the reporter protein under the control of the promoter of the protein of interest or (2) the scientist can genetically fuse the reporter to the DNA encoding the protein. These methods are usually more time consuming than IHC as various recombinant DNA constructs, gene delivery strategies, and/or transgenic strategies must be employed before an experiment can take place. However, this method overcomes some of the limitations of IHC. For example, reporter genes allow an investigator to determine the expression and location of a protein in living cells.

INVESTIGATING PROTEIN-PROTEIN INTERACTIONS

Almost all proteins function within the cell by interacting and/or forming a complex with other proteins. As the term *intracellular signaling* implies, proteins communicate with other proteins within the cell in a chain of events that serve as the basis of normal cellular physiology and metabolism. Therefore, a scientist might ask whether a specific protein interacts with a protein of interest. Alternatively, a scientist might want to know the identity of *all* proteins that interact with a protein of interest. The following techniques address these questions.

Coimmunoprecipitation (Co-IP)

One of the simplest methods to determine if a protein physically interacts with another protein is by performing a **coimmunoprecipitation** (co-IP). This technique is an extension of the IP assay described previously (Figure 14.7). A scientist lyses cells to extract proteins and then mixes the solution with beads linked to antibodies. An antibody directed toward one of the proteins in the complex immunoprecipitates the protein out of solution. Any proteins that are tightly associated with this protein will also precipitate out of solution and be released in the elution process (Figure 14.13). The eluted proteins can be separated and identified using an SDS-PAGE/western blot protocol with antibodies for the immunoprecipitated protein and potential binding partner (Figure 14.14). If the identities of the potential binding partners are unknown, they may be determined using a technique called **mass spectrometry** (Box 14.2).

Protein Affinity Chromatography

A scientist can use protein-affinity chromatography to isolate and identify proteins that physically interact with each other. To capture the interacting

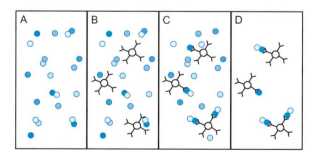

FIGURE 14.13 Coimmunoprecipitation. (A) Before an experiment begins, a protein of interest exists in a heterogenous sample bound to other proteins. **(B)** A scientist adds antibodies that recognize the protein to small beads and then adds the beads to the sample. **(C)** The antibodies bind to the proteins, causing them to precipitate out of solution and adhere to the beads. Any proteins that interact with the protein will also adhere to the beads. **(D)** The beads are removed from the sample, isolating the protein and any other proteins it interacts with. A scientist can then use other methods, such as a western blot, to identify the interacting proteins.

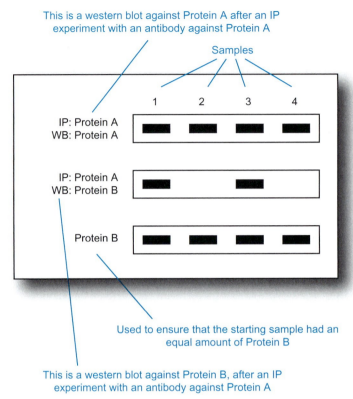

This is a western blot against Protein A after an IP experiment with an antibody against Protein A

Samples

1   2   3   4

IP: Protein A
WB: Protein A

IP: Protein A
WB: Protein B

Protein B

Used to ensure that the starting sample had an equal amount of Protein B

This is a western blot against Protein B, after an IP experiment with an antibody against Protein A

FIGURE 14.14 Typical data from a Co-IP experiment. This experiment represents a Co-IP followed by western blot.

proteins, a target molecule is attached to polymer beads within a column. These target molecules have specific interactions with the protein of interest that will capture the protein as it passes through the column. They can be ligands for a receptor, substrates for an enzyme, or antibodies for an antigen. Cellular proteins are washed through the column, and the proteins that interact with the target attach to the matrix, while all other proteins pass through. The proteins that interact with the target are eluted off the beads, and their identity can be determined using a western blot. Just as in the Co-IP assay, if the identities of the interacting proteins are unknown, they can be determined using mass spectrometry (Box 14.2).

Yeast Two-Hybrid Assay

A **yeast two-hybrid assay** investigates protein-protein interactions by exploiting an endogenous transcription system in yeast cells. In a normal yeast cell, a transcription factor called Gal4 binds to a promoter region called an upstream

BOX 14.2 Mass Spectrometry

Mass spectrometry (MS) is a technique used to identify and sequence peptides and proteins. It does this by determining the mass to charge ratio of proteins, peptides, and their fragments. Automated programs compare the calculated mass of protein and peptide fragments from a sample with the masses computed from protein databases. Thus, MS can identify components of an unknown mixture.

In *"de novo"* sequencing, MS can determine the amino acid sequence of an unknown peptide. By assembling many overlapping peptides, the whole sequence of a protein may be determined, though this is rarely done in practice. MS is often used to identify posttranslational modifications (PTMs), such as phosphorylation or acetylation, and the residues on which they reside. This technique is extremely sensitive and requires little material (sub-picomole amounts).

Determining the masses of proteins and peptides requires the sample to be in the gas phase, and two techniques have been developed to do this: **matrix-assisted laser desorption ionization** (MALDI) and **electrospray ionization** (ESI). In MALDI, peptides are mixed with a UV-absorbing organic acid and dried on a metal slide. The sample is then blasted with a UV laser; the matrix absorbs the laser light, causing it to go into the gas phase carrying the sample (often peptides), which become charged. In ESI, the sample is in an acidic solution that charges the sample. The charged sample travels through a capillary into a heated area where solvent molecules are lost due to the heat and the increasing charge on the droplets: as the charge density becomes higher and higher, the droplets explode, becoming finer and finer until only single molecules remain.

Once the molecules enter a gas phase, their mass-to-charge ratio can be measured. Many different types of detectors exist for this purpose; commonly coupled with MALDI is a **"time of flight"** (TOF) detector. In such an instrument, ionized peptides are accelerated in an electrical field and fly toward a detector. The time it takes each peptide to reach the detector is determined by its mass and charge, with large peptides moving slowly and highly charged peptides moving relatively quickly. Other detectors use electrical or magnetic fields to selectively filter or detect different mass to charge ratios.

Many molecules can have the same mass-to-charge ratio despite having very different chemical structures. For example, the peptides PEPTIDE and TIDEPEP are indistinguishable by their mass alone. This situation is common and occurs with proteins, nucleic acids, and organic molecules in general. In order to distinguish these types of molecules, they may be fragmented and the masses of their constituent components measured; with peptides, this usually involves smashing them with gas molecules such as argon or nitrogen, to cleave the peptide bonds. The result of this "collision induced dissociation" (CID) is overlapping fragments that reveal the original sequence of the peptide. In the preceding example, observing EPEP or PEPT would distinguish between the two possibilities. Ideally, collisions create a virtual ladder of fragments, each differing by only one amino acid. Practically, a perfect ladder is rarely observed, but enough detail is present to match a peptide to known protein sequences.

activating sequence (UAS). Gal4 is composed of a binding domain (BD), which binds to the UAS, as well as an activation domain (AD), which initiates transcription of a target gene.

In a yeast two-hybrid assay, a scientist uses recombinant DNA technology to divide the Gal4 protein into these two separate domains (Figure 14.15A). The Gal4 binding domain is fused to a protein of interest thought to interact with another protein. This protein can be considered the "bait" in that its role in the experiment is to attract other binding partners. The scientist fuses the Gal4 activation domain to a potential binding partner. This binding partner can be considered "prey" in that it could potentially bind with the bait. If the two proteins do not interact, the BD and AD fragments will be physically separated, and no transcription will occur. However, if the two proteins are able to bind,

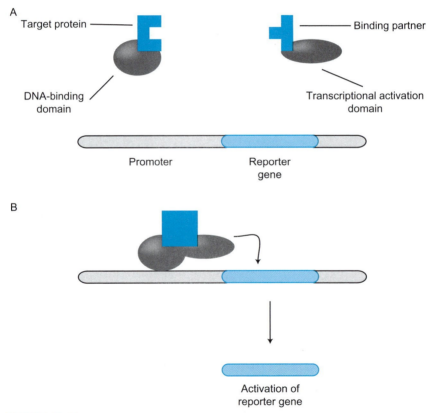

FIGURE 14.15 **Yeast two-hybrid assay.** (A) A scientist uses recombinant DNA technology to fuse the DNA binding domain of Gal4 to a protein of interest and the transcriptional activation domain of Gal4 to a potential binding partner. These constructs are introduced into yeast cells. If the two proteins do not interact, the DNA binding domain and transcriptional activation domain of Gal4 do not interact. However, **(B)** if the two proteins do interact, then the Gal4 complex is complete and can activate transcription of a reporter gene.

the Gal4 BD and AD fragments will become physically close enough to cause transcription of the downstream coding sequence (Figure 14.15B). If the gene adjacent to the UAS is a reporter gene, such as *lacZ*, the scientist will be able to identify the interaction between the two proteins.

A yeast two-hybrid assay is useful to test hypotheses that two proteins interact, but this assay can also be used as a screen to identify potential binding partners for a protein of interest (Figure 14.16). For example, a scientist can identify unknown "prey" for a protein of interest by attaching the activation domain to a large mixture of DNA fragments from a cDNA library (see Chapter 9). Individual ligation products are then introduced into yeast cells containing the target protein. If one of the members of the library encodes a protein that interacts with the target protein, the two Gal4 fragments will interact and activate transcription of the reporter. The scientist can then identify the yeast colony expressing the reporter, purify and sequence the DNA attached to the activation domain, and determine the identity of the binding protein. While this assay shows that the proteins *can* interact, it does not show that they interact *in vivo* or even in mammalian cells.

INVESTIGATING POSTTRANSLATIONAL MODIFICATIONS

A **posttranslational modification** (PTM) is a biochemical modification that occurs to one or more amino acids on a protein after the protein has been translated by a ribosome. One of the most commonly identified PTMs is the addition or removal of a phosphate group (Figure 14.17). Other common PTMs are listed in Table 14.3. These modifications often change the ultrastructural and functional properties of a protein, causing substantial downstream signaling effects. For example, the phosphorylation state of a protein often indicates whether it is functional or inactive. Thus, it would be informative to find proteins that can activate a protein of interest by adding a phosphate group or find complementary proteins that can inactivate a protein of interest by removing the phosphate group.

There are multiple questions a scientist may ask about a protein of interest regarding PTMs. For example, a scientist might want to know if a protein of interest receives a PTM. This question can be answered using mass spectrometry (Box 14.2) to analyze the molecular weights of protein fragments, identifying fragments with a PTM. Other questions include whether a specific enzyme is responsible for modifying a protein of interest, or how the PTM changes over time following a stimulus. These questions can be answered using the following techniques.

PTM-Specific Assays

PTM-specific assays are designed to test the hypothesis that a particular protein mediates the posttranslational modification of another protein. Each PTM has its own biochemical assay. Here, we describe a common type of assay: a kinase assay (Figure 14.18).

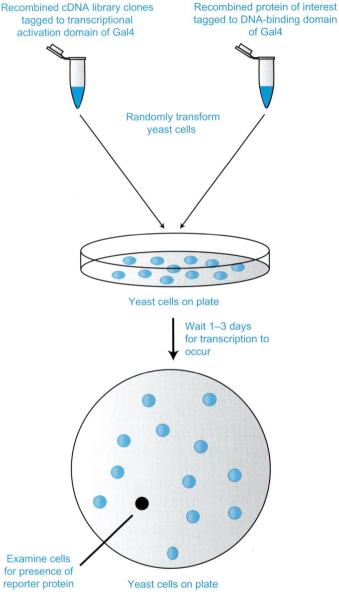

FIGURE 14.16 Using a yeast two-hybrid assay as a screen. Recombinant DNA technology is used to fuse thousands of genetic sequences from a cDNA library to the Gal4 transcriptional activator. These constructs, as well as a construct in which a protein of interest is fused to the Gal4 DNA binding domain, are randomly introduced into yeast cells. In some yeast cells, a target protein will interact with the protein of interest, and the Gal4 will activate a reporter gene. A scientist can visualize the expression of the reporter by examining the color of cell colonies grown on a plate. The scientist can then extract and sequence the DNA from these colonies to determine the identity of the protein-binding partner.

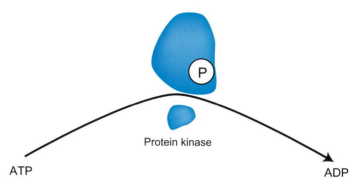

FIGURE 14.17 Phosphorylation. A protein kinase adds a phosphate group from ATP to a substrate protein.

TABLE 14.3 Common Posttranslational Modifications (most of these PTMs can be reversed by opposing enzymes)

Classification	Chemical Modification	Category of Enzyme Mediating the Modification
Phosphorylation	Addition of a phosphate group ($-PO_4$) to a residue of serine, tyrosine, threonine, or histidine	Protein kinase
Acetylation	Addition of an acetyl group ($-COCH_3$) at the N-terminus of a protein or a lysine residue	Acetyltransferase
Methylation	Addition of a methyl group ($-CH_3$) at a residue of lysine or arginine	Methyltransferase
Glycosylation	The addition of a glycosyl group to a residue of asparagine, hydroxylysine, serine, or threonine, resulting in a glycoprotein	(Multiple types of enzymes mediate this reaction)
Sulfation	Addition of a sulfate group (SO_4) to a tyrosine residue	Tyrosylprotein sulfotransferase
Farnesylation	Addition of a farnesyl group ($C_{15}H_{26}O$) to the C-terminus of a protein. Usually causes the protein to anchor to the cell membrane	Farnesyltransferase
Ubiquitination	Addition of a ubiquitin protein to a lysine residue. Usually targets proteins for proteasomal degradation.	Ubiquitin ligase

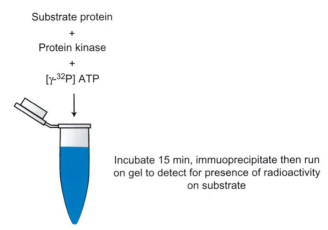

Substrate protein
+
Protein kinase
+
[γ-32P] ATP

Incubate 15 min, immuoprecipitate then run
on gel to detect for presence of radioactivity
on substrate

FIGURE 14.18 **A kinase assay.** A scientist mixes a protein, a protein kinase, and radioactive ATP into a reaction tube to determine if the kinase is able to phosphorylate the protein.

A **kinase assay** is used to determine whether one protein is capable of phosphorylating another protein. As mentioned previously, protein kinases are enzymes that catalyze the covalent transfer of phosphate from ATP to a substrate protein. A scientist can incorporate a radioactive label, such as $[^{32}P]$ orthophosphate, into molecules of ATP. This radioactive ATP acts as the phosphate donor for the protein kinase substrate. To perform this assay, the scientist combines the hypothetical protein kinase, the substrate protein, and radioactive ATP into a reaction tube. Next, the scientist isolates the substrate protein using immunoprecipitation and visualizes the presence of ^{32}P using autoradiography (Figure 14.19).

Other examples of PTM-specific assays include the methyltransferase or acetyltransferase assays, which are used to determine whether a protein is capable of adding a methyl or acetyl group, respectively, to a target protein. However, kinase assays are easily the most common PTM assay represented in the literature, as at least 2% of the genes in the mammalian genome code for protein kinases.

PTM-Specific Antibodies

It is often possible to produce an antibody to a protein of interest that is directed against a site on the protein that received a posttranslational modification (Figure 14.20). Such an antibody binds to its substrate protein only when the PTM is present. PTM-specific antibodies can be used in any of the previously described antibody-based techniques to assay the extent of modifications, or with spatially sensitive techniques such as IHC or IEM to determine the localization of modified proteins. Such analysis is especially informative when dealing with dynamic changes in protein localization and activation.

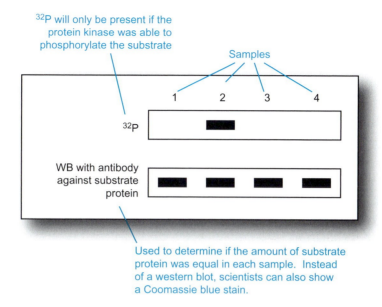

FIGURE 14.19 **Typical data from a kinase assay experiment.**

Antibody recognizes epitope on a phosphorylated protein,
but not in the absence of the phosphate group

FIGURE 14.20 **PTM-specific antibodies.**

By exposing a given sample to various molecular signals, such as growth factors, a scientist can assay any changes in localization as well as changes in activity corresponding to the modifications.

INVESTIGATING PROTEIN-DNA INTERACTIONS

The final step in many intracellular signaling pathways is the activation of a protein that directly affects gene transcription. Scientists often ask whether a specific protein of interest is able to bind to DNA and affect the expression of a specific target gene. Three techniques are commonly used to examine interactions between proteins and DNA: the electrophoretic mobility shift assay, chromatin immunoprecipitation, and the luciferase assay. Because these techniques answer slightly different questions about protein-DNA interactions,

BOX 14.3 Walkthrough of an Intracellular Signaling Experiment

Let's say that you work in a laboratory that studies a specific aspect of neural development: how a neuron's axon correctly extends to its postsynaptic target. The tip of an axon is referred to as the axonal growth cone, and it typically reaches its destination by responding to extracellular guidance cues produced by other cells. There are three general steps that occur to guide the growth cone: (1) an extracellular guidance cue is secreted to pattern the axon's position; (2) a receptor on the axon's surface recognizes that cue and translates it into an intracellular signaling cascade that will affect other proteins within the developing neuron; and (3) various downstream organelles receive and respond to the signal, including the cytoskeleton that must be reorganized to move the growth cone toward the site where the initial cue was detected. If you know the identity of the receptor and the extracellular cue, you might be interested in figuring out how the receptor translates the cue into a response. What is the intracellular signaling pathway that leads to the cytoskeletal changes necessary for the growth cone to turn toward the cue? Let's focus on identifying the next step of signaling that occurs after the receptor is activated: the activation of a second protein that binds to the intracellular domain of the receptor.

There are multiple approaches for finding proteins that interact with the receptor. For example, you could immunoprecipitate the receptor and then perform mass spectrometry to identify any other proteins that coprecipitated. However, this approach would not tell you if the two proteins interacted directly or were simply part of a larger complex of proteins. Another approach would be to perform a yeast two-hybrid assay. This technique is used to identify protein-binding partners that directly interact with the receptor. To perform this assay, you would begin by genetically engineering a construct that attaches the receptor's intracellular domain to the DNA-binding domain of a transcriptional activator, such as Gal4, to act as the bait. Then you would create or use a cDNA library to attach different cDNA molecules to the activation domain of Gal4 to act as the prey. Finally, you would introduce the bait and prey constructs into yeast cells that express a reporter gene, often *lacZ,* under the control of a promoter that is recognized by the transcriptional activator (*UAS* for the Gal4 activator). Thus, the bait will bind to the promoter but cannot initiate transcription of the reporter until a specific prey with the activation domain interacts with the bait to create the full activator. Then, the reporter gene will be transcribed and can be detected.

In order to detect yeast cells that express *lacZ,* you could grow thousands of cells on an agar plate that also contains X-gal, the substrate that causes a visible blue by-product to form in those cells. The final step would be to pick blue yeast colonies off the plate, extract and sequence the DNA of the "prey," and identify the protein that interacts with the receptor based on the cDNA sequence. This screen allows you to conclude that the prey you found *can* interact with the intracellular domain of the receptor. Most likely, you will have discovered a number of positive hits to sequence and should confirm the interaction's specificity to determine which would be a good candidate for a follow-up.

This yeast-two hybrid experiment is a good start, but you could learn even more about the receptor-protein interaction by attaching different functional regions of the receptor's intracellular end to use as bait. This would tell you with finer resolution exactly what part of the receptor is interacting with the prey or allow you to capture only prey that will interact with a specific region—say, the kinase domain—or a particular region you know to be functionally important. To convince yourself and other scientists of the interaction between the receptor and intracellular protein, you could perform additional, complementary experiments. For example, coimmunoprecipitation, as just described, is often used to verify yeast two-hybrid results. You could also use protein-affinity chromatography. In cell signaling studies, there is no such thing as too much evidence; figures often contain several different methods to validate the same result.

Now that you have a candidate (or several candidates, based on the results of the yeast two-hybrid experiments), you should follow up with other experiments to make sure that your candidate protein makes sense as a protein that mediates the intracellular response to a guidance cue. One question you might ask right away is whether the candidate protein is expressed in the right place (the axon growth cone) and at the right time (the embryonic or postnatal period during which the axon grows toward its target). If the interacting protein is not present in axonal growth cones or at a time when axons are growing out, the interaction may not be relevant to the growth cone turning behavior you're interested in, or the interaction may not actually occur. To examine protein expression, you could use immunohistochemistry to stain for the expression of the protein in histological preparations of the growth cone at different developmental timepoints. You could also collect tissue samples containing the growth cones from different timepoints and perform western blots to examine the expression of the protein at each time point.

If you have determined that a protein binds to the receptor, is expressed in the growth cone, and is expressed during the developmental period in which the growth cone migrates toward its target, you have laid the foundation for many future experiments. For example, you could ask whether the interaction between the receptor and the protein causes the protein to be phosphorylated. You could use techniques described in other chapters, such as RNA interference (RNAi—Chapter 12), to investigate the consequence of knocking down the expression of the intracellular protein. There are many future experiments that can and should be performed, and the more complementary lines of evidence you collect, the more you will know and the better your study will be received.

scientists often use them in concert to strengthen the evidence that a protein interacts with DNA and affects the expression of target genes.

Electrophoretic Mobility Shift Assay (EMSA)

An **electrophoretic mobility shift assay** (also known as a gel shift assay) is used to determine if a protein is able to directly interact with a short, specific sequence of DNA. Before the experiment begins, the investigator hybridizes two complementary DNA strands (about 30–40 bp in length) and labels the strands with a radioactive probe. The investigator also purifies the protein that is hypothesized to interact with the strand of DNA.

During the experiment, the radiolabeled DNA is run in several lanes of a polyacrylamide gel (Figure 14.21). The speed at which the DNA molecules move from one side of the gel to the other during electrophoresis is determined by their size. In one lane, the DNA is run by itself. In another lane, the DNA is run with the purified protein. If the protein interacts with the DNA strand, the size of the DNA-protein complex will be greater than the DNA strand alone, and therefore the band of DNA will be shifted upwards on the gel (Figure 14.22).

Several controls are necessary for a proper EMSA experiment: (1) in addition to the DNA sequence of interest, the investigator should run a slightly scrambled DNA sequence to ensure that the DNA-protein interaction is specific to the exact sequence. (2) The investigator should perform a "competitive" binding assay in which nonradiolabeled DNA is added to the reactions with the radiolabeled DNA. If there is sufficient nonradiolabeled DNA, the amount of protein bound to the radiolabeled DNA will decrease, and the size

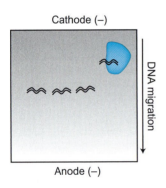

Cathode (−)

DNA migration

Anode (−)

FIGURE 14.21 Electrophoretic mobility shift assay (EMSA). A scientist radioactively labels short DNA sequences, mixes these sequences with purified proteins, and runs the mixture on a gel. Any DNA sequences that interact with a protein will migrate much slower on the gel.

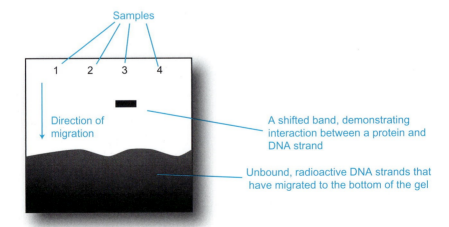

Samples

1   2   3   4

Direction of migration

A shifted band, demonstrating interaction between a protein and DNA strand

Unbound, radioactive DNA strands that have migrated to the bottom of the gel

FIGURE 14.22 Typical data from an EMSA experiment.

of the radiolabeled DNA band will be shifted back to normal. (3) An antibody to the purified protein should be added to a lane with the protein and labeled DNA. This will further increase the size of the DNA-protein complex and create a "supershifted" band, demonstrating that it is indeed the protein of interest that is causing the shift in the first place. (4) The investigator could run a non-specific protein known not to interact with DNA, such as a membrane-bound protein, to ensure that there is nothing "sticky" about the DNA strand in the EMSA reaction.

One of the strengths of an EMSA assay is that it demonstrates that a protein can directly bind to a sequence of DNA. One of the limitations is that all reactions take place in a test tube. Therefore, the only conclusions that can be reached using this assay is that a protein is capable of binding DNA, but not that it actually does so in a living cell. For this conclusion, it is necessary to perform a chromatin immunoprecipitation assay.

Chromatin Immunoprecipitation (ChIP)

Chromatin immunoprecipitation is used to determine if a protein interacts with a specific region of DNA. The advantage to the ChIP method over an EMSA assay is that the starting material comes from living cells, most often cultured cells, though fresh tissue is sometimes used. However, unlike an EMSA, it is impossible to determine if a protein interacts with a sequence of DNA directly or if it is part of a greater macromolecular complex.

The fundamental concept behind a ChIP assay is that before the experiment begins, a protein of interest is physically bound to DNA or is part of a DNA-protein complex in living cells (Figure 14.23). The scientist begins the experiment by quickly fixing the cells in a diluted formaldehyde solution so that DNA-protein complexes are **cross-linked** and remain intact. Next, the DNA is sonicated and sheared into approximately 600–1000 bp fragments. If the cross-linking is adequate, the proteins will still be bound to sonicated DNA. At this stage, the scientist immunoprecipitaes the DNA-protein complexes using an antibody against the protein of interest.

After the immunoprecipitation step, the purified proteins are "reverse cross-linked" so that the DNA fragments can be purified. If the scientist has a hypothesis about the identity of the bound DNA sequence, he or she can then determine the identity of the DNA fragments by performing a PCR reaction with primers for a specific region of DNA. The final result of a ChIP assay is usually an agarose gel revealing the end result of PCR reactions (Figure 14.24). Alternatively, quantitative real time PCR (qRT-PCR) can also be performed to better quantify the amount of PCR product in each lane.

What if a scientist doesn't have a hypothesis as to which DNA sequences are bound to a protein of interest? It is possible to screen all DNA sequences in the genomes of commonly used organisms by performing a **ChIP on chip** assay. In this assay, the DNA sequences obtained at the end of a ChIP experiment are

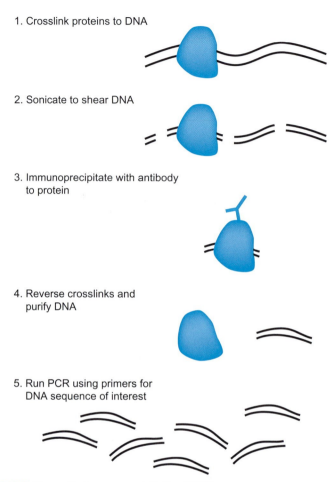

1. Crosslink proteins to DNA

2. Sonicate to shear DNA

3. Immunoprecipitate with antibody
   to protein

4. Reverse crosslinks and
   purify DNA

5. Run PCR using primers for
   DNA sequence of interest

FIGURE 14.23 Chromatin immunoprecipitation (ChIP).

run on a DNA microarray (Chapter 8) to determine all of the DNA sequences bound to the immunoprecipitated protein.

Several positive and negative controls are necessary for a proper ChIP experiment. An important positive control is an input lane containing PCR products from DNA obtained after sonication but prior to immunoprecipitation. This should result in a strong signal and ensures that the DNA of interest was properly obtained in the first place. Another important positive control is to perform an immunoprecipitation reaction with an antibody for a protein known to bind to any region of DNA, such as a histone protein. ChIP results using a histone antibody should therefore produce a strong DNA signal. Negative controls include performing an immunoprecipitation with nonspecific antibodies (random IgG is usually used) and using primers that amplify nonspecific DNA regions.

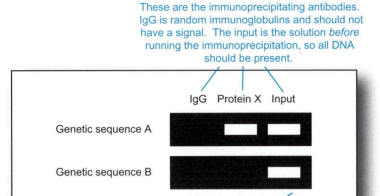

These are the immunoprecipitating antibodies. IgG is random immunoglobulins and should not have a signal. The input is the solution *before* running the immunoprecipitation, so all DNA should be present.

IgG   Protein X   Input

Genetic sequence A

Genetic sequence B

These are all PCR products, DNA run on an agarose gel

FIGURE 14.24 **Typical data from a ChIP experiment.** This data indicates that Protein X interacts with Sequence A but not Sequence B.

Luciferase Assay

A **luciferase assay** is used to determine if a protein can activate or repress the expression of a target gene. Unlike the ChIP or EMSA assays, which only assess the ability of a protein to interact with a region of DNA, a luciferase assay is able to establish a functional connection between the presence of the protein and the amount of gene product that is produced. This assay is unable to determine whether the protein directly interacts with DNA itself; the protein could indirectly affect transcription by activating or repressing a distinct protein that itself directly affects transcription.

Luciferase is an enzyme used for bioluminescence by various organisms in nature, most famously the firefly. The scientist produces a construct in which the regulatory region of a target gene is fused with the DNA coding sequence for luciferase (Figure 14.25). A separate DNA construct encodes the protein hypothesized to affect transcription. The scientist transfects a cell culture system, such as HEK 293T cells, with both constructs. If the protein is able to upregulate transcription of the target gene, the cells will express luciferase. If the protein downregulates transcription, the cells will express less luciferase than normal. The scientist examines the expression of luciferase about 2–3 days after the initial cell transfection. The cells are lysed, and the cell contents are placed in a reaction tube. If the proper substrate is added, luciferase will catalyze a reaction that produces light. This light is detected with a luminometer, a device that precisely quantifies how much light is produced in each reaction tube. The amount of light produced provides a quantitative measure of the affect of the protein on expression of the target gene.

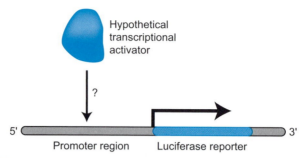

FIGURE 14.25 Luciferase assay. A scientist produces DNA constructs coding for a hypothetical transcriptional activator and a reporter sequence and introduces these constructs into cells in culture. If the protein is able to activate transcription, the cell will produce the luciferase reporter protein. The amount of luciferase produced can be quantified using a luminometer.

Due to variability in conditions within each well in a cell culture plate, a proper luciferase assay experiment should repeat each condition at least in triplicate. The scientist should compare luciferase expression between conditions in which (1) no protein is present, (2) protein is present, and (3) the regulatory sequence regulating luciferase expression is mutated so that the protein is no longer able to bind. If a scientist knows enough about the structure of a protein such that it is possible to mutate the DNA-binding domain or enhance the ability of a protein to affect transcription, such conditions would also strengthen the conclusions of a luciferase assay experiment.

CONCLUSION

The purpose of this chapter was to survey the common methods used to investigate protein expression and signaling. Although these methods are numerous, they essentially fall into four categories: (1) determining whether a protein is expressed in a particular population of cells and how much; (2) determining whether a protein interacts with other proteins; (3) determining the types of posttranslational modifications a protein can receive and which proteins cause these modifications; and (4) determining whether a protein is capable of interacting with DNA and/or regulating gene expression. Anyone outside the fields of biochemistry and cell signaling should not be intimidated by the huge variety of techniques. Focusing on the experimental questions rather than the methods can clarify the results and conclusions of an intracellular signaling study.

SUGGESTED READING AND REFERENCES

Books

Berg, J. M., Tymoczko, J. L., Stryer, L. (2007). *Biochemistry*, 6th ed. W. H. Freeman, NY.

Fields, R. D. (2008). *Beyond the Synapse: Cell-Cell Signaling in Synaptic Plasticity.* Cambridge University Press, Cambridge.

Wilson, K., & Walker, J. M. (2005). *Principles and Techniques of Biochemistry and Molecular Biology*, 6th ed. Cambridge University Press, NY.

Review Articles

Araujo, S. J. & Tear, G. (2003). Axon guidance mechanisms and molecules: lessons from invertebrates. *Nat Rev Neurosci* **4**, 910–922.

Greer, P. L. & Greenberg, M. E. (2008). From synapse to nucleus: calcium-dependent gene transcription in the control of synapse development and function. *Neuron* **59**, 846–860.

Guan, K. L. & Rao, Y. (2003). Signalling mechanisms mediating neuronal responses to guidance cues. *Nat Rev Neurosci* **4**, 941–956.

Tronson, N. C. & Taylor, J. R. (2007). Molecular mechanisms of memory reconsolidation. *Nat Rev Neurosci* **8**, 262–275.

Primary Research Articles — Interesting Examples from the Literature

Cowan, C. W., Shao, Y. R., Sahin, M., Shamah, S. M., Lin, M. Z., Greer, P. L., Gao, S., Griffith, E. C., Brugge, J. S. & Greenberg, M. E. (2005). Vav family GEFs link activated Ephs to endocytosis and axon guidance. *Neuron* **46**, 205–217.

Dolmetsch, R. E., Pajvani, U., Fife, K., Spotts, J. M. & Greenberg, M. E. (2001). Signaling to the nucleus by an L-type calcium channel-calmodulin complex through the MAP kinase pathway. *Science* **294**, 333–339.

Flavell, S. W., Cowan, C. W., Kim, T. K., Greer, P. L., Lin, Y., Paradis, S., Griffith, E. C., Hu, L. S., Chen, C. & Greenberg, M. E. (2006). Activity-dependent regulation of MEF2 transcription factors suppresses excitatory synapse number. *Science* **311**, 1008–1012.

Kim, M. S., Pak, Y. K., Jang, P. G., Namkoong, C., Choi, Y. S., Won, J. C., Kim, K. S., Kim, S. W., Kim, H. S., Park, J. Y., Kim, Y. B. & Lee, K. U. (2006). Role of hypothalamic Foxo1 in the regulation of food intake and energy homeostasis. *Nat Neurosci* **9**, 901–906.

Lee, S. K. & Pfaff, S. L. (2003). Synchronization of neurogenesis and motor neuron specification by direct coupling of bHLH and homeodomain transcription factors. *Neuron* **38**, 731–745.

Visel, A., Blow, M. J., Li, Z., Zhang, T., Akiyama, J. A., Holt, A., Plajzer-Frick, I., Shoukry, M., Wright, C., Chen, F., Afzal, V., Ren, B., Rubin, E. M. & Pennacchio, L. A. (2009). ChIP-seq accurately predicts tissue-specific activity of enhancers. *Nature* **457**, 854–858.

Protocols

Aparicio, O., Geisberg, J. V., Sekinger, E., Yang, A., Moqtaderi, Z. & Struhl, K. (2005). Chromatin immunoprecipitation for determining the association of proteins with specific genomic sequences in vivo. *Curr Protoc Mol Biol*, Chapter 21, Unit 21.3.

Golemis, E. A., Serebriiskii, I., Gyuris, J. & Brent, R. (2001). Interaction trap/two-hybrid system to identify interacting proteins. *Curr Protoc Neurosci*, Chapter 4, Unit 4.4.

Harlow, E., & Lane, D. (1999). *Using Antibodies: A Laboratory Manual*. Cold Spring Harbor Laboratory Press, Cold Spring Harbor, NY.

Pandey, A., Andersen, J. S. & Mann, M. (2000). Use of mass spectrometry to study signaling pathways. *Sci STKE* **2000**, PL1.

Walker, J. M. (2002). *The Protein Protocols Handbook*, 2nd ed. Humana Press, Totowa, NJ.

Glossary

Absolute recovery In microdialysis, the total amount of substance collected in the dialysate over a set amount of time.

Action potential A rapid, all-or-none depolarization of the cell membrane of a neuron that causes an electrochemical impulse to travel down the length of an axon. These impulses allow a neuron to communicate with other cells over great distances.

Acute culture A primary cell or tissue culture used for short-term experiments, usually lasting less than a day. For example, a slice culture for patch clamping experiments.

Adeno-associated virus (AAV) A type of virus used as a DNA delivery tool that infects dividing and postmitotic cells. The DNA integrates into the nucleus, permitting long-term gene expression.

Adenovirus A type of virus used as a DNA delivery tool that infects dividing and post-mitotic cells. The DNA does not integrate into the genome, so expression is often transient, lasting weeks or months.

Aequorin A chemiluminescent protein derived from the jellyfish *Aequorea victoria* that emits blue light when bound to calcium. This protein has been useful for imaging calcium activity in living cells.

Affinity chromatography A form of chromatography that uses small molecules fixed to a porous column to bind to proteins with a high affinity, such as a substrate binding to an enzyme.

Agonist A compound that can bind and activate an endogenous receptor, thus mimicking an endogenous ligand

Allatostatin receptor A ligand-gated receptor found in insects. In the presence of the insect hormone allatostatin, the receptor causes hyperpolarization of cells. Thus, this receptor can be used as a genetically coded tool to inhibit neural activity.

Amino acid A category of molecule consisting of amine and carboxyl functional groups, as well as a variable side chain. Used as the monomeric building blocks of proteins, as well as intermediates in metabolism.

Amperometry A form of voltammetry that uses an electrode to detect neurochemicals in living tissue. Unlike in other voltammetry techniques, the electrode is held at a specific, constant voltage.

Amplifier A device used in electrophysiology recordings to amplify the relatively weak electrical signals derived from a recording electrode.

Anisotropic Describes diffusion of substances that is not equal in all directions but instead tends to move randomly along a single axis. The opposite of **isotropic.**

Antagonist A compound that can bind to a receptor but does not cause activation. Therefore, it prevents the biological activity of an endogenous ligand.

Anterograde tracer A chemical probe that labels efferent axon tracts from the cell body through the axon to the presynaptic terminal.

Antibody Specialized protein used by an organism's immune system to detect and bind to foreign proteins. Used by neuroscientists to identify or purify proteins of interest.

Antigen A foreign substance recognized by an antibody.

Array tomography A technique that combines immunohistochemistry techniques with ultrathin, serial sections. This technique improves spatial resolution of traditional immunohistochemistry experiments and allows scientists to create highly detailed three-dimensional images of protein expression.

Artificial chromosome A large DNA molecule that contains the telomeric, centromeric, and replication origin sequences necessary for replication and preservation in yeast, bacterial, or phage. Useful to neuroscientists as a vector for cloning relatively large DNA fragments (50–350 kb).

Ataxia A neurological condition exhibited as a gross lack of coordination of muscle movements.

Ataxin A genetically encoded toxin that can be used to ablate specific neurons of interest.

Autoclave A large device used to sterilize tools, glassware, and reagents.

Autofluorescence The natural ability of some structures to fluoresce, even without the addition of exogenous fluorescent tags or dyes.

Autoradiography A technique that uses the pattern of decay emissions produced from a radioactive substance to form an image on an X-ray film. Used in neuroscience to detect radioactive probes.

Axial An imaginary plane that divides the brain and body into superior and inferior sections (top to bottom).

BAC See *Bacterial artificial chromosome.*

BAC construct/transgenic A recombinant DNA or recombineering construct that uses a bacterial artificial chromosome as the vector. Useful for creating transgenic animals with large, endogenous sequences of DNA.

Bacterial artificial chromosome A large (50–250 kb) DNA vector capable of replicating in bacteria. Useful for manipulating large sequences of DNA, such as in the generation of transgenic or knockout constructs.

Bacteriophage Any one of a number of viruses that infect bacteria. Sometimes simply referred to as "phage."

Barnes maze A maze used to test spatial learning and memory in rodents. The animal must learn the location of a drop box on a large circular table using spatial cues placed around the maze.

Base pair Describes a single nucleotide-nucleotide complementary match between the two strands of DNA. These base pairs are always hydrogen bonds between adenine-thymine or cytosine-guanine bases.

Basic Local Alignment Search Tool (BLAST) A free public access database used to identify and compare DNA and protein sequences.

Basophilic stain A histological stain used to visualize cell bodies.

Between-subjects study A study that compares two different groups of human subjects or animals. For instance, in functional imaging experiments, a between-subjects study may test the differences of old versus young, or male versus female participants.

Bimolecular fluorescence complementation (BiFC) A technique used to visualize the interaction between two proteins. Each protein is fused to half of a fluorescent protein. When the proteins interact, the two halves of the fluorescent protein can fold together to emit light and report the interaction.

Binary expression system An approach that uses two or more constructs to express a transgene in cells. Used to refine the spatial or temporal properties of transgene expression.

Biolistics A method of DNA delivery that uses a gene gun to physically shoot DNA-coated bullets into cells.

Biotin A small molecule with an extremely high affinity for another molecule called "avidin." Used to tag molecular probes so that future histochemical processing with avidin-bound probes can enhance a signal.

BLAST See *Basic Local Alignment Search Tool*.

Blastocyst An early embryonic structure that follows the fertilization of an egg.

Bleed-through A phenomenon in fluorescent microscopy that occurs when the signal of one fluorophore appears when a scientist uses a filter designed to detect the signal of a separate fluorophore.

Blocked design A strategy for presenting stimuli to human subjects in functional imaging experiments. Stimuli of the same category are presented grouped together in blocks. An alternative to an **event-related design**.

Blood oxygen level dependent (BOLD) effect The phenomenon in which the brain microvasculature increases the flow of oxygen-rich blood to neurons following an increase of neural activity.

Blunt ends Ends of fragments of DNA cut with restriction enzymes that result in no 5′ or 3′ overhangs; both strands of DNA terminate at the same complementary base pair. Opposite of **sticky ends**.

BOLD effect See *Blood oxygen level dependent effect*.

Brain atlas A tool that depicts brain anatomy in three-dimensional coordinates.

Bregma A landmark on the surface of the skull, defined as the intersection between the sagittal and coronal sutures.

Brightfield microscopy A type of light microscopy in which white light passes directly through or is reflected off a specimen.

Brodmann's areas Divisions of the cortex defined by the cytoarchitecture of the cortical layers. Developed by Korbinian Brodmann in the early twentieth century and still used today.

Calcium phosphate transfection A method of delivering DNA into cells by using chemical reactions that form calcium phosphate precipitates.

Cannula A narrow, cylindrical tube that can be implanted on the surface of the skull for long-term access to deep brain structures in a living animal.

Capsaicin receptor A genetically encoded, ligand-gated ion channel that depolarizes neurons in response to the capsaicin molecule. Can be used by neuroscientists to stimulate genetically defined populations of neurons.

cDNA "Complementary-DNA," a fragment of DNA produced from an RNA template following reaction with the reverse transcriptase enzyme.

cDNA library A collection of hundreds or thousands of DNA plasmids, each containing a unique cDNA molecule. Used to collect and store DNA sequences that were actively expressed in a sample prior to exposure with the reverse transcriptase enzyme.

Cell-attached mode A patch-clamp technique in which a glass micropipette forms a gigaseal with the membrane of a cell, but the membrane is not punctured.

Cell culture An *in vitro* approach in which a scientist grows and maintains cells under carefully controlled conditions outside of a living animal.

Cell fractionation A process in which a scientist centrifuges a sample multiple times at various speeds to divide the sample into different cellular components and organelles.

Cell signaling See *Intracellular signaling*.

Central dogma of molecular biology The overarching model of information flow in cells, briefly stated as "DNA codes for molecules of RNA, and molecules of RNA code for proteins."

Cerebral angiogram A structural brain imaging technique that uses X-rays and radio-opaque dyes to produce images of the brain's vasculature.

Channelrhodopsin-2 (ChR2) A genetically encoded cation channel capable of depolarizing a neuron with millisecond-timescale resolution in response to stimulation with blue light.

Chemical gene delivery The process by which a scientist uses a chemical reaction to deliver DNA into cells.

Chemical mutagenesis The use of chemical agents, such as EMS or ENU, to mutagenize hundreds or thousands of eggs/larvae for the purpose of performing a forward genetic screen.

Chemosensory jump assay A behavioral test in flies that exploits the tendency of flies to exhibit a startle response and jump when encountering a novel odor.

Chimeric mice Mice composed of patches of tissues from different genetic sources. An intermediate step in the procedure to create knockout or knockin mice.

ChIP on chip assay An assay in which chromatin immunoprecipitation is followed up with a microarray to determine the identity of all genetic sequences bound to a particular protein *in vitro* or *in vivo*.

Chromatin immunoprecipitation (ChIP) A technique used to determine whether a sequence of DNA physically interacts with a specific protein, usually a transcription factor or histone protein.

Chromatography A category of biochemistry techniques used to separate proteins on a column based on their size, charge, substrates, or other properties.

Chromogenic label A label on a probe that allows a scientist to perform a simple biochemical reaction that produces a colored byproduct. This byproduct can be visualized using simple light microscopy.

Chromophore-assisted laser inactivation (CALI) A method used to inhibit the activity of specific proteins in precise subcellular regions by tagging these proteins with a specialized dye molecule that becomes toxic after stimulation with laser light.

Chronic mild stress A series of mildly stressful events (periods of constant light, tilting an animal's cage, wetting the animal's bedding, etc.) that is used to create rodent models of depression and anxiety.

Circadian rhythms The regular, roughly 24-hour cycle of stereotyped biochemical, physiological, and behavioral processes.

Classical conditioning A paradigm of animal learning in which a scientist couples an initially neutral stimulus, such as a tone or light, with a salient stimulus, such as food or an electric shock. An animal eventually associates the neutral stimulus with the salient stimulus.

Cloning vector A DNA vector capable of storing DNA sequences for recombinant DNA techniques but lacking the necessary noncoding sequences that allow the DNA to be expressed in a host cell.

Co-culture system An *in vitro* culture system in which two different cultured cell types or explants are cultured together in the same chamber, allowing a scientist to examine the effect of one type of cell or tissue on another.

Codon A series of three nucleotides that code for a specific amino acid during translation.

Cognitive neuroscience The field of neuroscience dedicated to elucidating the neural basis of thought and perception.

Cohesive ends See *Sticky ends.*

Coimmunoprecipitation (Co-Ip) A method that determines whether a protein interacts with another protein.

Colorimetric label See *Chromogenic label.*

Competent cell A bacterial cell transiently capable of receiving DNA sequences through its cell wall.

Complementary DNA (cDNA) See *cDNA.*

Complementary strands Refers to two strands of DNA that have complementary base pairs, allowing the two strands to hybridize into a double-stranded molecule.

Complementation test A test used to determine if a mutant animal with an abnormal phenotype has a mutation in the same gene as a different mutant animal with the same abnormal phenotype.

Compound microscope A microscope in which two or more lenses are used in concert to enhance the magnification power.

Computerized tomography (CT)—also called Computerized Axial Tomography (CAT) A structural imaging technique in which several X-rays are taken from multiple orientations around a single plane to produce an image of the brain.

Condenser The part of a microscope that focuses light from a light source onto a specimen.

Conditional knockout An animal in which a gene is knocked out of the genome using gene targeting measures, but only in a specific cell type, and/or at a specific time.

Conditioned stimulus (CS) In classical conditioning, the neutral stimulus that eventually becomes associated with a salient stimulus.

Confluence The relative amount of space that a growing population of cultured cells takes up on the surface of a cell culture flask/plate.

Confocal microscope A specialized fluorescent microscope that can produce clear images of structures within specimens by selectively exciting and collecting emitted light from thin regions of the specimen.

Construct A recombinant DNA sequence capable of coding for a functional protein when properly expressed in host cells.

Construct validity Refers to a state in which an animal model and human model have the same underlying genetic or cellular mechanism that may result in a certain behavior, such as a transgenic mouse engineered to have the same mutation that causes Alzheimer's disease in humans.

Contextual fear conditioning A form of classical conditioning in which an animal associates a neutral stimulus with an aversive stimulus, such as a foot shock, and the animal subsequently displays a fearful behavior, such as freezing, when placed in the training chamber, even in the absence of the aversive stimulus.

Coomassie blue A blue dye used to stain proteins in SDS-PAGE gels.

Coronal plane A neuroanatomical plane that divides the brain into anterior and posterior sections (from ear to ear).

Cosmid A large (30–50 kb) DNA plasmid capable of replicating in bacteria. Useful for engineering very large DNA constructs.

Countercurrent apparatus A device used to fractionate a population of flies based on their behavioral responses to a sensory stimulus.

Cranial window An implant on the surface of a rodent's skull that allows for direct observation of fluorescent neurons *in vivo*.

Craniotomy A procedure used to remove a small section of the skull during a surgical procedure.

Cre/lox system A binary expression system used to control the spatial expression of a transgene.

Cre recombinase A protein derived from bacteriophage that recognizes lox sites. Cre excises any DNA sequences between two lox sites oriented in the same direction and inverts any DNA sequences between two lox sites oriented in inverted directions.

Cross-linking fixatives Chemical fixatives that create covalent chemical bonds between proteins in tissue and cells.

Cross-talk See *Bleed-through.*

Cryostat A device used to freeze and cut tissue specimects into 10–50 μm sections.

Cued fear conditioning A form of classical conditioning in which an animal associates a neutral stimulus with an aversive stimulus, such as a foot shock, and then displays freezing behaviors in the presence of the formerly neutral stimulus, even in the absence of foot shock.

Current (I) The rate of flow of electrical charge over time.

Current clamp An electrophysiological technique in which a scientist injects current into a cell and measures the change of voltage over time.

Cyclic voltammogram The end result of a fast-scan cyclic voltammetry experiment that plots the amount of measured current versus the applied voltage. This data is useful for detecting the presence of various neurochemicals in neural tissue.

Cyclotron A large machine that produces positron-emitting isotopes, such as the kind used in positron emission tomography (PET) experiments.

Cytokine Extracellular signaling molecules used in intercellular communication.

Darkfield microscopy A form of light microscopy that uses an oblique light source to illuminate a specimen from the side so that only scattered light enters the objective lens. This causes some organelles to stand out from a dark background.

Defensive marble burying assay A behavioral paradigm used to assay anxiety in a rodent. Marbles are placed in an animal's cage, and a scientist measures how many marbles an animal buries in the cage bedding over a fixed amount of time. Animals that are more anxious typically bury more marbles than nonanxious animals.

Dehydrating fixatives Histological fixatives that disrupt lipids and reduce the solubility of protein molecules, precipitating them out of the cytoplasmic and extracellular solutions.

Dependent variable The variable in an experiment that is measured and dependent upon the value of an independent variable.

Depolarization A decrease in the absolute value of a cell's membrane potential.

Dicer An enzyme that recognizes double-stranded RNA and cleaves it into fragments about 23 bp long (siRNA). This siRNA-dicer complex then recruits additional cellular proteins to form an RNA-induced silencing complex (RISC), which mediates the phenomenon of RNA interference (RNAi).

Differential interference contrast (DIC) microscopy A type of brightfield microscopy that uses optical modifications within the microscope to exaggerate changes in the light-scattering properties of cellular structures, producing a three-dimensional, textured appearance.

Diffuse optical imaging (DOI) A type of noninvasive optical imaging method in which light is reflected off neural structures through the scalp.

Diffusion The passive transport of molecules from a region of higher concentration to a region of lower concentration caused by random molecular interaction.

Diffusion MRI A technique used to image fiber tracts in the brain by measuring the anisotropic diffusion of water molecules along fiber bundles.

Diffusion tensor imaging (DTI) A diffusion MRI method that measures the anisotropic diffusion of water molecules along fiber tracts in the brain to image connections between different brain regions.

Digoxigenin A steroid molecule frequently used to tag molecular probes in histology experiments.

Diphtheria toxin receptor A genetically encoded receptor that will kill an expressing cell in the presence of the diphtheria toxin, which is normally harmless to mammalian neurons.

Dipstick assay See *Olfactory avoidance assay.*

Direct IHC A category of immunohistochemistry experiments in which the primary antibody is directly conjugated to a fluorescent or chromogenic label. This typically produces a weaker signal than indirect IHC.

Dissecting microscope See *Stereomicroscope.*

Dissociated cell culture A primary cell culture method in which tissue removed directly from an animal is separated into individual cells.

DNA ladder A sample of various DNA fragments of known sizes, run on an agarose gel next to experimental DNA samples to determine the size of fragments in the samples.

DNA library A comprehensive collection of hundreds or thousands of cloned DNA fragments from a biological sample, stored in expression vectors or artificial chromosomes.

Dominant negative A mutated protein that interferes with the function of the wild-type protein.

Double helix The "twisted ladder" shape of a double-stranded DNA molecule.

Doxycycline A drug that binds to the tetracycline transactivator protein (tTA) or reversed transactivator protein (rtTA) in Tet-off/Tet-on transgenic animals. This drug allows scientists to gain inducible control of the Tet-off/Tet-on system, allowing temporal control of transgene activity.

Electroencephalography (EEG) A noninvasive technique used to measure the gross electrical activity of the brain.

Electrolytic lesion A physical lesion in the brain caused by injection of current.

Electromyogram A technique used to record the electrical potential generated by muscle cells and therefore to record motor activity.

Electron microscope tomography See *Electron tomography.*

Electron microscope (EM) A microscope that uses a beam of electrons rather than a beam of photons to image a specimen, greatly enhancing the resolving power.

Electron tomography (ET) A procedure used to collect and combine several transmission electron microscope images from the same specimen, allowing scientists to image the three-dimensional ultrastructure of organelles and macromolecules.

Electrophoretic mobility shift assay (EMSA) A technique used to determine if a protein is capable of directly binding to a short sequence of DNA.

Electrophysiology A branch of neuroscience and neuroscience methodology that studies the electrical activity of whole brain regions, single cells, and single ion channels.

Electroporation The process of using an electric field to deliver DNA constructs into cells.

Electrospray ionization (ESI) A technique used to cause a sample to enter a gaseous phase for mass spectrometry experiments.

Elevated plus maze A behavioral assay used to assay anxiety in rodents.

Embedding The process of surrounding a brain or tissue section with a substance that infiltrates and forms a hard shell around the tissue.

Embryonic lethal A phenotype caused by a transgenic or knockout construct that causes the death of an animal during gestation.

Embryonic stem (ES) cell A pluripotent cell capable of giving rise to any tissue in an organism.

Emission filter A filter on a fluorescent microscope that blocks extraneous wavelengths of light, including the light used to illuminate the specimen, but allows emitted light wavelengths to pass through to a detector.

Ensembl A public genomic database, useful for looking up genomic sequences and genetic information for a variety of organisms.

Enzyme-linked immunosorbent assay (ELISA) A technique used to quantify the amount of protein in a sample.

Epifluorescent microscopy Standard fluorescent microscopy. Specimens labeled with fluorescent probes are illuminated by light of the excitation wavelength. The specimen is then viewed using a second filter that is opaque to the excitation wavelength but transmits the longer wavelength of the emitted light.

Epitope The specific region of an antigen where an antibody binds.

ER/tamoxifen system A system used in genetic engineering experiments to allow inducible gene activity/suppression at specific timepoints. The estrogen receptor (ER) is sequestered in the cytoplasm. In the presence of tamoxifen, ER translocates to the nucleus. In genetic engineering experiments, ER can be fused to other useful proteins, such as Cre recombinase, to allow for inducible activation of Cre activity when a scientist adds tamoxifen.

Estrogen receptor (ER) An endogenous receptor that resides in the cytoplasm until binding with its ligand, at which point it translocates to the nucleus.

Ethology The study of natural animal behaviors.

Event-related design A strategy for presenting stimuli to human subjects in functional imaging experiments in which stimuli are presented as isolated, individual events of short duration. An alternative to a **blocked design**.

Excitation filter A filter on a fluorescent microscope that allows only light in a specific range of wavelengths to pass through to an objective to illuminate a specimen.

Excitatory postsynaptic potential (EPSP) A localized potential change in a small region of a neuron that causes the membrane potential to depolarize, bringing the membrane potential closer to the threshold to generate an action potential.

Exon A sequence of an RNA molecule that remains in the final mRNA sequence after splicing occurs.

Explant culture A primary cell culture method in which intact chunks of tissue are removed from a living organism and cultured in an *in vitro* environment.

Expression vector A DNA vector containing all of the necessary noncoding sequences necessary for a host cell to express a coding DNA sequence as a protein.

Extracellular recording An electrophysiological recording in which the recording electrode is placed outside a neuron in the extracellular environment.

Eye coil A wire loop implanted around the outer circumference of a monkey's eye. Used to ensure that the monkey correctly fixes its gaze during a visual task.

Face validity A concept that establishes the validity of an animal model by demonstrating that the animal's behavior is similar to the analogous human behavior.

Fast-scan cyclic voltammetry A technique used to measure neurochemicals in neural tissue.

Feeding acceptance assay A *Drosophila* assay that examines taste preferences. Slightly starved flies are offered a choice of appetitive, aversive, or neutral stimuli, each dyed a different color. The amount of each stimulus can be scored by examining the color of the fly's stomach.

Fiber stain A histological stain used to mark the presence of fiber tracts in the nervous system.

Fixation The process of using chemical methods to preserve, stabilize, and strengthen a biological specimen for subsequent histological procedures and microscopic analysis.

Flat skull position The placement of an animal on a stereotaxic frame such that the top of the skull is perfectly level, at the same height along the rostral-caudal axis.

Flight simulators Specialized chambers that suspend a fly from a thin pin and project visual stimuli on a screen, depicting the external, visual environment. Used to determine the effect of visual stimuli on flying behavior.

Flippase recognition target (Frt) A 34 bp sequence recognized by the flippase recombinase (Flp) enzyme.

Flippase recombinase (Flp) A protein that recognizes flippase recognition target (Frt) sites. Flp excises any DNA sequences between Frt sites oriented in the same direction, and inverts any DNA sequences between two Frt sites oriented in inverted directions.

Floxed Term used to describe any DNA sequence flanked by two lox sites.

Flp/Frt system A binary expression system used to control the spatial expression of a transgene.

Fluorescence recovery after photobleaching (FRAP) A protein visualization technique in which a scientist uses strong laser illumination to bleach fluorescently tagged proteins in a particular region of a cell and then measures the timecourse of fluorescence that returns to the bleached region.

Fluorescence resonance energy transfer (FRET) A technique used to visualize if and when two proteins interact.

Fluorescent _in situ_ hybridization (FISH) A technique used to examine the location of specific DNA sequences within chromosomes, commonly used to detect chromosomal abnormalities.

Fluorodeoxyglucose (FDG) A radioactive form of glucose used to image functional brain activity in positron emission tomography experiments.

Fluorophore A molecule that has the ability to absorb light at a specific wavelength and then emit light at a different, typically longer wavelength.

FM dyes Lipophilic dyes that fluoresce when bound to a membrane. They are particularly useful for reporting the exocytosis and endocytosis events that occur during synaptic vesicle recycling and can stain nerve terminals in an activity-dependent manner.

Footprint pattern assay A behavioral assay used to examine a rodent's motor coordination and balance by observing its footprint patterns.

Forced swim test A behavioral experiment used to assay depression by placing an animal in a narrow chamber of water and measuring how long it takes for the animal to stop trying to escape.

Formalin assay An assay used to examine an animal's sensitivity to noxious chemical stimuli by observing its response to a small injection of the noxious chemical formalin.

Förster resonance energy transfer (FRET) See *Fluorescence resonance energy transfer.*

Forward genetics A method used to identify genes that contribute to a phenotype. Scientists generate random mutations in eggs/larvae and then examine the phenotypes of the offspring.

Founder Individuals in the first generation of a transgenic or genetically targeted animal.

Fractional intensity change The change in fluorescent signal during calcium or voltage sensitive dye imaging compared to baseline fluorescence. Denoted as $\Delta F/F$ or $\Delta I/I$.

Freezing microtome A microtome with a cooling unit that allows a tissue sample to be frozen during sectioning.

Functional brain imaging Techniques used to measure neural activity in the central nervous system without physically penetrating the skull.

Functional magnetic resonance imaging (fMRI) A technology that uses the principles of magnetic resonance imaging (MRI) to indirectly measure neural activity in the brain over time.

Gal4 A transcription factor derived from yeast that binds to a specialized upstream activation sequence (UAS) to activate the transcription of a gene.

Gal4/UAS system A binary expression system that exploits a transcription system in yeast cells to control the spatial expression of a transgene. Widely used with invertebrate and zebrafish model organisms.

Gal80 A protein derived from yeast that inhibits Gal4 transcription factors.

Gel electrophoresis The process of using an electrical field to force DNA through an agarose gel. Used to isolate, identify, and characterize the properties of DNA fragments.

Gel-filtration chromatography A chromatography technique that separates proteins on the basis of their size.

Gel shift assay See *Electrophoretic mobility shift assay.*

Geller-Seifter conflict test A behavioral paradigm that assays anxiety in rodents. A rodent is trained to press a lever in order to receive a food reward. In subsequent experiments, the lever is paired with an aversive foot shock. Anxious animals tend to press the lever significantly less to receive food than nonanxious animals.

Gene The fundamental unit of heredity.

Gene gun A device used in biolistic experiments to physically shoot DNA-coated bullets onto tissue specimens.

Gene targeting The process of incorporating a DNA construct into a specific locus in an animal's genome.

Genome The complete set of genes or genetic material present in a cell or organism.

Genomic DNA library A collection of thousands of plasmids that each contain fragments of an animal's genome.

Genotype The genetic constitution of an animal.

Germ cells Cells that pass on genetic information to an organism's offspring (sperm in males and eggs in females).

GFP See *Green fluorescent protein.*

Golgi stain A classic histological stain used to label a small subset of neurons in their entirety.

Graded potential See *Localized potential.*

Green fluorescent protein (GFP) A protein derived from jellyfish, capable of absorbing wavelengths of blue light and emitting wavelengths of green light. Ubiquitously used in the biosciences as a reporter protein.

Growth factors Proteins produced by certain cells as cell communication molecules that stimulate the growth and differentiation of other cells.

Gyrus A ridge on the cerebral cortex.

Halorhodopsin (NpHR) A genetically encoded transmembrane pump that hyperpolarizes neurons upon stimulation with yellow light.

Hanging wire assay A behavioral assay used to measure neuromuscular deficits in rodents.

Hargreaves assay A behavioral assay used to measure pain in rodents by aiming a high-intensity beam of light at the hind paw and measuring the time it takes for the rodent to withdraw its paw.

Hemoglobin The protein that transports oxygen throughout the bloodstream.

Herpes simplex virus A neurotropic virus that can be used to deliver DNA to some cell types and also serve as a multisynaptic tracer.

Heterologous expression systems A dissociated cell culture system that can easily be transfected with DNA.

Holding potential The set voltage that is held constant in a voltage clamp experiment.

Homologous recombination A natural phenomenon in which DNA regions with strong sequence similarity can exchange genetic material.

Homology arms Regions of homology that flank a genetic-targeting construct to exploit the phenomenon of homologous recombination in the production of a knockout or knockin mouse.

Hot plate assay A behavioral paradigm used to assay nociception in rodent models. A rodent is enclosed on a heated plate and observed for licking and paw withdrawl.

Hyperpolarization Any change in the membrane potential of a neuron that causes a net increase in the absolute value of the membrane potential.

I/V curve A plot of the voltage across a cell membrane compared with the current that flows through the ion channels of the membrane.

Immediate early gene (IEG) A gene that tends to be actively transcribed during periods of high neural activity.

Immersion The process of fixing small brains or even entire animals by immersing them in fixative solutions.

Immortalized cell line A cell line manipulated so that it will continually divide and multiply.

Immunoblot See *Western blot*.

Immunocytochemistry (ICC) A histological technique that uses antibodies to stain proteins in cells.

Immunofluorescence (IF) A histological technique that uses antibodies to stain proteins with fluorescent reagents.

Immunohistochemistry (IHC) A histological technique that uses antibodies to stain proteins in tissues.

Immunopanning A technique used to purify certain cell types. A scientist coats the bottom of a plate with antibodies that recognize cell surface markers on the outside of specific cells. When heterogeneous populations of cells are added to the plate, the scientist can purify the cells of interest by allowing the cells to bind to the bottom of the plate and then wash off the undesired, unbound cells.

Immunoprecipitation (IP) A protein purification method that uses antibodies to purify specific proteins out of solution.

In silico **screen** A method of identifying genes or proteins of interest by searching public genome and bioinformatics databases.

In situ **hybridization *(ISH)*** A histological method used to label the location of mRNA transcripts.

In vitro Any process that takes place in a controlled environment outside a living organism, such as a cell culture dish or test tube.

In vivo Any process that takes place in a whole, living organism.

Independent variable The experimental variable that is intentionally manipulated by the researcher and is hypothesized to cause a change in the dependent variable.

Index of refraction A measure of how much the speed of light is reduced in a specific medium, such as air, water, or oils.

Indirect IHC A method of performing an immunohistochemistry experiment in which a primary antibody binds to an antigen and then a secondary antibody binds to the primary antibody. The secondary antibody is conjugated to a fluorescent or chromogenic tag that allows for subsequent visualization.

Induced pluripotent stem (iPS) cell A somatic stem cell that has been manipulated into reverting back to a pluripotent stem cell state, capable of giving rise to multiple cell types.

Infection The process of delivering DNA into cells using viruses.

Inhibitory postsynaptic potential (IPSP) A localized potential change in a small region of a neuron that causes the membrane potential to hyperpolarize, bringing the membrane potential farther from the threshold to generate an action potential.

Inner cell mass A group of cells within the blastocyst that give rise to all cells and tissues in an organism.

Insertional mutagenesis The process of causing genetic mutations by insertion of mobile genetic elements called transposons into the genomes of offspring. These transposons insert at random locations in the genome, occasionally disrupting an endogenous gene.

Inside-out mode A patch clamp recording method in which a tiny patch of the membrane is isolated and oriented such that the intracellular surface is exposed to the medium and the extracellular surface resides on the inside of the glass pipette.

Institutional Animal Care and Use Committee (IACUC) An academic institution's internal review committee that approves and inspects all procedures using vertebrate animals.

Institutional Review Board (IRB) An academic institution's internal review committee that approves and inspects all procedures that use human subjects.

Internal ribosome entry site (IRES) A sequence within a DNA construct (and subsequently, within a transcribed mRNA strand) that allows a ribosome to start translation. Thus, an IRES allows a single mRNA molecule to code for two separate functional proteins.

Intracellular recording An electrophysiology recording technique in which an electrode is gently placed inside a cell.

Intracellular signaling The biochemical pathways that take place within a cell that affect the cell's physiology and metabolism.

Intracerebroventricular (i.c.v.) injection The injection of a solution directly into the ventricular system of the brain.

Intraperitoneal (i.p.) injection The injection of a solution into the peritoneum (body cavity) of an animal.

Intron A sequence of an RNA molecule that is cut out of a final mRNA sequence during RNA splicing.

Inverted microscope A microscope in which the objective lens is placed beneath the sample and the light source and condenser are placed above the sample.

Ion-exchange chromatography The process of using chromatography to separate proteins based on their charge.

Irradiation mutagenesis The use of high-intensity UV light to mutagenize hundreds or thousands of eggs/larvae for the purpose of performing a forward genetic screen.

Isofluorane A gaseous anesthetic agent used in mammals.

Isotropic The diffusion of a substance in all directions within a medium. The opposite of **anisotropic.**

Ketamine A chemical anesthetic agent that acts by inhibiting NMDA and HCN1 ion channels. Often used in concert with **Xylazine**.

Kinase assay An assay used to determine if a protein kinase is capable of phosphorylating a substrate protein or protein fragment.

Knockdown The disruption of mRNA transcript expression, thus causing a decrease in functional protein products.

Knockin The use of gene targeting procedures to add a functional gene or gene sequence to a specific location in a genome.

Knockout The use of gene targeting procedures to remove a functional sequence from the genome, thus knocking out a gene of interest.

LacZ A bacterial gene that encodes the enzyme β-galactosidase. This gene is often used as a reporter gene, as β-galactosidase can react with the substrate X-gal to produce a dark blue byproduct.

Lambda Landmark on the surface of the skull, defined as the intersection between the lines of best fit through the sagittal and lambdoid sutures.

Learned helplessness A paradigm used to create rodent models of depression. A scientist exposes an animal to aversive stimuli at random intervals. Theoretically, this treatment creates a condition in which the animal experiences a lack of control, causing it to show symptoms of behavioral despair.

Lectins Proteins that exhibit extremely high binding affinities for sugars, many of which can be used as transsynaptic neural tracers.

Lentivirus A retrovirus that can be used to deliver DNA sequences to both dividing and postmitotic cells and integrate these sequences in the genome of the host cell.

Linkage analysis A strategy to map a gene's location in the genome by identifying the position of the novel gene in relation to the location of known genes.

Lipofection A chemical transfection method that uses a lipid complex to deliver DNA to cells.

Liposomes Tiny vesicular structures composed of a sphere of phospholipids that have the same composition as the plasma membrane.

Local field potential An electrophysiological measurement of the sum of all dendritic synaptic activity within a volume of tissue.

Localized potential A local change in membrane potential caused by the activity of individual ion channels. These potentials include excitatory postsynaptic potentials (EPSPs) and inhibitory postsynaptic potentials (IPSPs).

Lox site A 34 bp DNA sequence recognized by the Cre recombinase enzyme.

Luciferase A genetically encoded bioluminescent protein cloned from various species of fireflies.

Luciferase assay A technique used to determine whether a protein is able to interact with a genetic regulatory sequence and activate the transcription of a gene.

Magnetic resonance imaging (MRI) A structural imaging technique that uses the magnetic properties of neural tissue to non-invasively produce highly detailed structural images of the brain and body.

Magnetoencephalography (MEG) A functional imaging technique that measures changes in magnetic fields on the surface of the scalp produced by changes in underlying patterns of neural electrical activity.

Magnification A critical parameter in microscopy that measures how much larger the sample appears compared to its actual size.

Mass spectrometry A technique used to identify proteins and/or peptide sequences by determining the mass to charge ratio of proteins, peptides, and their fragments.

Match to sample task A memory assay that requires the subject to choose the stimulus that was seen previously. Used as a working memory assay by adding a delay between the sample and the test stimuli.

Matrix-assisted laser desorption ionization (MALDI) A mass spectrometry method of vaporizing proteins into the gas phase by blasting them with a UV laser so that their mass to charge ratio can be measured.

Membrane potential The voltage across the cell membrane generated by differential ionic charges on the intracellular versus extracellular sides of the membrane.

Microarray A tool used to measure the expression of thousands of genes between one or more tissue samples. It consists of a grid (array) of known nucleic acid sequences spotted on a slide so that cDNA from a biological sample can be hybridized with the sequences to detect whether any of the known sequences are present in the sample.

Microdialysis A method to sample neurochemicals from brain extracellular fluid based on diffusion of chemicals from the extracellular fluid into a microdialysis probe's semipermeable membrane.

Microelectrode A small (micrometer scale) metal or glass probe used to measure electrical signals from cells or tissue.

Microinjection The process of injecting small volumes of a solution (usually containing DNA) into cells with a thin needle.

Microiontophoresis The process of using a small electrical current to infuse substances out of a glass electrode into the brain.

Microscope An instrument that manipulates the trajectory of light rays so that small objects appear larger.

Microstimulation The process of sending electrical currents through an electrode to induce changes in membrane potential.

Microtome A piece of equipment with a sharp knife to cut medium thickness (25–100 μm) sections of tissue.

Midsagittal A view or section of the brain that perfectly divides the left and right hemispheres.

Mixed design A type of fMRI experimental task paradigm that uses elements of both blocked and event-related designs.

MNI template A standardized three dimensional coordinate map of the human brain produced by the Montreal Neurological Institute. Based on the average of hundreds of individual MRI brain scans and matched to landmarks in Talairach space.

Molecular cloning The process of identifying, isolating, and making copies of a particular DNA fragment through recombinant DNA technology.

Monoclonal antibody An antibody made by a single clone of cells that only recognizes a single epitope, or specific region of a protein. Contrast with **polyclonal antibody**.

Morpholinos Stable, synthetic 22–25 bp antisense oligonucleotide analogues that complement an RNA sequence to block proper mRNA translation as an alternative to RNAi.

Morris water maze A spatial learning and memory task for rodents that involves a hidden platform submerged in opaque water whose location can be recognized through distant visual cues.

Multielectrode array (MEA) A group of individual electrodes or tetrodes arranged into one electrical recording unit to record extracellular activity from multiple cells at the same time.

Multiple cloning site A region in a DNA vector with a variety of recognition sequences for different restriction enzymes that make it simple to insert a foreign DNA sequence into the vector.

Multipotent The ability of some cells to give rise to all types of cells found in a particular tissue—for example, the ability of a neural stem cell to give rise to neurons, astrocytes, and oligodendrocytes.

Myelin The fatty insulating material surrounding axons produced by glial cells that provide insulation to neurons.

Near-infrared spectroscopy (NIRS) A noninvasive optical brain imaging method that records neural activity-induced changes in the reflectance properties of near-infrared light as it is reflected off the scalp.

Neural progenitor A dividing cell with a limited ability to give rise to another dividing cell but that has the ability to generate either a neuron, astrocyte, or oligodendrocyte.

Neural stem cell (NSC) A multipotent, self-renewing cell that divides and gives rise to neurons, astrocytes, or oligodendrocytes, as well as other neural stem cells.

Neurosphere A ball of cells generated from a dividing neural progenitor or neural stem cell that indicates an ability to proliferate.

Neurosphere assay See *Primary neurosphere assay.*

Neuroethology The study of the neural basis of an animal's natural behaviors.

Neuron doctrine The theory advocated by Ramón y Cajal that discrete neurons are the basic structural and functional units of the nervous system.

Nissl stain A basophilic stain that highlights RNA (the "Nissl substance") in cells. Cresyl violet is a commonly used Nissl stain.

Nociception The detection of a noxious stimulus, typically perceived as pain.

Nomarski microscopy See *Differential interference contrast (DIC) microscopy.*

Nonmatch to sample task A memory assay that requires the subject to choose the stimulus that was *not* previously seen. This task can be used as an assay of working memory by adding a delay between the sample and the test stimuli. Contrast with **match to sample task**.

Nonratiometric dye An organic chemical whose fluorescence intensity varies directly with its ability to bind a substance such as calcium. Contrast with **ratiometric dyes**.

Northern blot A nucleic acid hybridization technique used to detect the presence and relative concentration of mRNA in a sample.

Novel object recognition A learning and memory task in which a rodent will spend more time exploring an object that has not previously been presented, demonstrating memory for the previously presented object.

Nucleotide A molecule composed of a phosphate group, a nitrogenous base (adenine, thymine, cytosine, guanine, or uracil), and a sugar molecule (2'-deoxyribose for DNA, ribose for RNA) that combines with other nucleotides to form a DNA or RNA polymer.

Numerical aperture (NA) A measure of the light-collecting ability of a microscope objective lens. The NA depends on the angle by which light enters the objective and the medium light must pass through. Higher NA objectives collect more light and consequently have better resolving power.

Objective lens The lens in a microscope that gathers and focuses light from the specimen in a microscope. This lens is typically placed adjacent to the specimen.

Ocular lens A lens, typically 10× magnification, located in the eyepiece of a compound microscope.

Ohm's law The relationship between voltage (V), current (I), and resistance (R) in an electrical circuit, expressed as. $V = I \times R$.

Olfactory avoidance assay A *Drosophila* chemosensory assay in which a scientist presents an odor on a stick in a fly chamber and measures the distance a fly maintains from the odor. Also called a *dipstick assay*.

Olfactory jump response See *Chemosensory jump assay.*

Open field test A rodent assay used to measure exploratory behavior in a large open chamber. More anxious animals stay near the walls of the chamber and do not explore the center area.

Operant conditioning The use of consequences (a positive reward or negative punishment) in response to a specific behavior in an attempt to influence that behavior. For example, a rodent learns to press a lever to receive a pleasurable food reward.

Optical imaging A brain imaging technique that measures changes in light reflectance from the surface of the brain due to changes in blood flow and metabolism caused by neural activity.

Optodes See *Optrodes*.

Optrodes "Optical electrodes" that record changes in light reflectance in optical brain imaging techniques.

Organotypic slice culture A slice of brain tissue kept in culture conditions for multiple days to weeks. Such cultures require an air/liquid interface to properly regulate gas exchange throughout the slice.

Oscilloscope An instrument that displays the membrane potential over time.

Outside-out mode A patch-clamp technique that exposes the extracellular side of an excised membrane patch to the bath solution and the intracellular side to the inside of the recording pipette.

P element A transposable genetic element used in *Drosophila* that is able to change locations in the genome when the P element transposase enzyme is also present.

Packaging cells An immortalized cell line that is transfected with the necessary recombinant DNA sequences to make viral particles. The endogenous cell machinery produces virus, which is eventually harvested in the extracellular media.

Particle-mediated gene transfer The process of introducing DNA into cells by physically shooting DNA-coated pellets through the cell membrane.

Passaging The process of transferring a fraction of cells into a new container to provide space for the cells to continue to divide.

Patch-clamp techniques A set of electrophysiological methods in which a glass electrode forms a tight seal with a patch of membrane, allowing the study of single or multiple ion channels.

Pavlovian conditioning See *Classical conditioning*.

PCR See *Polymerase chain reaction*.

Perforated patch A category of patch-clamp technique in which a chemical is added to the recording pipette to cause small holes to form in the membrane. This is useful for making the contents of the glass pipette continuous with the cell but without the disadvantage of cytoplasmic contents leaking into the pipette.

Perfusion A method used to circulate a chemical fixative through the cardiovascular system so that the brain is thoroughly preserved.

Peri-stimulus time histogram (PSTH) A graph of the number of action potentials recorded with respect to the time a stimulus was presented.

Phage See *Bacteriophage*.

Phase-contrast microscopy An optical method of enhancing contrast in a light microscope to amplify small differences in the index of refraction of different cellular structures.

Phenotype An observable trait or set of traits in an animal.

Photoactivation One of two processes involving the use of light to alter a chemical property. In the context of a caged molecule, photoactivation releases an active compound by uncaging the molecule. In the context of photoactivatable fluorescent proteins, photoactivation changes the fluorescence properties of a fluorophore, usually causing it to fluoresce.

Photobleaching The phenomenon in which fluorescence intensity emitted from a fluorophore decreases over time as it is continuously exposed to light.

Photoconversion The process of delivering a pulse of light to a fluorophore whose emission spectra changes from one color to another.

Phototaxis The tendency of an organism to move toward light and away from dark.

Phototoxicity The phenomenon in which illumination leads to the death of cells expressing a fluorophore, typically because illumination of the fluorophore generates free-radicals.

Physical gene delivery A method of introducing foreign DNA into cells by physically disrupting the cell's membrane. Includes microinjection, electroporation, and particle-mediated gene transfer (biolistics).

Plasmids Independently replicating accessory chromosomes in bacteria that exist as circles of DNA. Commonly used as vectors for molecular cloning.

Pluripotent The ability of a cell to generate any type of cell, including more pluripotent cells. Contrast with **multipotent** and **unipotent**.

Polyclonal antibody A collection of antibodies that each recognizes a distinct epitope, or region, of the same protein. Contrast with **monoclonal antibody**.

Polymerase chain reaction (PCR) A biochemical reaction that uses controlled heating and cooling in the presence of DNA synthesizing enzymes to exponentially amplify a small DNA fragment.

Porsolt test See *Forced swim test.*

Position effects The phenomenon of transgenes having different expression patterns depending on where the transgene randomly integrates in the genome.

Positional cloning The process of using molecular cloning technology to map the location of a gene in the genome.

Positron An antimatter counterpart of an electron. When a positron contacts an electron, an annihilation event occurs that generates gamma photons.

Positron emission tomography (PET) A brain imaging technique in which positron-emitting isotopes generate annihilation events that can be detected by a gamma detector. Depending on the positron-emitting isotope, PET can detect neural activity or the location of specific receptors.

Positron-emitting isotope An unstable version of a molecule that emits positrons as it decays, allowing it to be detected in positron emission tomography (PET).

Post-translational modification (PTM) A modification to a protein, usually the addition or subtraction of a functional chemical group, after it has been translated by a ribosome. For example, the addition or removal of a phosphate group.

Potential See *Membrane potential*.

Precess The circular movement of the axis of a spinning body around another axis due to a torque. For example, the circular movement of a proton.

Predictive validity An aspect of an animal model's validity that is fulfilled if treatments used in human patients have the same effect on the animal model. For example, rodent models of depression have predictive validity if human antidepressants relieve behavioral measures of depression in rodents.

Prepulse inhibition (PPI) The attenuation of a startle response caused by the presentation of a weaker sensory stimulus before the presentation of the startle-causing stimulus. Often used in animal models of schizophrenia because human patients with schizophrenia display deficits in PPI.

Primary antibody The antibody that recognizes and binds to a specific antigen.

Primary cell culture The culture of cells extracted directly from a tissue of interest. Can be in the form of dissociated cells, explants, or slices.

Primary neurosphere assay An assay of cultured cells for the ability to form neurospheres to examine whether they are proliferative. Cannot distinguish between progenitor cells and stem cells. Compare to **secondary neurosphere assay**.

Primer A short, single-stranded oligonucleotide used in PCR to hybridize to a template strand in the region flanking the region to be amplified.

Proboscis extension response (PER) A reflex in *Drosophila* that can be used to test the responses of specific gustatory receptors. A scientist applies taste ligands to gustatory receptor neurons on the leg or proboscis and examines proboscis extension.

Progenitor A cell that divides, but with a limited capacity for self-renewal.

Promoter A regulatory region of DNA that controls the spatial and temporal expression of a gene under its control.

Pronucleus One of the nuclei present in a recently fertilized egg—either the male sperm nucleus or the female egg nucleus.

Protein A biological macromolecule composed of one or more sequences of amino acids. Proteins serve as the structural and enzymatic molecules that regulate virtually every process within cells.

Protein kinase A protein that adds a phosphate group to another protein.

Pseudopregnant mouse A female mouse that has been mated with a sterile male mouse so she produces the proper hormones to act as a foster mom.

Pulse chase labeling A process in which a labeled probe (the pulse) is briefly injected into animals or added to cultured cells, then washed away and replaced by unlabeled molecules (the chase). By following the changes in localization of the labeled probe, different protein trafficking pathways can be observed.

qRT-PCR See *Quantitative real time PCR*.

Quantitative real time PCR (qRT-PCR) A method to both amplify and quantify a specific DNA fragment (especially a cDNA fragment produced from mRNA).

Quantum dot Semiconductor nanocrystals that can be used as bright fluorophores.

Radial arm maze A behavioral test used to assay spatial learning and memory in rodents. It utilizes a maze consisting of an array of arms radiating out from a central starting point, with only some arms containing food. A rodent is trained to traverse down only one arm and not others.

Radiofrequency (RF) pulse A pulse of electromagnetic energy used in MRI to excite hydrogen protons in a human (or rarely, an animal) subject. After the RF pulse is switched off, the protons relax, generating the MR signal used to form images of the brain.

Radioimmunoassay (RIA) A sensitive method for measuring extremely low concentrations of proteins.

Raster plot A graph that displays each action potential (or membrane potential over a threshold) as a dot or tick mark in relation to a certain time period. Often, the Y axis represents categories of stimuli while the X axis represents time.

Ratiometric dye An organic molecule that is either excited at or emits at a slightly different wavelength when bound to a substance, such as calcium. Thus, it is the ratio of fluorescence intensity at either different excitation or emission wavelengths that reveals changes in the substance concentration. In contrast with nonratiometric dyes, ratiometric dyes allow corrections for changes in background fluorescence and artifacts.

Recognition site See *Restriction site*.

Recombineering A method of creating recombinant DNA molecules using homologous recombination to avoid restriction enzyme digests and ligation reactions.

Reduction One of the guidelines for animal welfare stating that scientists must use the minimum number of animals required to obtain statistically significant data.

Refinement One of the guidelines for animal welfare stating that scientists must minimize distress and pain and enhance animal well-being.

Refractive index See *Index of refraction*.

Region-of-interest (ROI) analysis Method of fMRI data analysis that divides the brain into discrete regions that can be compared for significant differences in signal intensity.

Relative recovery In microdialysis, the relative concentration of a substance in the collected dialysate from the probed brain region compared to concentration in the perfusion solution.

Replacement One of the guidelines for animal welfare stating that scientists must do as much research without using animals as possible, including using computational methods and cell culture methods. Also applies to choice of a "less-sentient" animal species, such as choosing to study invertebrates over vertebrates or mice rather than monkeys.

Reporter gene/protein A visible, nonendogenous gene or protein that is controlled under the same promoter as a gene of interest to report the expression pattern of that gene. Alternatively, a reporter gene is fused to a gene of interest to report the subcellular localization of a reporter fusion protein.

Resident intruder assay A social behavior assay in rodents that measures territorial behavior in males by adding an "intruder" animal into the cage of a "resident" animal and measuring specific aggressive behaviors.

Resistance (R) An electrical property that restricts the movement of charge for example, across a membrane.

Resolution In microscopy, the minimum distance by which two points can be separated and still distinguished as two separate points.

Resolving power See *Resolution.*

Resting potential The voltage across the cell membrane of a cell at rest, generated by differential buildup of ionic charges on the intracellular vs. extracellular sides of the membrane. Neurons typically have a resting potential of about $-70\,mV$.

Restriction digest The process of using restriction enzymes to cut DNA into fragments.

Restriction endonuclease See *Restriction enzyme.*

Restriction enzyme An enzyme that recognizes and cuts specific sequences of DNA.

Restriction mapping The process of constructing a map of the locations of various recognition sites on a piece of DNA by using restriction digests and separating the digested fragments by gel electrophoresis.

Restriction site An approximately 4–8 unique base pair sequence that a particular restriction enzyme recognizes and cuts.

Retrograde tracer A chemical probe that labels an axon path by being transported from the synaptic terminal back up to the cell body, revealing what areas of the brain project to the location where the tracer was deposited.

Retrovirus A type of virus that uses RNA rather than DNA as its genetic material.

Reversal potential The membrane voltage at which there is no overall flow of ions across the membrane. Determining this potential can be used to identify the ion species that passes through a particular channel.

Reverse genetics The directed mutagenesis of a specific gene so that a scientist can determine if it is necessary for a certain phenotype to occur. Contrast with **forward genetics**.

Reverse microdialysis The use of a microdialysis probe to introduce chemical substances into a brain region.

Reverse transcriptase The enzyme that uses RNA as a template for creating strands of DNA, called complementary DNA (cDNA).

Reverse transcription The process of creating complementary DNA (cDNA) from RNA templates using reverse transcriptase.

Reverse transcription PCR (RT-PCR) A PCR reaction that amplifies a cDNA fragment generated from reverse transcription of an RNA fragment.

Ribosome Macromolecular complexes of RNA and protein that translate mRNA into a protein by assembling amino acids into polypeptide chains.

RNA-induced silencing complex (RISC) A complex of proteins that uses siRNA as a template for finding a complementary mRNA to degrade.

RNA interference (RNAi) The process of silencing gene expression through endogenous cellular mechanisms that degrade mRNA.

RNA polymerase The enzyme that binds to and transcribes DNA templates to produce RNA.

RNA splicing The removal of introns and joining of exons by a spliceosome of a newly transcribed mRNA before it leaves the nucleus for translation.

Rotarod A rodent behavioral assay of muscle coordination and balance. Consists of a rotating rod on which rodents must walk and maintain their balance. This device can also be used to examine motor learning, as animals will improve after multiple trials of walking on the rotarod.

RT-PCR See *Reverse transcription PCR*.

Run In a brain imaging experiment, one complete scan of the brain, during which a volume of information is collected.

Sagittal plane The plane dividing the brain into left and right portions such that the brain is complete top-to-bottom and rostral-to-caudal.

Sanger dideoxy chain termination method A technique for determining the sequence of a DNA molecule.

Scanning electron microscope (SEM) A microscope that uses electron interactions at the surface of a specimen to generate an image of the specimen's surface that appears three-dimensional.

SDS-PAGE See *Sodium dodecylsulfate polyacrylamide gel electrophoresis*.

Secondary neurosphere assay An assay that measures whether cells taken from a primary neurosphere are able to form new neurospheres, indicating the cell is still proliferative.

Sectioning The process of physically or optically creating thin sections of a specimen to gain greater access and/or visibility internal structures.

Selection cassette A sequence within a genetic targeting construct that encodes genes that can be used to identify cells that have properly incorporated the targeting construct. Typically, this cassette includes a positive selection gene, such as neomycin resistance gene, that can be used to kill cells that have not incorporated the construct. It also typically includes a negative selection gene, such as thymidine kinase, that can be used to kill cells that have improperly incorporated the targeting construct.

Session In a brain imaging experiment, the scheduled time when an actual experiment is conducted.

Short hairpin RNA (shRNA) A strand of RNA containing complementary sequences at either end that cause it to fold onto itself, producing a hairpin shape.

shRNA See *Short hairpin RNA*.

Signal transduction The conversion of an extracellular cue, such as a ligand binding to a receptor, into a signal that begins an intracellular signaling cascade.

Single-proton emission computerized tomography (SPECT) A functional brain imaging technique in which radioactive probes incorporated into the blood generate high-energy photons that can be detected using a gamma camera placed around a subject's head.

siRNA See *Small interfering RNA*.

Slice culture A primary tissue culture technique in which sections of brain tissue are sliced using a vibratome and then placed in a culture medium. These slices can be used acutely for short-term experiments, or organotypically, for multiple days.

Small interfering RNA (siRNA) Short (19–23 bp) double-stranded RNA fragments used as a template in the RNAi pathway to target mRNA molecules for degradation.

Sodium dodecylsulfate polyacrylamide gel electrophoresis (SDS-PAGE) A technique used to separate proteins based on their size and folding properties. Often used while performing a western blot.

Somatic cell A differentiated cell that is not part of the germline (eggs or sperm), such as skin or blood cells.

Southern blot A nucleic acid hybridization technique that can identify the presence of a specific sequence of DNA.

Spatial resolution The minimum size that can be resolved to distinguish individual components of an organ or tissue. Compare to **temporal resolution**.

Spike Another name for an action potential.

Spike sorting Differentiating action potentials recorded in an extracellular electrophysiology experiment and assigning them to different neurons based on their pattern, shape of spiking activity, and signal size.

Spliceosome The macromolecular complex that catalyzes RNA splicing.

Splitting See *Passaging.*

Startle response assay A behavioral assay that measures the ability of an animal to exhibit a startle response, such as an eye blink or sudden muscle contraction, in response to the presentation of an unexpected sensory stimulus. This assay can be used to indicate whether sensory processing is intact.

Stem cell A pluripotent cell that has an unlimited ability to renew itself.

Stereomicroscope A microscope that magnifies small objects and provides a three-dimensional perspective for examining the surface of brains or large neural structures. Often used for dissections, surgeries, or examining electrodes and implants.

Stereotaxic instrument A piece of equipment that stabilizes an animal's head and allows precise positioning for injections or implants.

Stereotaxic surgery A survivable procedure in which a scientist physically penetrates the brain of an animal under anesthesia for the injection of substances or implantation of devices.

Sterile Aseptic and free from microorganisms. Important condition for performing surgeries on live animals or culturing cells and tissue outside of an organism.

Sterile field A dedicated aseptic surface where sterile instruments can be placed when not in use.

Sterilizer A small device filled with heated glass beads used to sterilize tools and fine instruments when autoclave sterilization is inconvenient, such as between surgeries.

Sticky end Single-stranded DNA overhang generated by some restriction enzyme digests. Sticky ends can be matched to a complementary sticky end produced by the same restriction enzyme on a different sample of DNA. Compare to **blunt end**.

Structural brain imaging Techniques that are used to resolve brain anatomy in a living subject without physically penetrating the skull. Examples include cerebral angiography, CT scans, and magnetic resonance imaging (MRI).

Subcloning The process of copying or moving a fragment of DNA from one recombinant DNA construct into a new vector to create a new recombinant DNA construct.

Sucrose preference test A behavioral measure of anhedonia in rodents. A scientist places a bottle of plain tap water and a bottle containing sweet water in a rodent's cage. Normal rodents prefer sweet water, but rodent models of depression do not.

Sulci The indentations created by the folds in the brain.

Suture Either (a) a seam between different parts of the skull or (b) thread used after a surgery to close a wound.

SynaptopHlorin A pH-sensitive fluorescent protein variant of GFP that is used to examine synaptic vesicle recycling and release. It does not fluoresce when present inside synaptic vesicles, but does fluoresce when the vesicle fuses during exocytosis.

Systems neuroscience The branch of neuroscience that examines the coordinated activity of neural ensembles or the firing properties of individual neurons that generate behavior and cognition.

T1 In MRI and fMRI, the time (in milliseconds) required for a certain percentage of protons that had been excited by radiofrequency pulses to realign in the longitudinal direction.

T1-weighted image An MRI image formed based on T1 signal intensity.

T2 In MRI and fMRI, the time (in milliseconds) required for a certain percentage of protons that had been excited by radiofrequency pulses to relax in the transversal direction.

T2-weighted image An MRI image formed based on T2 signal intensity.

T-maze A behavioral apparatus with two arms oriented in opposite directions, giving it a T shape. A *Drosophila* T-maze contains an additional loading arm to move flies into the choice point between the T. This maze can be used to test sensory preferences as well as learning.

Tail flick assay A rodent assay for nociception in which painful heat or cold is applied to the tail, causing the rodent to rapidly move its tail out of the way.

Tail suspension assay An assay for depression in which a rodent is held by the tail and the amount of time the animal struggles is measured. Rodent models of depression stop struggling sooner than normal rodents but struggle for a normal amount of time if given antidepressant medication.

Talairach space A coordinate system based on the stereotaxic measurements of a single postmortem brain used to normalize MRI data so that anatomical comparisons can be made among different brains.

Tamoxifen A molecule that is able to bind to the estrogen receptor, causing its translocation from the cytoplasm to the nucleus. Useful to exert temporal control over the Cre/lox system when Cre recombinase is fused to the estrogen receptor.

Task paradigm A strategy for presenting stimuli to subjects during an experiment. In human brain imaging experiments, these paradigms are typically classified as a blocked design, event-related design, or mixed design.

Temporal delay In functional brain imaging techniques, the time delay between presentation of the stimulus and measurement of neural activity due to the time it takes for blood to flow into an active region.

Temporal resolution The ability to distinguish discrete neural events over time.

Tesla Units of measurement for magnetic field strength. Conventional MRI scanners create external magnetic fields at 1.5–3 T.

Tet-off/Tet-on system An inducible promoter system that allows transgene expression to be either turned off (Tet-off) or turned on (Tet-on) through the use of an antibiotic called doxycycline.

Tetrode A bundle of four microelectrodes used to perform extracellular electrophysiological recordings, often *in vivo* recordings in awake, behaving rodents.

Tetrodotoxin (TTX) A toxin produced by specific species of fish, such as the puffer-fish, that binds pores of voltage-gated sodium channels to block action potentials. The purified chemical is often used as a pharmacological agent to block action potentials or sodium channel current.

Thermocycler A machine that controls the heating and cooling of samples, most often used for PCR.

Time of flight (TOF) detector In mass spectrometry, an instrument that measures the mass to charge ratio of molecules in the gas phase. Often combined with MALDI.

Total internal reflection fluorescence (TIRF) microscopy A form of microscopy used to image molecules and/or events that occur at the cell surface. A TIRF microscope uses a rapidly decaying evanescent wave of fluorescence excitation to restrict imaging to a thin (~100 nm) region of contact between a specimen and the surface it is on (usually a glass coverslip).

Tracer A chemical that can be injected and transported along neural processes to highlight axonal paths and connections.

Transcranial magnetic stimulation (TMS) A noninvasive method used to manipulate neural activity in the brain. A coil is positioned adjacent to a subject's head. This coil generates magnetic field pulses which induce electrical activity in superficial brain structures. Depending on the specific sequence and strength of pulses, TMS can reversibly activate or inactivate brain regions.

Transcription The process of synthesizing RNA from a DNA template, performed by the enzyme RNA polymerase.

Transcription factor A DNA-binding protein that recognizes a specific genetic regulatory element and regulates transcription by recruiting or blocking RNA polymerase.

Transcriptional start site The location within a gene where RNA polymerase begins transcription.

Transduction The process of using a nonreplicating viral vector to deliver foreign DNA into a cell.

Transfection The process of delivering DNA to cells using nonviral methods.

Transgene A foreign gene expressed in an organism that does not normally express the gene.

Transgenic organism An organism that carries foreign DNA that it does not normally express.

Translation The process by which a ribosome synthesizes a chain of amino acids based on an mRNA template.

Transmission electron microscopy (TEM) A microscope that transmits an electron beam through a thin specimen to generate an image based on differences in electron density of the material. Produces extremely high-resolution images of cellular ultrastructure.

Transposable element See *Transposon*.

Transposase An enzyme that acts on specific DNA sequences at the end of a transposon to disconnect the sequence from flanking DNA and insert it into a new target DNA site.

Transposition The natural process of a transposon moving to different positions within the genome.

Transposon A genetic sequence capable of translocating to different positions within the genome.

Transsynaptic tracer A tracer capable of passing through synapses useful for labeling multisynaptic connections.

Trophoblast Cells in a blastocyst that form the outer spherical shell and eventually develop into the placenta.

Two-photon microscope A specialized form of fluorescent microscopy that sends two relatively low energy pulses of laser light that can add up to excite fluorophores in a restricted section of a specimen. Regions beyond the restricted section do not receive enough energy for excitation, greatly reducing out-of-focus fluorescence.

UAS See *Upstream activation sequence.*

Unconditioned stimulus (US) In classical conditioning, the salient stimulus that eventually becomes associated with a previously neutral stimulus.

Unipotent A cell that is capable of dividing but giving rise to only a single cell type.

Upright microscope A standard light microscope in which the objective lens is above the specimen and the light source and condenser are beneath the specimen.

Upstream activation sequence (UAS) A strong promoter derived from yeast that is activated by the Gal4 transcription factor.

Validity A subjective measure of how relevant an animal model is toward understanding humans.

Vector A carrying vehicle composed of DNA that can hold an isolated DNA sequence of interest for cloning and recombinant DNA experiments.

Vertical pole test A behavioral paradigm used to assay motor coordination and balance in rodents.

Vibratome A device that uses a vibrating blade to section a fresh, unfrozen specimen into tissue slices that can be kept alive.

Viral gene delivery The delivery of DNA constructs into cells *in vivo* or *in vitro* using one of many available viral vectors.

Visual cliff An assay used to examine normal visual function in rodents and even human infants.

Voltage (V) An electromotive force that exists across a membrane.

Voltage clamp An electrophysiology technique that allows a scientist to hold the membrane potential at a constant voltage to measure currents generated by ions moving across the membrane.

Voltage sensitive dye imaging (VSDI) A cell or tissue visualization technique that uses dyes that shift their absorption or emission fluorescence based on the membrane potential, thus allowing scientists to visualize changes in membrane potential over time.

Voltammetry A method used to measure the presence and relative amounts of neurochemicals in living tissue.

Volume In brain imaging experiments, a complete scan of the brain composed of multiple, individual slices.

Von Frey assay A behavioral test used to assay noxious mechanical and pinch stimuli in rodents.

Voxel A three-dimensional unit of space, often used in terms of brain space during whole brain imaging experiments.

Voxelwise analysis A strategy for analyzing data in an fMRI experiment, in which each voxel of data is examined for significant changes in signal intensity.

Western blot A method used to detect the expression of a protein in a biological sample.

Whole-cell mode A patch clamp recording technique in which a scientist places a glass pipette adjacent to a cell membrane, forms a tight gigaseal with the membrane, and provides enough suction so that the membrane patch is removed, allowing the interior of the pipette to become continuous with the cytoplasm of the cell.

Whole-mount preparations A histological preparation in which a chunk of tissue or an entire small animal brain is mounted onto a slide for analysis.

Wide-field fluorescent microscopy See *Epifluorescent microscopy.*

Within-subjects study A study that compares the effects of two different stimuli within the same subject pool.

Worm tracker Computer software that can automatically track locomotor behavior in *C. elegans.*

X-ray A structural visualization technique used to image parts of the body with natural contrast, such as a bone in tissue. This technique is inadequate by itself to visualize individual structures within the brain, but serves as the basis of cerebral angiogram and computerized tomography technology.

Xylazine A chemical sedative and analgesic, often used in combination with ketamine to anesthetize small rodents.

Yeast artificial chromosome (YAC) A large (100–2000 kb) DNA vector capable of replicating in yeast. Useful for manipulating large sequences of DNA, such as in the generation of transgenic or knockout constructs.

Yeast two-hybrid assay A method that takes advantage of yeast transcriptional machinery in order to determine whether two proteins directly interact with each other.

Index